HANDBOOK
OF
ENERGY
ENGINEERING

HANDBOOK
OF
ENERGY
ENGINEERING
Third Edition

Albert Thumann, P.E., C.E.M.
D. Paul Mehta, Ph.D.

Published by
THE FAIRMONT PRESS, INC.
700 Indian Trail
Lilburn, GA 30247

Library of Congress Cataloging-in-Publication Data

Thumann, Albert.
    Handbook of energy engineering / Albert Thumann, D. Paul Mehta. --
3rd ed.
    p. cm.
    Includes bibliographical references and index.
    ISBN 0-88173-202-8
    1. Power resources--Handbooks, manuals, etc. 2. Power
(Mechanics)--Handbooks, manuals, etc. I. Mehta, D. Paul, 1940-
II. Title.
TJ163.2.T49   1994   621.042--dc20   94-37714
                                                              CIP

Published by The Fairmont Press, Inc.
700 Indian Trail
Lilburn, GA 30247

Printed in the United States of America

10 9 8 7 6 5 4 3 2 1

ISBN 0-88173-202-8   FP

ISBN 0-13-227687-9   PH

While every effort is made to provide dependable information, the publisher, authors, and
editors cannot be held responsible for any errors or omissions.

Distributed by Prentice Hall PTR
Prentice-Hall, Inc.
A Simon & Schuster Company
Englewood Cliffs, NJ 07632

Prentice-Hall International (UK) Limited, London
Prentice-Hall of Australia Pty. Limited, Sydney
Prentice-Hall Canada Inc., Toronto
Prentice-Hall Hispanoamericana, S.A., Mexico
Prentice-Hall of India Private Limited, New Delhi
Prentice-Hall of Japan, Inc., Tokyo
Simon & Schuster Asia Pte. Ltd., Singapore
Editora Prentice-Hall do Brasil, Ltda., Rio de Janeiro

Contents

1
Fundamentals of Energy Engineering

INTRODUCTION

Energy engineering is a profession which applies scientific knowledge for the improvement of the overall use of energy. It combines the skills of engineering with the knowledge of energy problems. The energy engineer must be able to identify problems in the use of energy, creatively design solutions, and implement the process.

In order to develop economic and socially acceptable ways to use available resources for the benefit of mankind, the energy engineer searches for a better way to combine a broad base of knowledge with experience.

Energy engineering requires a "system approach" and is multidisciplinary in nature. Thus an energy engineer must have both engineering skills in, for example, electrical, mechanical and process engineering, as well as good management knowledge.

Problems such as the high cost of energy, the depletion of resources, and the degradation of the environment are not transitory but will plague mankind for decades to come. The energy engineering profession is addressing these and other problems in a systematic and cost-effective manner.

ENERGY CONVERSION FACTORS

In order to communicate energy engineering goals and to analyze the literature in the field, it is important to understand the conversion factors used in energy engineering and how they are applied.

Each fuel has a heating value, expressed in terms of the British thermal unit, Btu. The Btu is the heat required to raise the temperature of one pound of water 1°F. Table 1-1 illustrates the heating values of various fuels. To compare efficiencies of various fuels, it is best to convert fuel usage in terms of Btu's. Table 1-2 illustrates conversions used in energy engineering calculations.

When comparing costs of fuels, the term "cents or dollars per therm" (100,000 Btu) is commonly used.

The chemical energy stored in fuel is sometimes expressed as Higher Heating Value (HHV) or Lower Heating Value (LHV).

Table 1-1. Heating Values for Various Fuels

Fuel	Average Heating Value
Fuel Oil	
Kerosene	134,000 Btu/gal.
No. 2 Burner Fuel Oil	140,000 Btu/gal.
No. 4 Heavy Fuel Oil	144,000 Btu/gal.
No. 5 Heavy Fuel Oil	150,000 Btu/gal.
No. 6 Heavy Fuel Oil 2.7% sulfur	152,000 Btu/gal.
No. 6 Heavy Fuel Oil 0.3% sulfur	143,800 Btu/gal.
Coal	
Anthracite	13,900 Btu/lb
Bituminous	14,000 Btu/lb
Sub-bituminous	12,600 Btu/lb
Lignite	11,000 Btu/lb
Gas	
Natural	1,000 Btu/cu ft
Liquefied butane	103,300 Btu/gal.
Liquefied propane	91,600 Btu/gal.

Source: Brick & Clay Record, October 1972.

Table 1-2. List of Conversion Factors

1 U.S. barrel	= 42 U.S. gallons
1 atmosphere	= 14.7 pounds per square inch absolute (psia)
1 atmosphere	= 760 mm (29.92 in.) mercury with density of 13.6 grams per cubic centimeter
1 pound per square inch	= 2.04 inches head of mercury
	= 2.31 feet head of water
1 inch head of water	= 5.20 pounds per square foot
1 foot head of water	= 0.433 pound per square inch
1 British thermal unit (Btu)	= heat required to raise the temperature of 1 pound of water by 1°F
1 therm	= 100,000 Btu
1 kilowatt (kW)	= 1.341 horsepower (hp)
1 kilowatt-hour (kWh)	= 1.34 horsepower-hour
1 horsepower (hp)	= 0.746 kilowatt (kW)
1 horsepower-hour	= 0.746 kilowatt hour (kWh)
1 horsepower-hour	= 2545 Btu
1 kilowatt-hour (kWh)	= 3412 Btu

To generate 1 kilowatt-hour (kWh) requires 10,000 Btu of fuel burned by average utility

1 ton of refrigeration	= 12,000 Btu per hr

1 ton of refrigeration requires about 1 kW (or 1.341 hp) in commercial air conditioning
1 standard cubic foot is at standard conditions of 60°F and 14.7 psia

1 degree day	= 65°F minus mean temperature of the day, $^\circ$F
1 year	= 8760 hours
1 year	= 365 days
1 MMBtu	= 1 million Btu
1 kW	= 1000 watts
1 trillion barrels	= 1×10^{12} barrels
1 KSCF	= 1000 standard cubic feet
1 Quad	= one quadrillion (10^{15}) Btu's

Note: In these conversions, inches and feet of water are measured at 62°F (16.7°C), and inches and millimeters of mercury at 32°F (0°C).

The higher heating value is obtained by burning a small sample of fuel in an oxygen environment and recording the heat transferred to the water sample surrounding it. This test includes the latent heat of vaporization of the condensed vapor.

The lower heating value subtracts the latent heat of vaporization since this energy is usually unavailable in practice.

European countries usually use lower heating values for fuels while in the United States higher heating values are used. Thus a heating device tested in Europe will be approximately 10% more efficient than if tested in the United States due to the standard in heating values assumed.

Example Problem 1-1

The supply of fuel oil is projected at 11.5 million barrels per day. What is the supply in Btu per year, assuming an average fuel oil value of 140,000 Btu/gallon?

Answer

Fuel supply = 140,000 Btu/gallon X 11.5 X 10^6 barrels/day
X 42 gallons/barrel X 365 days/year = 24.7 X
10^{15} Btu/year.

Example Problem 1-2

The warehouse of a plant is required to be minimally heated at night. Two methods of heating are being considered. One method is electric heaters. The second is to operate the oil-fired boiler (using No. 2 fuel oil) which keeps the radiators warm all night. Electricity costs, excluding demand charges, is 10 cents per kilowatt hour; No. 2 fuel oil costs $1.00 per gallon. Assume the boiler is 70% efficient and exclude the costs of running pumps and fans to distribute heat. What is the relative cost to heat the building using electricity or the oil-fired boiler?

Answer

To compare fuel costs, a common base of $ per therm will be used.
Electricity Cost
$$1 \text{ kWh} = 3412 \text{ Btu}$$

$$\text{Cost} = \frac{100,000}{3412} \text{ X } .10 = \$2.93/\text{therm}$$

No. 2 Fuel Oil Cost
$$\text{No. 2 fuel oil} = 140,000 \text{ Btu/gallon}$$

$$\text{Cost} = \frac{100,000}{140,000} \text{ X } \$1.00 = 71 \text{ cents/therm}$$

Taking into account an efficiency of combustion of 0.7, the relative cost becomes

$$\text{Cost} = \frac{71}{0.7} = \$1.02/\text{therm}$$

CODES, STANDARDS AND LEGISLATION

The flurry of activity in energy legislation brought about by the war in the Gulf will dramatically impact on consumption to the year 2000.

On October 24, 1992 the Energy Policy Act of 1992 was signed into law. The Act contains a wide ranging set of provisions to increase the efficiency of energy usage, develop new fuels, increase production of energy and stimulate new electric generating capacity.

This section will highlight some of the classic laws, codes and standards which are impacting the energy industry and the new rules of the game.

State Codes
More than three quarters of the states have adopted ASHRAE Standard 90-80 as a basis for their energy efficiency standard for new building design.

ASHRAE Standard 90-80 has been updated into two new standards:

ASHRAE 90.1-1989 Energy Efficient Design of New Buildings Except New Low-Rise Residential Buildings

ASHRAE 90.2 Energy Efficient Design of New Low Rise Residential Building

The American Building Officials (CABO) has upgraded their Model Energy Code (MEC). The MEC is a widely accepted model energy standard that is developed in an open process involving code officials, builders, manufacturers, government agencies and researchers.

The Energy Policy Act of 1992 requires states to establish minimum energy efficiency standards for buildings, and to consider minimum residential codes based on current voluntary codes.

FEDERAL LEGISLATION

National Energy Conservation Policy Act (NECPA)—1978. This act is probably the most important legislation passed.

MATCHING GRANTS PROGRAM
FOR SCHOOLS AND HOSPITALS

Background

The Matching Grants Program for Schools and Hospitals is a voluntary program to assist these non-profit institutions in saving energy.

Authorized by Part 1, Title III of the National Energy Conservation Policy Act of 1978 (NECPA) and administered by the Department of Energy's Institutional Conservation Programs (ICP) through state energy offices, the ICP provides grants for detailed energy analyses and for installation of energy-saving capital improvements to eligible institutions. Participation in the grant program requires a 50 percent match of funds from recipient institutions, except in hardship cases.

Program Activities

NECPA divided the program into two main phases: Phase I, Preliminary Energy Audits and Energy Audits, and Phase II, Technical Assistance and Energy Conservation Measures. Funding for Phase I activities has expired; however, the program still administers the Technical Assistance/Energy Conservation Measure phase.

The Preliminary Energy Audit (PEA) was a data-gathering activity to determine basic information about the number, size, type, and rate of energy use in eligible buildings and the actions taken to conserve energy. The Energy Audit (EA) consisted of an on-site visit which confirmed and expanded upon information gathered in the PEA and provided further information to establish a "building profile."

The Technical Assistance (TA) Audit consists of detailed engineering analyses and reports specific costs, payback periods, and projected energy savings possible from the purchase and installation of devices or systems, or from modification to building operations and the building shell. Examples of this type of activity would be the installation of storm windows, insulation, solar hot water or space heating systems, or automatic setback devices.

Energy Conservation Measure (ECM) grants provide for the actual purchase and installation of energy measures recommended as a result of the TA Audit.

Funding — Schools and Hospitals

The ICP has completed its eighth funding cycle, the previous seven having been conducted between fiscal years 1979 and 1985. More than $700 million has been distributed in the first eight cycles to institutions to help finance conservation improvements; these funds covered 57,000 buildings through 20,378 grants in the 50 states.

Multi-Year Strategy

In late 1984, the ICP initiated a strategy development process to identify ways of compounding the impact of the ICP grant program by systematically incorporating more non-federal resources and expertise in institutional energy efficiency. During 1985, through a process of broad-based research, analysis, coordination, and consultation, a strategy was developed, which was approved by the Assistant Secretary for Conservation and Renewable Energy on September 4, 1985.

The strategy addresses the three most important energy-related needs of institutions, as identified during the course of interviews and focus group meetings: financing, expertise, and information. To respond to these needs, specific actions were developed and organized in two components, one conducted at the state and local level, the other conducted by the national program office.

State/Local Component. Seed money is provided to state energy offices (SEO's) in a two-tiered effort to (1) conduct innovative developmental projects (Tier 1) and (2) implement proven Tier 1-type projects through multiplier grants (Tier 2). Tier 1 and 2 projects will be administered by DOE field offices.

ICP awarded 12 Tier 1 grants in September 1986 for SEO's to conduct innovative developmental projects in the generic areas of technology transfer and financial counseling coordination, association networking and heroes programs, and alternate energy efficiency financing coordination. Tier 1 grants are competitive, 1-year grants with no match requirements, whose features include

- leveraged non-federal resources;
- transferability;
- high potential for self-sufficiency; and
- high potential for evaluation.

Any SEO wishing to implement Tier 1 type projects in their states would be eligible for a Their 2 multiplier grant beginning is FY 1988, subject to availability of funds. These are maximum 2-year matching grants for up to $50,000 to implement proven projects.

National Component. Beginning in 1986, the national program office is conducting enhanced information development/networking activities and demonstration projects in partnership with national associations, state and local governments, federal programs, and others. These cooperative activities focus primarily on providing products and services such as

- energy management and financing case studies;
- information and training in operations and maintenance techniques and the economics of energy management; and
- demonstrations of alternate financing methods and other innovative approaches to institutional energy efficiency.

INDOOR AIR QUALITY (IAQ) STANDARDS[1]

Indoor Air Quality (IAQ) is an emerging issue of concern to building managers, operators, and designers. Recent research has shown that indoor air is often less clean than outdoor air and Federal legislation has been proposed to establish programs to deal with this issue on a national level. This, like the asbestos issue, will have an impact on building design and operations. Americans today spend long hours inside buildings, and building operators, managers and designers must be aware of potential IAQ problems and how they can be avoided.

IAQ problems, sometimes termed "Sick Building Syndrome," have become an acknowledged health and comfort problem. Buildings are characterized as "sick" when occupants complain of acute symptoms such as headache, eye, nose and throat irritation, dizziness, nausea, sensitivity to odors and difficulty in concentrating. The complaints may become more clinically defined so that an occupant may develop an actual "building-related illness" that is believed to be related to IAQ problems.

---

[1] Source: Indoor Air Quality: Problems & Cures, M. Black & W. Robertson, Presented at 13th World Energy Engineering Congress.

The most effective means to deal with an IAQ problem is to remove or minimize the pollutant source, when feasible. If not, dilution and filtration may be effective.

Dilution (increased ventilation) is to admit more outside air to the building. ASHRAE's 1981 standard recommended 5 CFM/person outside air in an office environment. The new ASHRAE ventilation standard, 62-1989, now requires 20 CFM/person for offices if the prescriptive approach is used.

Increased ventilation will have an impact on building energy consumption. However, this cost need not be severe. If an airside economizer cycle is employed and the HVAC system is controlled to respond to IAQ loads as well as thermal loads, 20 CFM/person need not be adhered to and the economizer hours will help attain air quality goals with energy savings at the same time. Incidentally, it was the energy cost of treating outside air that led to the 1981 standard. The superseded 1973 standard recommended 15-25 CFM/person.

REGULATIONS & STANDARDS IMPACTING CFCs

For many years, chlorofluorocarbons (CFCs) have been used in air conditioning and refrigeration systems designed for long-term use. However, because CFCs are implicated in the depletion of the earth's ozone layer, regulations will require the complete phase out of the production of new CFCs by the turn of the century. Many companies, like Du Pont, are developing alternative refrigerants to replace CFCs. The need for these alternatives will become even greater as regulatory cutbacks cause continuing CFC shortages.

Sometime in the future, air conditioning and refrigeration systems designed to operate with CFCs will need to be retrofitted (where possible) to operate with alternative refrigerants so that these systems can remain in use for their intended service life.

Du Pont and other companies plan to commercialize their series of alternatives—hydrochlorofluorocarbon (HCFC) and hydrofluorocarbon (HFC) compounds—in a 3- to 5-year time period beginning in late 1990. This commercialization schedule will depend on favorable toxicology results, successful market development and regulatory acceptance of the candidate alternatives. See Table 1-3.

Table 1-3. Candidate Alternatives for CFCs in Existing Cooling Systems

CFC	Alternative	Potential Retrofit Applications
CFC-11	HCFC-123	Water and brine chillers; process cooling
CFC-12	HFC-134a or Ternary Blends	Auto air conditioning; medium temperature commercial food display and transportation equipment; refrigerators/freezers; dehumidifiers; ice makers; water fountains
CFC-114	HCFC-124	Water and brine chillers
R-502	HFC-125; HCFC-22	Low-temperature commercial food equipment

The Montreal Protocol which is being implemented by the United Nations Environment Program (UNEP) is a worldwide approach to the phaseout of CFCs. The initial provisions required a 50% phaseout in CFC consumption by mid 1998.

Production of chlorofluorocarbons (CFCs) will be phased out by the year 2000 as a result of amendments to the Montreal Protocol.

The new agreement was approved in June 1990 by representatives of 58 countries that ratified the original 1987 Protocol.

The amended Protocol will limit CFC production and consumption to 80%of 1986 levels by 1993, 50% by 1995, and 15% by 1997. Total phaseout will be required by January 1, 2000. Recycled CFCs are not included in the ban.

In a non-binding resolution, Protocol parties agreed to restrict use of hydrochlorofluorocarbon (HCFC) substitutes to applications currently using CFCs and to phase out production of HCFCs no later than 2040.

The new agreement also establishes a $240 million fund to provide technical and financial assistance to help developing countries adopt alternative technologies. The U.S. will contribute 25% of total funding.

CLEAN AIR ACT AMENDMENT

On November 15, 1990, the new Clean Air Act (CAA) was signed by President Bush. The legislation includes a section entitled Stratospheric Ozone Protection (Title VI). This section contains extraordinarily comprehensive regulations for the production and use of CFC's, Halons, carbontetrachloride, methyl chloroform and HCFC and CFC substances. These regulations will be phased in over the next 40 years and they will impact every industry that currently uses CFC's.

The seriousness of the ozone depletion is such that as new findings are obtained, there is tremendous political and scientific pressure placed on CFC end-users to phase out use of CFC's. This has resulted in the U.S., under the signature of President Bush in February 1992, to have accelerated the phase out of CFC's to December 31, 1995.

MAJOR PROVISION DEALING WITH CFC'S

1) Phaseout Schedules:

- CFC's and halons (so called "Class I" substances) will be phased out in steps until a total phaseout occurs. Currently the law states that this phaseout will occur on January 1, 2000. However, as stated above, the more likely date will be December 31, 1995 for TOTAL phase out of CFC's.

- HCFC's (so called "Class II" substances) are regulated as follows:
    — Production frozen and use limited to refrigeration equipment on January 1, 2015.
    — Use in new refrigeration equipment is allowed until January 1, 2020.
    — Only service of in-place refrigeration equipment is allowed after January 1, 2020 and until the HCFC production ban becomes effective on January 1, 2030.

The reader must note that, as is the case with CFC's, there is a tremendous pressure being placed on industry to phase out HCFC's well before the current 2015 to 2030 time frame. It is currently estimated that phase out of "long lived" HCFC's (such as refrigerant R-22) may occur as early as 2000 or 2005. It is likely, however, that "short lived" HCFC's such as R-123 may not be phased out until the 2015-2030 time frame as discussed above.

The reader is also advised to carefully consider both the "alternate" refrigerants entering the market place *and* the alternate technologies that are available. Alternate refrigerants come in the form of HCFC's, and HFC's. HFC's have the attractive attribute of having no impact on the ozone layer (and correspondingly are not named in the Clean Air Act). Alternative technologies include absorption and ammonia refrigeration (established technologies since the early 1900's), as well as desiccant cooling.

2) Taxes on CFC's

The reader should also note that a tax placed on each pound of CFC's produced has been in effect since 1990. This tax is currently slated to increase to $5.00 per pound by 1999. However, given the 1995 phase-out date for CFC's, it is anticipated that "market forces" will be a much more significant concern than will be these taxes.

REGULATORY AND LEGISLATIVE ISSUES IMPACTING COGENERATION & INDEPENDENT POWER PRODUCTIONS[2]

Federal, state and local regulations must be addressed when considering any cogeneration project. This section will provide an overview of the federal regulations that have the most significant impact on cogeneration facilities.

Federal Power Act

The Federal Power Act asserts the federal government's policy toward competition and anti-competitive activities in the electric power industry. It identifies the Federal Energy Regulatory Commission (FERC) as the agency with primary jurisdiction to prevent undesirable anti-competitive behavior with respect to electric power generation. In addition, it provides cogenerators and small power producers with a judicial means to overcome obstacles put in place by electric utilities.

Public Utility Regulatory Policies Act (PURPA)

This legislation was part of the 1978 National Energy Act and has had perhaps the most significant effect on the development of cogeneration and other forms of alternative energy production in the past decade

---

[2]Source: *Georgia Cogeneration Handbook*, published by The Governor's Office of Energy Resources.

Certain provisions of PURPA also apply to the exchange of electric power between utilities and cogenerators.

PURPA provides a number of benefits to those cogenerators who can become Qualifying Facilities (QFs) under the act. Specifically, PURPA

- Requires utilities to purchase the power made available by cogenerators at reasonable buy-back rates. These rates are typically based on the utilities' cost.

- Guarantees the cogenerator or small power producer inter-connection with the electric grid and the availability of back-up service from the utility.

- Dictates that the supplemental power requirements of the cogenerator must be provided for at a reasonable cost.

- Exempts cogenerators and small power producers from fed-eral and state utility regulations and the associated reporting requirements of these regulating bodies.

In order to assure a facility the benefits of PURPA, a cogenera-tor must become a Qualifying Facility. To achieve Qualifying Status, a cogenerator must generate electricity and useful thermal energy from a single fuel source. In addition, a cogeneration facility must be less than 50% owned by an electric utility or an electric utility hold-ing company. Finally, the plant must meet the minimum annual operating efficiency standard established by FERC when using oil or natural gas as the principal fuel source. The standard is that the useful electric power output plus one half of the useful thermal output of the facility must be no less than 42.5% of the total oil or natural gas energy input. The minimum efficiency standard increases to 45% if the useful thermal energy is less than 15% of the total energy output of the plant.

FERC has provided two methods for achieving certification as a Qualifying Facility:

1. Self-certification
2. FERC-certification

Self-certification may be obtained by writing to FERC and pro-viding basic information about a cogeneration plant. Specifically, you should provide the name and address of the facility, the primary fuel type, the electric power output, and the percentage of owner-ship, if any, by any utilities. You should also include a description of the cogeneration system that is sufficiently detailed to show that the facility will meet the minimum operating efficiency standard.

FERC certification offers the advantage of future regulatory protection. In addition, FERC certification may be required in order to obtain financing for your project or to prevent utility challenges to your Qualifying Facility status. To obtain FERC certification, a more detailed description of the system must be provided in order to demonstrate that your facility meets all of the efficiency standards.

Natural Gas Policy Act (NGPA)

The major objective of this legislation was to create a deregulated national market for natural gas. It provides for incremental pricing of higher cost natural gas supplies to industrial customers who use gas, and it allows the cost of natural gas to fluctuate with the cost of fuel oil. Cogenerators classified as Qualifying Facilities under PURPA are exempt from the incremental pricing schedule established for indus-trial customers.

**Resource Conservation and
Recovery Act of 1976 (RCRA)**

This act requires that disposal of non-hazardous solid waste be handled in a sanitary landfill instead of an open dump. It affects only cogenerators with biomass and coal-fired plants. This legisla-tion has had little, if any, impact on oil and natural gas cogeneration projects.

**PUBLIC UTILITY
HOLDING COMPANY ACT OF 1935[3]**

The Public Utility Holding Company Act of 1935 (the 35 Act) authorizes the Securities and Exchange Commission (SEC) to regu-late certain utility "holding companies" and their subsidiaries in a wide range of corporate transactions.

---

[3]Source: Cogeneration Institute of the Association of Energy Engineers.

The Energy Policy Act of 1992 (the "Energy Bill") creates a new class of wholesale-only electric generators—"exempt wholesale generators" (EWGs)—which are exempt from the Public Utility Holding Company Act (PUHCA). The Energy Bill will dramatically enhance competition in U.S. wholesale electric generation markets, including broader participation by subsidiaries of electric utilities and holding companies. The Bill will also open up foreign markets by exempting companies from PUHCA with respect to retail sales as well as wholesale sales.

Under the Bill, EWG status can only be obtained from FERC, on a case-by-case basis, and not from the SEC. Since FERC retains jurisdiction under the Federal Power Act (FPA) over the rates and certain financial transactions of EWGs, the Bill effectively centralizes regulation of EWGs at FERC. However, states must approve EWGs consisting of wholly or partially rate-based assets and EWG sales to utility affiliates, as well as the prudency of utility purchases of wholesale power. (The SEC maintains certain jurisdiction over registered holding companies [RHCs] and non-EWGs.)

Unlike "Qualifying facilities" (QFs) under the Public Utility Regulatory Policies Act (PURPA), which partially deregulated the generation sector, EWG status will not be subject to restrictions relating to type of fuel, maximum size, technology or permissible utility ownership. EWGs will remain subject to regulation of wholesale power rates by the Federal Energy Regulatory Commission (FERC) under the Federal Power Act (FPA) and to state regulation, as further discussed below. However, the PUHCA exemption will enable EWGs to obtain project financing without the necessity for sponsors to relinquish control in order to avoid PUHCA's regulation of upstream owners.

The Bill's implementation will be affected by a series of federal and state agency proceedings, although EWG transactions can and will proceed in the interim. FERC is expected to promptly initiate rulemakings regarding eligibility for EWG status, transmission access and "full cost" transmission pricing. States will consider the competitive, financial and reliability impacts of purchased power from EWGs and other suppliers on state-regulated utilities and procedures for pre-approval of purchased power. And the SEC will initiate rulemakings on RHC investments in EWGs and foreign utility companies.

OPPORTUNITIES IN THE SPOT MARKET[4]

Basics of the Spot Market

A whole new method of contracting has emerged in the natural gas industry through the spot market. The market has developed because the Natural Gas Policy Act of 1978 (NGPA) guaranteed some rights for end-users and marketers in the purchasing and transporting of natural gas. It also put natural gas supplies into a more competitive position with deregulation of several categories.

The Federal Energy Regulatory Commission (FERC) provided additional rulings that facilitated the growth of the spot market. These rulings included provisions for the Special Marketing Programs in 1983 (Order 2346) and Order 436 in 1985, which encouraged the natural gas pipelines to transport gas for end-users through blanket certificates.

The change in the structure of markets in the natural gas industry has been immense in terms of both volumes and the participants in the market. By year-end 1986, almost 40 percent of the interstate gas supply was being transported on a carriage basis. Not only were end-users participating in contract carriage, but local distribution companies (LDCs) were accounting for about one half of the spot volumes on interstate pipelines.

The "spot market" or "direct purchase" market refers to the purchase of gas supplies directly from the producer by a marketer, end-user or LDC. (The term "spot gas" is often used synonymously with "best efforts gas," "interruptible gas," "direct purchase gas" and "self-help gas.") This type of arrangement cannot be called "new" because the pipelines have always sold some supplies directly to end-users.

The new market differs from the past arrangements in terms of the frequency in contracting and the volumes involved in such contracts. Another characteristic of the spot market is that contracts are short-term, usually only 30 days, and on an interruptible basis. The interruptible nature of spot market supplies is an important key to understanding the operation of the spot market and the costs of dealing in it. On both the production and transportation sides, all activities in transportation or purchasing supplies are on a "best efforts" basis. This means that when a cold snap comes, the direct purchaser may not get delivery on his contracts because of producer shutdowns, pipeline capacity and operational problems or a combination

---

[4]Source: *New Opportunities for Purchasing Natural Gas*, Fairmont Press, Atlanta, GA.

of these problems. The "best efforts" approach to dealing can also lead to problems in transporting supplies when demand is high and capacity limited.

FERC's Order No. 436

The impetus for interstate pipeline carriage came with FERC's Order No. 436, which provided more flexibility in pricing and transporting natural gas. In passing the 1986 ruling, FERC was attempting to get out of the day-to-day operations of the market and into more generic rule making. More significantly, FERC was trying to get interstate pipelines out of the merchant business into the transportation business—a step requiring a major restructuring of contracting in the gas industry.

FERC has expressed an intent to create a more competitive market so that prices would signal adjustments in the markets. The belief is that direct sales ties between producers and end-users will facilitate market adjustments without regulatory requirements clouding the market. As more gas is deregulated, FERC reasoned that natural gas prices will respond to the demand: Lower prices would assist in clearing excess supplies; then as markets tightened, prices would rise drawing further investment into supply development.

FERC Order 436, which instituted the last major changes of pipeline services, was struck down in court and eventually was replaced by FERC Order 500.

FERC Order No. 636

Order 636 required significant "Restructuring" in interstate pipeline services, starting in the Fall of 1993. The original Order 636:
- Separates (unbundles) pipeline gas sales from transportation
- Provides open access to pipeline storage
- Allows for "no notice" transportation service
- Requires access to upstream pipeline capacity
- Uses bulletin boards to disseminate information
- Provides for a "capacity release" program to temporarily sell firm transportation capacity
- Pregrants a pipeline the right to abandon gas sales
- Bases rates on straight fixed variable (SFV) design
- Passes through 100% of transition costs in fixed monthly charges to firm transport customers

FERC Order No. 636A

Order 636A makes several relatively minor changes in the original order and provides a great deal of written defense of the original order's terms. The key changes are:

- Concessions on transport and sales rates for a pipeline's traditional "small sales" customers (like municipalities).
- The option to "release" (sell) firm capacity for less than one month— *without posting it on a bulletin board system or bidding*
- Greater flexibility in designing special transportation rates (i.e., offpeak service) while still requiring overall adherence to the straight fixed variable rate design.
- Recovery of 10% of the transition costs from the interruptible transportation customers (Part 284)

Court action is still likely on the order. Further, each pipeline will submit its own unique tariff to comply with the Order. As a result, additional changes and variations are likely to occur.

HIGHLIGHTS OF ENERGY POLICY ACT OF 1992

Energy Efficiency Provisions
<u>Buildings</u>
- Requires states to establish minimum commercial building energy codes and to consider minimum residential codes based on current voluntary codes.
- Ties the availability of federal mortgage assistance for new residential buildings to compliance with minimum efficiency codes.
- Requires the development of voluntary home energy rating guidelines and the promotion of energy-efficient mortgages.
- Establishes a federal energy-efficient mortgage pilot program in five states to demonstrate the financing of efficiency improvements in existing homes.

<u>Utilities</u>
- Requires states to consider new regulatory standards that would: require utilities to undertake integrated resource planning; allow efficiency programs to be at least as profitable as new supply options;

and encourage improvements in supply system efficiency.
- Requires integrated resource planning by the Tennessee Valley Authority and the customers of the Western Area Power Administration.

Equipment Standards
- Establishes efficiency standards for: commercial heating and air-conditioning equipment; large electric motors; and lamps.
- Gives the private sector an opportunity to establish voluntary efficiency information/labeling programs for windows, office equipment and luminaires, or the Dept. of Energy will establish such programs.
- Establishes maximum flow rates for showerheads, faucets and other plumbing products.

Industry
- Establishes an industrial energy efficiency grant program to encourage industrial associations to begin or strengthen their efficiency programs, including energy reporting and efficiency targets.
- Establishes a program of grants to states to encourage states and utilities to cooperate with local industries to assess industrial efficiency opportunities, and to finance cost-effective efficiency improvements.
- Requires the Dept. of Energy to develop voluntary guidelines for industrial energy audits and for the proper level of insulation in industrial equipment.

Federal Energy Management
- Updates and expands the existing federal energy management program to establish new goals and procedures, and to integrate the efforts of the several federal agencies.
- Requires the Dept. of Energy to develop new regulations for the use of energy performance contracts with which the federal government can tap private-sector funding for federal efficiency improvements.

Provisions for Fuels, Alternative Energy & the Environment
- *Renewable Energy.* Establishes a program for providing federal support on a competitive basis for renewable energy technologies. Expands program to promote export of these renewable energy technologies to emerging markets in developing countries.
- *Alternative Fuels.* Gives Dept. of Energy authority to require a private

and municipal alternative fuel fleet program starting in 1998. Provides for a federal alternative fuel fleet program with a phased-in acquisition schedule; also provides for state fleet program for large fleets in large cities.

- *Electric Vehicles.* Establishes comprehensive program for the research and development, infrastructure promotion, and vehicle demonstration for electric motor vehicles.

- *Electricity.* Removes obstacles to wholesale power competition in the Public Utilities Holding Company Act by allowing both utilities and non-utilities to form exempt wholesale generators without triggering the PUHCA restrictions.

- *Global Climate Change.* Directs the Energy Information Administration to establish a baseline inventory of greenhouse gas emissions and establishes a program for the voluntary reporting of those emissions. Directs the Dept. of Energy to prepare a report analyzing the strategies for mitigating global climate change and to develop a least-cost energy strategy for reducing the generation of greenhouse gases.

- *Research and Development.* Directs the Dept. of Energy to undertake research and development on a wide range of energy technologies, including: energy efficiency technologies, natural gas end-use products, renewable energy resources, heating and cooling products, and electric vehicles.

2

Energy Economic Analysis

To justify the energy investment cost, a knowledge of life-cycle costing is required.

The life-cycle cost analysis evaluates the total owning and operating cost. It takes into account the "time value" of money and can incorporate fuel cost escalation into the economic model. This approach is also used to evaluate competitive projects. In other words, the life-cycle cost analysis considers the cost over the life of the system rather than just the first cost.

THE TIME VALUE OF MONEY CONCEPT

To compare energy utilization alternatives, it is necessary to convert all cash flow for each measure to an equivalent base. The life-cycle cost analysis takes into account the "time value" of money; thus a dollar in hand today is more valuable than one received at some time in the future. This is why a time value must be placed on all cash flows into and out of the company.

DEVELOPING CASH FLOW MODELS

The cash flow model assumes that cash flows occur at discrete points in time as lump sums and that interest is computed and payable at discrete points in time.

To develop a cash flow model which illustrates the effect of "compounding" of interest payments, the cash flow model is developed as follows:

End of Year 1: $P + i(P) = (1 + i)P$
       Year 2: $(1 + i) P + (1 + i) Pi = (1 + i)P [(1 + i)]$
                 $= (1 + i)^2 P$
       Year 3: $(1 + i)^3 P$
       Year n $(1 + i)^n P$ or $F = (1 + i)^n P$
       Where P = present sum
               i = interest rate earned at the end of each interest period
               n = number of interest periods
               F = future value

$(1 + i)^n$ is referred to as the "Single Payment Compound Amount" factor (F/P) and is tabulated for various values of i and n in Tables 15-1 through 15-8.

The cash flow model can also be used to find the present value of a future sum F.

$$P = \left(\frac{1}{(1 + i)^n}\right) \times F$$

Cash flow models can be developed for a variety of other types of cash now as illustrated in Figure 2-1.

To develop the Cash Flow Model for the "Uniform Series Compound Amount" factor, the following cash flow diagram is drawn.

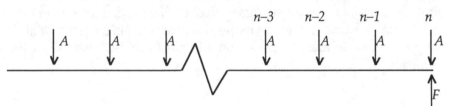

Where A is a Uniform Series of year-end payments and F is the future sum of A payments for n interest periods.

The A dollars deposited at the end of the nth period earn no interest and, therefore, contribute A dollars to the fund. The A dollars deposited at the end of the $(n - 1)$ period earn interest for 1 year and will, therefore, contribute $A (1 + i)$ dollars to the fund. The A dollars deposited at the end of the $(n - 2)$ period earn interest for 2 years and will, therefore, contribute

$A(1+i)^2$. These years of earned interest in the contributions will continue to increase in this manner, and the A deposited at the end of the first period will have earned interest for $(n-1)$ periods. The total in the fund F is, thus, equal to $A + A(1+i)+A(1+i)^2 + A(1+i)3 + A(1+i)4 +...+ A(1+i)^{n-2} + A(1+i)^{n-1}$. Factoring out A,

(1) $F = A [1+ (1 + i) + (1 + i)2+ ... +(1 + i)n^{-2} + (1 + i)n^{-1}]$
      Multiplying both sides of this equation by $(1 = i)$;

(2) $(1 + i)F = A [(1 + i)+(1+i)2 + (1+i)3 + ... +(1 + i)n^{-1} + (1 + i)n]$
      Subtracting equation (1) from (2):

$(1 + i)F - F = A [(1 + i) + (1 + i)2 + (1 + i)3$
$\qquad +(1 + i)n^-1+(1 + i)n]-A [1 + (1 + i)$
$\qquad + (1 + i)2 +_{,,} . + (1 + i)n^-2 + (1 + i)n^{-1}]$
$iF = A [(1 + i)n^{-1}]$

$$F = A \left| \frac{(1+i)^n-1}{i} \right|$$

Interest factors are seldom calculated. They can be determined from computer programs, and interest tables included in Chapter 15, Appendix. Each factor is defined when the number of periods (n) and interest rate (i) are specified. In the case of the Gradient Present Worth factor the escalation rate must also be stated.

The three most commonly used methods in life-cycle costing are the annual cost, present worth and rate-of-return analysis.

In the present worth method a minimum rate of return (i) is stipulated. All future expenditures are converted to present values using the interest factors. The alternative with lowest effective first cost is the most desirable.

A similar procedure is implemented in the annual cost method. The difference is that the first cost is converted to an annual expenditure. The alternative with lowest effective annual cost is the most desirable.

In the rate-of-return method, a trial-and-error procedure is usually required. Interpolation from the interest tables can determine what rate of return (i) will give an interest factor which will make the overall cash flow

Single Payment Compound Amount - F/P

The *F/P* factor is the future value of one dollar in "*n*" periods at interest of "*i*" percent.

$$F/P = (1+i)^n \qquad (2\text{-}1)$$

Single Payment Present Worth - P/F

The *P/F* factor is the present worth of one dollar, in "*n*" periods from now at interest of "*i*" percent.

$$P/F = \frac{1}{(1+i)^n} \qquad (2\text{-}2)$$

Uniform Series Compound Amount - F/A

The *F/A* factor is the future value of a uniform series of one dollar deposits.

$$F/A = \frac{(1+i)^n - 1}{1} \qquad (2\text{-}3)$$

Sinking Fund Payment - A/F

The *A/F* factor is the uniform series of deposits whose future value is one dollar.

$$A/F = \frac{i}{(1+i)^n - 1} \qquad (2\text{-}4)$$

Uniform Series Present Worth - P/A

The *P/A* factor is the present value of uniform series of one dollar deposits.

Capital Recovery - A/P

The *A/P* factor is the uniform series of deposits whose present value is one dollar.

$$P/A = \frac{(1+i)^n - 1}{i\,(+i)^n} \qquad (2\text{-}5)$$

$$A/P = \frac{i\,(1+i)^n}{(1+i)^n - 1} \qquad (2\text{-}6)$$

Gradient Present Worth - GPW

The GPW factor is the present value of a gradient series.

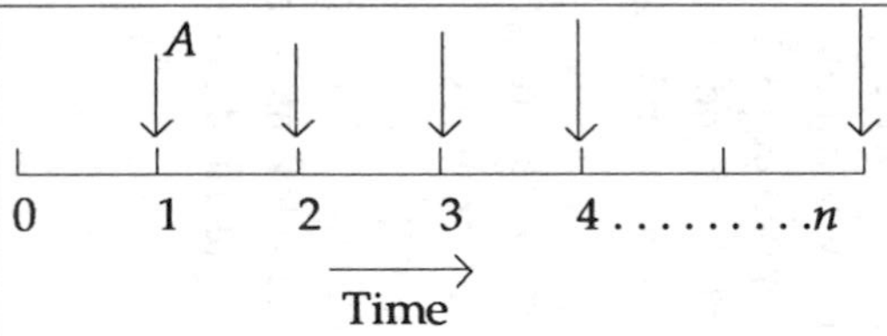

$$GPW = P/A = \frac{\frac{1+e}{1+i}\left[1 - \left(\frac{1+e}{1+i}\right)^n\right]}{1 - \frac{1+e}{1+i}} \qquad (2\text{-}7)$$

NOTES

where

P is the present worth (occurs at the beginning of the interest period).
F is the future worth (occurs at the end).
n is the number of periods that the interest is compounded.
i is the interest rate or desired rate of return.
A is the uniform series of deposits (occurs at the end of the interest period).
e is the escalation rate

Figure 2-1. Interest Factors

balance. The rate-of-return analysis gives a good indication of the overall ranking of independent alternates.

The effect of escalation in fuel costs can influence greatly the final decision. When an annual cost grows at a steady rate it may be treated as a gradient and the Gradient Present Worth factor can be used.

Special appreciation is given to Rudolph R. Yaneck and Dr. Robert Brown for use of their specially designed interest and escalation tables used in this text.

When life-cycle costing is used to compare several alternatives, the differences between costs are important. For example, if one alternate forces additional maintenance or an operating expense to occur, then these factors as well as energy costs need to be included. Remember, what was previously spent for the item to be replaced is irrelevant. The only factor to be considered is whether the new cost can be justified based on projected savings over its useful life.

PAYBACK ANALYSIS

The simple payback analysis is sometimes used instead of the methods previously outlined. The simple payback is defined as initial investment divided by annual savings after taxes. The simple payback method does not take into account the effect of interest or escalation rate.

Since the payback period is relatively simple to calculate, and due to the fact managers wish to recover their investment as rapidly as possible, the payback method is frequently used.

It should be used in conjunction with other decision-making tools. When used by itself as the principal criterion, it may result in choosing less profitable investments which yield high initial returns for short periods as compared with more profitable investments which provide profits over longer periods of time.

Example Problem 2-1
An electrical energy audit indicates electrical motor consumption is 4×10^6 kWh per year. By upgrading the motor spares with high efficiency motors, a 10% savings can be realized. The additional cost for these motors is estimated at $80,000. Assuming an 8¢ per kWh energy charge and 20-year life, is the expenditure justified based on a minimum rate of

return of 20% before taxes? Solve the problem using the present worth, annual cost, and rate-of-return methods.

Analysis

Present Worth Method

	Alternate 1 Present Method	Alternate 2 Use High Efficiency Motor Spares
(1) First Cost (P)	–	$80,000
(2) Annual Cost (A)	$4 \times 106 \times .08$	$.9 \times \$320,000$
	= $320,000	= $288,000
USPW (Table 15-4)	4.87	4.87
(3) $A \times 4.87 =$	$1,558,400	$1,402,560
Present Worth	$1,558,400	$1,482,560
(1) + (3)		Choose Alternate with Lowest Present Worth Cost

Annual Cost Method

	Alternate I	Alternate 2
(1) First Cost (P)	–	$80,000
(2) Annual Cost (A)	$320,000	$288,000
CR (Table 15-4)	.2	.2
(3) $P \times .2$	–	$16,000
Annual Cost	$320,000	$304,000
(2) + (3)		Choose Alternate with Lowest Annual Cost

Rate of Return Method

$$P = (\$320,000 - \$288,000)$$

$$P/A = \frac{80,000}{32,000} = 2.5$$

What value of i will make $P/A = 2.5$? $i = 40\%$ (Table 15-7).

Example Problem 2-2

Show the effect of 10% escalation on the rate-of-return analysis given the

Energy equipment investment	=	$20,000
After tax savings	=	$ 2,600
Equipment life (n)	=	15 years

Analysis
   Without escalation

$$CR = \frac{A}{P} = \frac{2{,}600}{20{,}000} = .13$$

From Table 15-1, the rate of return in 10%.
   With 10% escalation assumed:

$$GPW = \frac{P}{A} = \frac{20{,}000}{2{,}600} = 7.69$$

From Table 15-11, the rate of return is 21%.

   Thus we see that taking into account a modest escalation rate can dramatically affect the justification of the project.

TAX CONSIDERATIONS

Depreciation
   Depreciation affects the "accounting procedure" for determining profits and losses and the income tax of a company. In other words, for tax purposes the expenditure for an asset such as a pump or motor cannot be fully expensed in its first year. The original investment must be charged off for tax purposes over the useful life of the asset. A company wishes to expense an item as quickly as possible.
   The Internal Revenue Service allows several methods for determining the annual depreciation rate.
   Straight-Line Depreciation: The simplest method is referred to as a straight-line depreciation and is defined as

$$D = \frac{P - L}{n} \hspace{3cm} Formula\ (2\text{-}8)$$

Where
   D is the annual depreciation rate
   L is the value of equipment at the end of its useful life, commonly
       referred to as salvage value

n is the life of the equipment which is determined by Internal
      Revenue Service Guidelines
P is the initial expenditure.

Sum-of-Years Digits: Another method is referred to as the sum-of
years digits. In this method the depreciation rate is determined by finding
the sum of digits using the following formula:

$$N = n \, \frac{n+1}{2}$$

Formula (2-9)

Where n is the life of equipment.

Each year's depreciation rate is determined as follows:

First year
$$D = \frac{n}{N} \, (P - L)$$
Formula (2-10)

Second year
$$D = \frac{n-1}{N} \, (P - L)$$
Formula (2-11)

n year
$$D = \frac{1}{N} \, (P - L)$$
Formula (2-12)

Declining-Balance Depreciation: The declining-balance method allows
for larger depreciation charges in the early years, which is sometimes
referred to as fast write-off.

The rate is calculated by taking a constant percentage of the declining
undepreciated balance. The most common method used to calculate the declining balance is to predetermine the depreciation rate. Under certain circumstances a rate equal to 200% of the straight-line depreciation rate may
be used. Under other circumstances the rate is limited to 1-1/2 or 1-1/4
times as great as straight-line depreciation. In this method the salvage value
or undepreciated book value is established once the depreciation rate is pre-
established.

To calculate the undepreciated book value, Formula 2-13 is used:

$$D = 1 - \left(\frac{L}{P}\right)^{1/N}$$

Formula (2-13)

Where
    D is the annual depreciation rate
    L is the salvage value
    P is the first cost

Example Problem 2-3

Calculate the depreciation rate using the straight-line, sum-of-years digit, and declining-balance methods.
    Salvage value is 0.
    $n = 5$ years
    $P = 150,000$
    For declining balance use a 200% rate.

Straight-Line Method.

$$D = \frac{P - L}{n} = \frac{150,000}{5} = \$30,000 \text{ per year}$$

Sum-of-Years Digits

$$N = \frac{n(n + 1)}{2} = \frac{5(6)}{2} = 15$$

$$D_1 = \frac{n}{N}(P) = \frac{5}{15}(150,000) = 50,000$$

N	P
1 =	$54,000
2 =	40,000
3 =	30,000
4 =	20,000
5 =	10,000

$$D = 2 \times 20\% = 40\%$$

Declining-Balance Method
(Straight-Line Depreciation Rate = 20%)

Year	*Undepreciated Balance At Beginning of Year*	*Depreciation Charge*
1	150,000	60,000
2	90,000	36,000
3	54,000	21,600
4	32,400	12,960
5	19,440	7,776
	TOTAL	138,336

Undepreciated Book Value (150,000 − 138,336) = $11,664

Cogeneration Equipment Depreciation

Most cogeneration equipment is depreciated over a 15- or 20-year period, depending on the particular type of equipment involved, using the 150% declining balance method switching to straight-line to maximize deductions. Gas and combustion turbine equipment used to produce electricity for sale is depreciated over a 15-year period. Equipment used in the steam power production of electricity for sale (including combustion turbines operated in combined cycle with steam units), as well as assets used to produce steam for sale, are normally depreciated over a 20-year period.

However, most electric and steam generation equipment owned by a taxpayer and producing electric or thermal energy for use by the taxpayer in its industrial process and plant activity, and not ordinarily for sale to others, is depreciated over a 15-year period. Electrical and steam transmission and distribution equipment will be depreciated over a 20-year period at the same 150 percent declining balance rate.

Energy Efficiency Equipment
and Real Property Depreciation

Energy conservation equipment, still classified as real property, is depreciated on a straight line basis over a recovery period. Equipment installed in connection with residential real property qualifies for a 27-1/2-year period, while equipment placed in nonresidential facilities is subject to a 31-1/2-year period. Other real property assets are depreciated over the above period, depending on their residential or nonresidential character.

After-Tax Analysis

Tax-deductible expenses such as maintenance, energy, operating costs, insurance and property taxes reduce the income subject to taxes.

For the after-tax life-cycle cost analysis and payback analysis, the actual incurred annual savings is given as follows:

$$AS = (1 - I) E + ID \qquad \qquad Formula\ (2\text{-}14)$$

Where

AS = yearly annual after-tax savings (excluding effect of tax credit)

E = yearly annual energy savings (difference between original expenses and expenses after modification)

D = annual depreciation rate

I = income tax bracket

Formula 2-14 takes into account that the yearly annual energy savings is partially offset by additional taxes which must be paid due to reduced operating expenses. On the other hand, the depreciation allowance reduces taxes directly.

To compute a rate of return which accounts for taxes, depreciation, escalation, and tax credits, a cash-flow analysis is usually required. This method analyzes all transactions including first and operating costs. To determine the after-tax rate of return, a trial and error or computer analysis is required.

The Present Worth factors tables in Chapter 15, Appendix, can be used for this analysis. All money is converted to the present assuming an interest rate. The summation of all present dollars should equal zero when the correct interest rate is selected, as illustrated in Figure 2-2.

This analysis can be made assuming a fuel escalation rate by using the Gradient Present Worth interest of the Present Worth Factor.

Example Problem 2-4

Comment on the after-tax rate of return for the installation of a heat-recovery system given the following:

- First Cost $100,000
- Year Savings   36,363
- Straight-line depreciation life and equipment life of 5 years

- Income tax bracket 34%

Year	1 Investment	2 Tax Credit	3 After Tax Savings (AS)	4 Single Payment Present Worth Factor	(2 + 3) x 4 Present Worth
0	–P				–P
1		+TC	AS_1	$SPPW_1$	$+P_1$
2			AS_2	$SPPW_2$	P_2
3			AS_3	$SPPW_3$	P_3
4			AS_4	$SPPW_4$	P_4
Total					ΣP

$$AS = (1 - I)\, E + ID$$
Trial & Error Solution:
Correct i when $\Sigma P = 0$

Figure 2-2. Cash Flow Rate of Return Analysis

Analysis

$D = 100,000/5 = 20,000$

$AS = (1 - I)\,E + ID = .66(36,363) + .34(20,000) = \$30,800$

First Trial $i = 20\%$

Investment	After Tax Savings	SPPW 20%	PW
0–100,000			–100,000
1	30,800	.833	25,656
2	30,800	.694	21,375
3	30,800	.578	17,802
4	30,800	.482	14,845
5	30,800	.401	12,350
			$\Sigma - 7,972$

Since summation is negative a higher present worth factor is required. Next try is 15%.

Investment	After Tax Savings	SPPW 15%	PW
0–100,000			–100,000
1	30,800	.869	+ 26,765
2	30,800	.756	+ 23,284
3	30,800	.657	+ 20,235
4	30,800	.571	+ 17,586
5	30,800	.497	+ <u>15,307</u>
			+ 3,177

Since rate of return is bracketed, linear interpolation will be used.

$$\frac{3177 + 7972}{-5} = \frac{3177 - 0}{15 - i\%}$$

$$i = \frac{3177}{2229.6} + 15 = 16.4\%$$

Impact of Fuel Inflation on Life-Cycle Costing

As illustrated by Problem 2-2, a modest estimate of fuel inflation has a major impact on improving the rate of return on investment of the project. The problem facing the energy engineer is how to forecast what the future of energy costs will be. All too often no fuel inflation is considered because of the difficulty of projecting the future. In making projections the following guidelines may be helpful:

- Is there a rate increase that can be forecast based on new nuclear generating capacity? In locations such as Georgia, California, and Arizona electric rates will rise at a faster rate due to commissioning of new nuclear plants and rate increases approved by the Public Service Commission of that state.
- What has been the historical rate increase for the facility? Even with fluctuations there are likely to be trends to follow.
- What events on a national or international level would impact on your costs? New state taxes, new production quotas by OPEC and other factors will affect your fuel prices.
- What do the experts say? Energy economists, forecasting services, and your local utility projections all should be taken into account.

The rate of return on investment becomes more attractive when life-cycle costs are taken into account. Tables 15-9 through 15-12 can be used to show the impact of fuel inflation on the decision-making process.

Example Problem 2-5

Develop a set of curves that indicate the capital that can be invested to give a rate of return of 15% after taxes for each $1000 saved for the following conditions:

1. The effect of escalation is not considered.
2. A 5% fuel escalation is considered.
3. A 10% fuel escalation is considered.
4. A 14% fuel escalation is considered.
5. A 20% fuel escalation is considered.

Calculate for 5-, 10-, 15-, 20-year life.

Assume straight-line depreciation over useful life, 34% income tax bracket, and no tax credit.

Answer

$$AS = (1 - I)\,E + ID$$
$$I = 0.34,\ E = \$1000$$
$$AS = 660 + \frac{0.34P}{N}$$

Thus, the after-tax savings (AS) are composed of two components. The first component is a uniform series of $660 escalating at e percent/year. The second component is a uniform series of $0.34P/N$.

Each component is treated individually and converted to present day values using the GPW factor and the USPW factor, respectively. The sum of these two present worth factors must equal P. In the case of no escalation, the formula is

$$P = 660\,P/A + \frac{0.34P}{N}\,P/A$$

In the case of escalation

$$P = 660\,GPW + \frac{0.34P}{N}\,P/A$$

Since there is only one unknown, the formulas can be readily solved. The results are indicated below.

	$N = 5$ $\$P$	$N = 10$ $\$P$	$N = 15$ $\$P$	$N = 20$ $\$P$
$e = 0$	2869	4000	4459	4648
$e = 10\%$	3753	6292	8165	9618
$e = 14\%$	4170	7598	10,676	13,567
$e = 20\%$	4871	10,146	16,353	23,918

Figure 2-3 illustrates the effects of escalation. This figure can be used as a quick way to determine after-tax economics of energy utilization expenditures.

Example Problem 2-6

It is desired to have an after-tax savings of 15%. Comment on the investment that can be justified if it is assumed that the fuel rate escalation should not be considered and the annual energy savings is $2000 with an equipment economic life of 10 years.

Comment on the above, assuming a 10% fuel escalation.

Answer

From Figure 2-3, for each $1000 energy savings, an investment of $3600 is justified or $8000 for a $2000 savings for which no fuel increase is accounted.

With a 10% fuel escalation rate on investment of $6300 justified for each $1000 energy savings, $12,600 can be justified for $2000 savings. Thus, a 57% higher expenditure is economically justifiable and will yield the same after tax rate of return of 15% when a fuel escalation of 10% is considered.

Figure 2-3. Effects of Escalation
On Investment Requirements

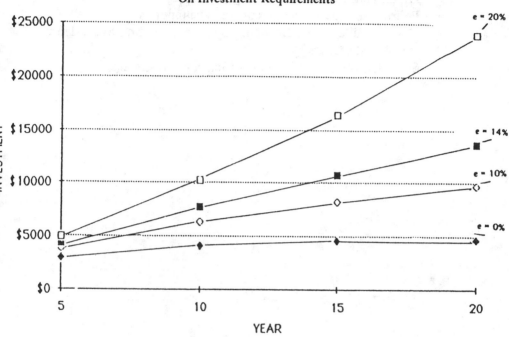

Note: Maximum investment in order to attain a 15% after-tax rate of return on investment for annual savings of $1000.

COMPUTER ANALYSIS

The Alliance to Save Energy, 1925 K Street, NW, Suite 206, Washington, DC 20006, has introduced an investment analysis software package, ENVEST, which costs only $75 for 5-1/4-inch disc and $85 for 3-1/2-inch hard drive disc and includes a 170-page user manual and 30 days of telephone support. The program can be run on an IBM PC, PCXT, PCAT with 256K ram. The program enables the user to

• Generate spreadsheets and graphs showing the yearly cash flows from any energy-related investment.

- Compute payback, internal rate of return, and other important investment measures.
- Experiment with differing energy price projections.
- Perform sensitivity analysis on key assumptions.
- Compare alternative financing options, including loans, leases, and shared savings.
- Store data on over 100 energy efficiency investments.

3

Energy Auditing and Accounting

TYPES OF ENERGY AUDITS

The simplest definition for an energy audit is as follows: An energy audit serves the purpose of identifying where a building or plant facility uses energy and identifies energy conservation opportunities.

There is a direct relationship to the cost of the audit (amount of data collected and analyzed) and the number of energy conservation opportunities to be found. Thus, a first distinction is the cost of the audit which determines the type of audit to be performed.

The second distinction is the type of facility. For example, a building audit may emphasize the building envelope, lighting, heating, and ventilation requirements. On the other hand, an audit of an industrial plant emphasizes the process requirements.

Most energy audits fall into three categories or types, namely, Walk-Through, Mini-Audit, or Maxi-Audit.

Walk-Through—This type of audit is the least costly and identifies preliminary energy savings. A visual inspection of the facility is made to determine maintenance and operation energy saving opportunities plus collection of information to determine the need for a more detailed analysis.

Mini-Audit—This type of audit requires tests and measurements to quantify energy uses and losses and determine the economics for changes.

Maxi-Audit—This type of audit goes one step further than the mini-audit. It contains an evaluation of how much energy is used for each function such as lighting, process, etc. It also requires a model analysis, such as a computer simulation, to determine energy use patterns and predictions on a year-round basis, taking into account such variables as weather data.

As noted in the audit definition, there are two essential parts, namely, data acquisition and data analysis.

Data Acquisition

This phase requires the accumulation of utility bills, establishing a baseline to provide historical documentation and a survey of the facility.

All energy flows should be accounted for; thus all "energy in" should equal "energy out." This is referred to as an energy balance.

All energy costs should be determined for each fuel type. The energy survey is essential. Instrumentation commonly used in conducting a survey is discussed at the conclusion of the chapter.

Data Analysis

As a result of knowing how energy is used, a complete list of "Energy Conservation Opportunities" (ECOs) will be generated. The life-cycle costing techniques presented in Chapter 2 will be used to determine which alternative should be given priority.

A very important phase of the overall program is to continuously monitor the facility even after the ECOs have been implemented. Documentation of the cost avoidance or savings is essential to the audit.

Remember, in order to have a continuous ongoing program, *individuals must be made accountable* for energy use. As part of the audit, recommendations should be made as to where to add "root" or submetering.

ENERGY USE PROFILES

The energy audit process for a building emphasizes building envelope, heating and ventilation, air conditioning, plus lighting functions. For an industrial facility the energy audit approach includes process consideration. Figures 3-1 through 3-3 illustrate how energy is used for a typical industrial plant. It is important to account for total consumption, cost, and how energy is used for each commodity such as steam, water, air and natural gas. This procedure is required to develop the appropriate energy conservation strategy.

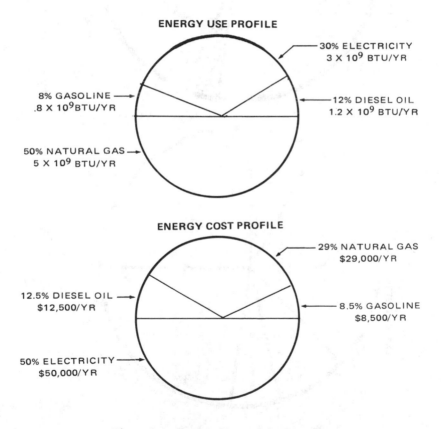

Figure 3-1. Energy Use and Cost Profile

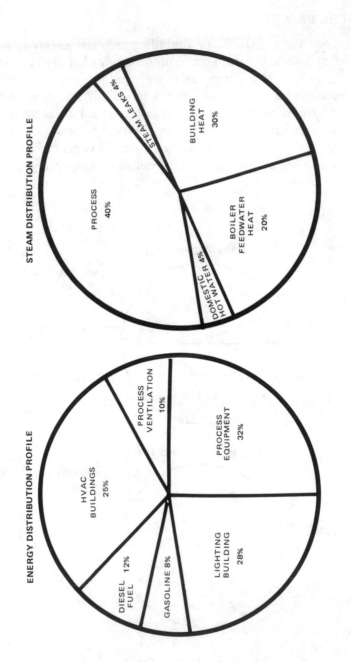

Figure 3-2. Energy Profile by Function

The top portion of Figure 3-1 illustrates how much energy is used by fuel type and its relative percentage. The pie chart below shows how much is spent for each fuel type. Using a pie-chart presentation or nodal flow diagram can be very helpful in visualizing how energy is being used.

Figure 3-2, on the other hand, shows how much of the energy is used for each function such as lighting, process, and building heating and ventilation. Pie charts similar to the right-hand side of the figure should be made for each category such as air, steam, electricity, water and natural gas.

Figure 3-3 illustrates an alternate representation for the steam distribution profile.

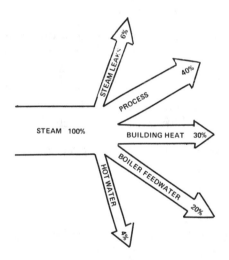

Figure 3-3. Steam Distribution Nodal Diagram

Several audits are required to construct the energy use profiles, such as:

Envelope Audit—This audit surveys the building envelope for losses or gains due to leaks, building construction, doors, glass, lack of insulation, etc.

Functional Audit—This audit determines the amount of energy required for a particular function and identifies energy conservation opportunities. Functional audits include:

- Heating, ventilation and air conditioning
- Building
- Lighting
- Domestic hot water
- Air distribution

Process Audit—This audit determines the amount of energy required for each process function and identifies energy conservation opportunities. Process functional audits include:

- Process machinery
- Heating, ventilation and air-conditioning process
- Heat treatment
- Furnaces

Transportation Audit—This audit determines the amount of energy required for forklift trucks, cars, vehicles, trucks, etc.

Utility Audit—This audit analyzes the monthly, daily or yearly energy usage for each utility.

ENERGY USERS

Energy use profiles for several end-users are summarized in Tables 3-1 through 3-11.

Table 3-1. Energy Use in Apartment Buildings

	Range (%)	Norms (%)
Environmental Control	50 to 80	70
Lighting and Wall Receptacles	10 to 20	15
Hot Water	2 to 5	3
Special Functions		
Laundry, Swimming Pool, Restaurants,		
Parking, Elevators, Security Lighting	5 to 20	10

Table 3-2. Energy Use in Bakeries

Housekeeping Energy	Percent
Space Heating	21.5
Air Conditioning	1.6
Lighting	1.4
Domestic Hot Water	1.8
TOTAL	26.3
Process Energy	Percent
Baking Ovens	49.0
Pan Washing	10.6
Mixers	4.1
Freezers	3.3
Cooking	2.0
Fryers	1.8
Proof Boxes	1.8
Other Processes	1.1
TOTAL	73.7

Data are for a 27,000-square-foot bakery in
Washington, D.C.

Table 3-3. Energy Use in Die Casting Plants

Housekeeping Energy	Percent
Space Heating	24
Air Conditioning	2
Lighting	2
Domestic Hot Water	2
TOTAL	30
Process Energy	Percent
Melting Hearth	30
Quiet Pool	20
Molding Machines	10
Air Compressors	5
Other Processes	5
TOTAL	70

Table 3-4. Energy Use in Hospital Buildings

	Range (%)	Norms (%)
Environmental Control	40 to 65	58
Lighting and Wall Receptacles	10 to 20	15
Laundry	8 to 15	12
Food Service, Kitchen Operations	5 to 10	7
Medical Equipment, Sterilization, Incinerator, Parking, Elevators, Security Lighting	5 to 15	8

Table 3-5. Energy Use in Hotels and Motels

	Range (%)	Norms (%)
Space Heating	45 to 70	60
Lighting	5 to 15	11
Air Conditioning	3 to 15	10
Refrigeration	0 to 10	4
Special Functions	5 to 20	15
Laundry, Kitchen, Restaurant,		
Swimming Pool, Garage,		
Security Lighting, Hot Water		

Table 3-6. Energy Use in Retail Stores

	Range (%)	Norms (%)
HVAC	20 to 50	30
Lighting	40 to 75	60
Special Functions	5 to 20	10
Elevators, General Power, Parking		
Security Lighting, Hot Water		

Table 3-7. Energy Use in Restaurants

	Table Restaurant Norms (%)	Fast Food Restaurant Norms (%)
HVAC	32	36
Lighting	8	26
Special Functions		
Food Preparation	45	27
Food Storage	2	6
Sanitation	12	1
Other	1	4

Table 3-8. Energy Use in Schools

	Range (%)	Norms (%)
Environmental Control	45 to 80	65
Lighting and Wall Receptacles	10 to 20	15
Food Service	5 to 10	7
Hot Water	2 to 5	3
Special Functions	0 to 20	10

Table 3-9. Energy Use in Transportation Terminals

	Range (%)	Norms (%)
Space Heating	50 to 75	60
Lighting	5 to 25	15
Air Conditioning	5 to 25	15
Special Functions	3 to 20	10
Elevators, General Power,		
Parking, Security Lighting,		
Hot Water		

Table 3-10. Energy Use in Warehouses and Storage Facilities

(Vehicles Not Included)	Range (%)	Norms (%) *
Space Heating	45 to 80	67
Air Conditioning	3 to 10	6
Lighting	4 to 12	7
Refrigeration	0 to 40	12
Special Functions	5 to 15	8
Elevators, General Power,		
Parking, Security Lighting,		
Hot Water		

* Norms for a warehouse or storage facility are strongly dependent on the products and their specific requirements for temperature and humidity control.

Table 3-11. Comparative Energy Use by System

		Heating & Ventilation	Cooling & Ventilation	Lighting	Power & Process	Domestic Hot Water
Schools	A	4	3	1	5	—
	B	1	4	2	5	3
	C	1	4	2	5	3
Colleges	A	5	2	1	4	3
	B	1	3	2	5	4
	C	1	5	2	4	3
Office Bldg.	A	3	1	2	4	5
	B	1	3	2	4	5
	C	1	3	2	4	5
Commercial Stores	A	3	1	2	4	5
	B	2	3	1	4	5
	C	1	3	2	4	5
Religious Bldg.	A	3	2	1	4	5
	B	1	3	2	4	5
	C	1	3	2	4	5
Hospitals	A	4	1	2	5	3
	B	1	3	4	5	2
	C	1	5	3	4	2

Climatic Zone A: Fewer than 2500 degree days
Climate Zone B: 2500—5500 degree days
Climate Zone C: 5500—9500 degree days

Source: Guidelines For Saving Energy In Existing Buildings ECM—1

Note: Numbers indicate energy consumption relative to each other
(1) greatest consumption
(5) least consumption

ENERGY USE IN BUILDINGS[1]

More than 36% of the nation's primary energy was consumed in providing services to residential and commercial buildings in 1985 (26.9 quads). The residential sector consumed 15.3 quads; the commercial sector (including public buildings) consumed 11.6 quads.

The largest end use is space heating in both residential (40%) and commercial (34%) buildings. Water heating is second in residences (17%), followed by refrigerators (9%), air conditioning (7%), lighting (7%), and kitchen ranges and ovens (6%). In commercial buildings, lighting is second (25%), followed by ventilation (12%), air conditioning (10%) and water heating (6%). (See Table 3-12.)

Table 3-12. Building Sector Energy Use

BUILDING SECTOR ENERGY USE BY END USE 1985-2010 FORECAST		
	1985	2010
Space Heating	38%	34%
Air Conditioning and Ventilation	15	16
Lighting	15	15
Water Heating	12	12
Refrigeration	7	5
Other	13	18

The forecast changes as the type of energy used becomes more significant. The analysis estimates a major increase in the use of electricity by the building sector, increasing from 17.0 quads in 1985 to 26.8 quads in 2010. These figures represent the source (primary) energy used to generate electricity.

During this period, natural gas use is forecast to increase slightly from 7.0 quads to 7.2 quads. The use of solar and renewable energy is expected to more than double from 1.2 quads to 2.9 quads; the use of oil is expected to remain flat at 2.6 quads.

[1] Source: Energy Conservation Goals for Buildings, A Report to the Congress of The United States, May 1988.

THE ENERGY SURVEY

As part of the data acquisition phase, a detailed survey should be conducted. The various types of instrumentation commonly used in the survey are discussed in this section.

Infrared Equipment

Some companies may have the wrong impression that infrared equipment can meet most of their instrumentation needs.

The primary use of infrared equipment in an energy utilization program is to detect building or equipment losses. Thus it is just one of the many options available.

Several energy managers find infrared in use in their plant prior to the energy utilization program. Infrared equipment, in many instances, was purchased by the electrical department and used to detect electrical hot spots.

Infrared energy is an invisible part of the electromagnetic spectrum. It exists naturally and can be measured by remote heat-sensing equipment. Within the last four years lightweight portable infrared systems became available to help determine energy losses. Differences in the infrared emissions from the surface of objects cause color variations to appear on the scanner. The hotter the object, the more infrared radiated. With the aid of an isotherm circuit, the intensity of these radiation levels can be accurately measured and quantified. In essence the infrared scanning device is a diagnostic tool which can be used to determine building heat losses. Equipment costs range from $400 to $25,000.

An overview energy scan of the plant can be made through an aerial survey using infrared equipment. Several companies offer aerial scan services starting at $1500. Aerial scans can determine underground stream pipe leaks, hot gas discharges, leaks, etc.

Since IR detection and measurement equipment have gained increased importance in the energy audit process, a summary of the fundamentals is reviewed in this section.

The visible portion of the spectrum runs from .4 to .75 micrometers (μm). The infrared or thermal radiation begins at this point and extends to approximately 1000 μm. Objects such as people, plants, or buildings will emit radiation with wavelengths around 10 μm. (See Figure 3-4.)

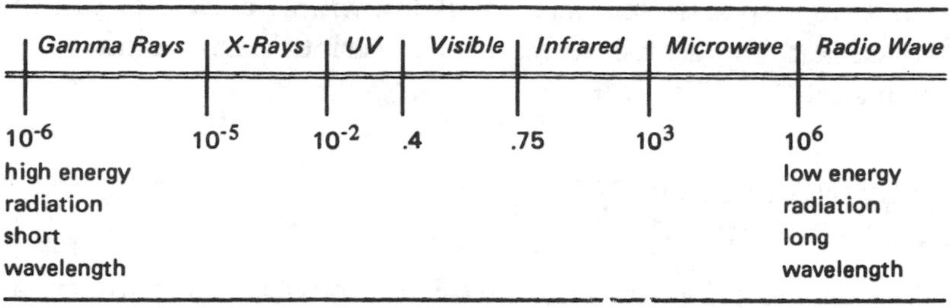

Gamma Rays	X-Rays	UV	Visible	Infrared	Microwave	Radio Wave

10^{-6} 10^{-5} 10^{-2} .4   .75 10^{3} 10^{6}

high energy   low energy
radiation   radiation
short   long
wavelength   wavelength

Figure 3-4. Electromagnetic Spectrum

Infrared instruments are required to detect and measure the thermal radiation. To calibrate the instrument, a special "black body" radiator is used. A black body radiator absorbs all the radiation that impinges on it and has an absorbing efficiency or emissivity of 1.

The accuracy of temperature measurements by infrared instruments depends on the three processes which are responsible for an object acting like a black body. These processes—absorbed, reflected, and transmitted radiation—are responsible for the total radiation reaching an infrared scanner.

The real temperature of the object is dependent only upon its emitted radiation.

Corrections to apparent temperatures are made by knowing the emissivity of an object at a specified temperature.

The heart of the infrared instrument is the infrared detector. The detector absorbs infrared energy and converts it into electrical voltage or current. The two principal types of detectors are the thermal and photo type. The thermal detector generally requires a given period of time to develop an image on photographic film. The photo detectors are more sensitive and have a higher response time. Television-like displays on a cathode ray tube permit studies of dynamic thermal events on moving objects in real time.

There are various ways of displaying signals produced by infrared detectors. One way is by use of an isotherm contour. The lightest areas of the picture represent the warmest areas of the subject, and the darkest areas represent the coolest portions. These instruments can show thermal variations of less than 0.1°C and can cover a range of −30°C to over 2000°C.

The isotherm can be calibrated by means of a black body radiator so that a specific temperature is known. The scanner can then be moved and the temperatures of the various parts of the subject can be made.

MEASURING ELECTRICAL
SYSTEM PERFORMANCE

The ammeter, voltmeter, wattmeter, power factor meter, and footcandle meter are usually required to do an electrical survey. These instruments are described below.

Ammeter and Voltmeter

To measure electrical currents, ammeters are used. For most audits, alternating currents are measured. Ammeters used in audits are portable and are designed to be easily attached and removed.

There are many brands and styles of snap-on ammeters commonly available that can read up to 1000 amperes continuously. This range can be extended to 4000 amperes continuously for some models with an accessory step-down current transformer.

The snap-on ammeters can be either indicating or recording with a printout. After attachment, the recording ammeter can keep recording current variations for as long as a full month on one roll of recording paper. This allows the study of current variations in a conductor for extended periods without constant operator attention.

The ammeter supplies a direct measurement of electrical current, which is one of the parameters needed to calculate electrical energy. The second parameter required to calculate energy is voltage, and it is measured by a voltmeter.

Several types of electrical meters can read the voltage or current. A voltmeter measures the difference in electrical potential between two points in an electrical circuit.

In series with the probes are the galvanometer and a fixed resistance (which determine the voltage scale). The current through this fixed resistance circuit is then proportional to the voltage, and the galvanometer deflects in proportion to the voltage.

The voltage drops measured in many instances are fairly constant and need only be performed once. If there are appreciable fluctuations, additional readings or the use of a recording voltmeter may be indicated.

Most voltages measured in practice are under 600 volts and there are many portable voltmeter/ammeter clamp-ons available for this and lower ranges.

Wattmeter and Power Factor Meter

The portable wattmeter can be used to indicate by direct reading electrical energy in watts. It can also be calculated by measuring voltage, current and the angle between them (power factor angle).

The basic wattmeter consists of three voltage probes and a snap-on current coil which feeds the wattmeter movement.

The typical operating limits are 300 kilowatts, 650 volts, and 600 amperes. It can be used on both one- and three-phase circuits.

The portable power factor meter is primarily a three-phase instrument. One of its three voltage probes is attached to each conductor phase and a snap-on jaw is placed about one of the phases. By disconnecting the wattmeter circuitry, it will directly read the power factor of the circuit to which it is attached.

It can measure power factor over a range of 1.0 leading to 1.0 lagging with "ampacities" up to 1500 amperes at 600 volts. This range covers the large bulk of the applications found in light industry and commerce.

The power factor is a basic parameter whose value must be known to calculate electric energy usage. Diagnostically, it is a useful instrument to determine the sources of poor power factor in a facility.

Portable digital kWh and kW demand units are now available.

Digital read-outs of energy usage in both kWh and kW demand or in dollars and cents, including instantaneous usage, accumulated usage, projected usage for a particular billing period, alarms when over-target levels are desired for usage, and control-outputs for load-shedding and cycling are possible.

Continuous displays or intermittent alternating displays are available at the touch of a button for any information needed such as the cost of operating a production machine for one shift, one hour or one week.

Footcandle Meter

Footcandle meters measure illumination in units of footcandles through a light-sensitive barrier layer of cells contained within them. They are usually pocket-size and portable and are meant to be used as field instruments to survey levels of illumination. Footcandle meters differ from conventional photographic lightmeters in that they are color and cosine corrected.

TEMPERATURE MEASUREMENTS

To maximize system performance, knowledge of the temperature of a fluid, surface, etc. is essential. Several types of temperature devices are described in this section.

Thermometer

There are many types of thermometers that can be used in an energy audit. The choice of what to use is usually dictated by cost, durability, and application.

For air-conditioning, ventilation and hot-water service applications (temperature ranges 50°F to 250°F), a multipurpose portable battery-operated thermometer is used. Three separate probes are usually provided to measure liquid, air or surface temperatures.

For boiler and oven stacks (1000°F) a dial thermometer is used. Thermocouples are used for measurements above 1000°F.

Surface Pyrometer

Surface pyrometers are instruments which measure the temperature of surfaces. They are somewhat more complex than other temperature instruments because their probe must make intimate contact with the surface being measured.

Surface pyrometers are of immense help in assessing heat losses through walls and also for testing steam traps.

They may be divided into two classes: low-temperature (up to 250°F) and high-temperature (up to 600°F to 700°F). The low-temperature unit is usually part of the multipurpose thermometer kit. The high-temperature unit is more specialized but needed for evaluating fired units and general steam service.

There are also noncontact surface pyrometers which measure infrared radiation from surfaces in terms of temperature. These are suitable for general work and also for measuring surfaces which are visually but not physically accessible.

A more specialized instrument is the optical pyrometer. This is for high-temperature work (above 1500°F) because it measures the temperature of bodies which are incandescent because of their temperature.

Psychrometer

A psychrometer is an instrument which measures relative humidity based on the relation of the dry-bulb temperature and the wet-bulb temperature.

Relative humidity is of prime importance in HVAC and drying operations. Recording psychrometers are also available. Above 200°F humidity studies constitute a specialized field of endeavor.

Portable Electronic Thermometer

The portable electronic thermometer is an adaptable temperature measurement tool. The battery-powered basic instrument, when housed in a carrying case, is suitable for laboratory or industrial use.

A pocket-size digital, battery-operated thermometer is especially convenient for spot checks or where a number of rapid readings of process temperatures need to be taken.

Thermocouple Probe

No matter what sort of indicating instrument is employed, the thermocouple used should be carefully selected to match the application and properly positioned if a representative temperature is to be measured. The same care is needed for all sensing devices—thermocouple, bimetals, resistance elements, fluid expansion, and vapor pressure bulbs.

Suction Pyrometer

Errors arise if a normal sheathed thermocouple is used to measure gas temperatures, especially high ones. The suction pyrometer overcomes these by shielding the thermocouple from wall radiation and drawing gases over it at high velocity to ensure good convective heat transfer. The thermocouple thus produces a reading which approaches the true temperature at the sampling point rather than a temperature between that of the walls and the gases.

MEASURING COMBUSTION SYSTEMS

To maximize combustion efficiency, it is necessary to know the composition of the flue gas. By obtaining a good air-fuel ratio, substantial energy will be saved.

Combustion Tester

Combustion testing consists of determining the concentrations of the products of combustion in a stack gas. The products of combustion usually considered are carbon dioxide and carbon monoxide. Oxygen is tested to assure proper excess air levels.

The definitive test for these constituents is an Orsat apparatus. This test consists of taking a measured volume of stack gas and measuring successive volumes after intimate contact with selective absorbing solutions. The reduction in volume after each absorption is the measure of each constituent.

The Orsat has a number of disadvantages. The main ones are that it requires considerable time to set up and use and that its operator must have a good degree of dexterity and be in constant practice.

Instead of an Orsat, there are portable and easy to use absorbing instruments which can easily determine the concentrations of the constituents of interest on an individual basis. Setup and operating times are minimal and just about anyone can learn to use them.

The typical range of concentrations are CO_2, 0-20%; O_2, 0-21%; and CO, 0-0.5%. The CO_2 or O_2 content, along with knowledge of flue gas temperature and fuel type, allows the flue gas loss to be determined off standard charts.

Boiler Test Kit

The boiler test kit contains the following:

CO_2 Gas analyzer
O_2 Gas analyzer
Inclined monometer
CO Gas analyzer.

The purpose of the components of the kit is to help evaluate fireside boiler operation. Good combustion usually means high carbon dioxide (CO_2), low oxygen (O_2), and little or no trace of carbon monoxide (CO).

Gas Analyzers

The gas analyzers are usually of the Fyrite type. The Fyrite type differs from the Orsat apparatus in that it is more limited in application and less accurate. The chief advantages of the Fyrite are that it is simple and easy to use and is inexpensive. This device is used many times in an energy audit. Three readings using the Fyrite analyzer should be made and the results averaged.

Draft Gauge

The draft gauge is used to measure pressure. It can be the pocket type or the inclined monometer type.

Smoke Tester

To measure combustion completeness the smoke detector is used. Smoke is unburned carbon, which wastes fuel, causes air pollution, and fouls heat-exchanger surfaces. To use the instrument, a measured volume of flue gas is drawn through filter paper with the probe. The smoke spot is compared visually with a standard scale and a measure of smoke density is determined.

Combustion Analyzer

The combustion electronic analyzer permits fast, close adjustments. The unit contains digital displays. A standard sampler assembly with probe allows for stack measurements through a single stack or breaching hole.

MEASURING HEATING, VENTILATION
AND AIR-CONDITIONING (HVAC)
SYSTEM PERFORMANCE

Air Velocity Measurement

The following suggests the preference, suitability, and approximate costs of particular equipment:

- *Smoke pellets*—limited use but very low cost. Considered to be useful if engineering staff has experience in handling.
- *Anemometer* (deflecting vane)—good indication of air movement with acceptable order of accuracy. Considered useful (approximately $50).
- *Anemometer* (revolving vane)—good indicator of air movement with acceptable accuracy. However, easily subject to damage. Considered useful (approximately $100).
- *Pitot tube*—a standard air measurement device with good levels of accuracy. Considered essential. Can be purchased in various lengths—12" about $20, 48" about $35. Must be used with a monometer. These vary considerably in cost, but could be on the order of $20 to $60.
- *Impact tube*—usually packaged air flow meter kits, complete with various jets for testing ducts, grills, open areas, etc. These units are convenient to use and of sufficient accuracy. The costs vary around $150 to $300, and therefore this order of cost could only be justified for a large system.
- *Heated thermocouple*—these units are sensitive and accurate but costly. A typical cost would be about $500 and can only be justified for regular use in a large plant.
- *Hot wire anemometer*—not recommended. Too costly and too complex.

Temperature Measurement

The temperature devices most commonly used are as follows:
- *Glass thermometers*—considered to be the most useful to temperature measuring instruments—accurate and convenient but fragile. Cost runs from $5 each for 12" long mercury in glass. Engineers should have a selection of various ranges.

- *Resistance thermometers*—considered to be very useful for A/C testing. Accuracy is good and they are reliable and convenient to use. Suitable units can be purchased from $150 up, some with a selection of several temperature ranges.
- *Thermocouples*—similar to resistance thermocouple but do not require battery power source. Chrome-Alum or iron types are the most useful and have satisfactory accuracy and repeatability. Costs start from $50 up.
- *Bimetallic thermometers*—considered unsuitable.
- *Pressure bulb thermometers*—more suitable for permanent installation. Accurate and reasonable in cost—$40 up.
- *Optical pyrometers*—only suitable for furnace settings and therefore limited in use. Cost from $300 up.
- *Radiation pyrometers*—limited in use for A/C work and costs from $500 up.
- *Indicating crayons*—limited in use and not considered suitable for A/C testing—costs around $2/crayon.
- *Thermographs*—use for recording room or space temperature; gives a chart indicating variations over a 12- or 168-hour period. Reasonably accurate. Low cost at around $30 to $60. (Spring-wound drive.)

Pressure Measurement (Absolute and Differential)

Common devices used for measuring pressure in HVAC applications (accuracy, range, application, and limitations are discussed in relation to HVAC work) are as follows:

- *Absolute pressure manometer*—not really suited to HVAC test work.
- *Diaphragm*—not really suited to HVAC test work.
- *Barometer (Hg manometer)*—not really suited to HVAC test work.
- *Micromanometer*—not usually portable, but suitable for fixed measurement of pressure differentials across filter, coils, etc. Cost around $30 up.
- *Draft gauges*—can be portable and used for either direct pressure or pressure differential. From $30 up.
- *Manometers*—can be portable. Used for direct pressure reading and with pitot tubes for air flows. Very useful. Costs from $20 up.

- *Swing Vane gauges*—can be portable. Usually used for air flow. Costs about $30.
- *Bourdon tube gauges*—very useful for measuring all forms of system fluid pressures from 5 psi up. Costs vary greatly, from $10 up. Special types for refrigeration plants.

Humidity Measurement

The data given below indicate the type of instruments available for humidity measurement. The following indicates equipment suitable for HVAC applications:

- *Psychrometers*—basically these are wet and dry bulb thermometers. They can be fixed on a portable stand or mounted in a frame with a handle for revolving in air. Costs are low ($10 to $30) and they are convenient to use.
- *Dewpoint hygrometers*—not considered suitable for HVAC test work.
- *Dimensional change*—device usually consists of a "hair," which changes in length proportionally with humidity changes. Not usually portable, fragile, and only suitable for limited temperature and humidity ranges.
- *Electrical conductivity*—can be compact and portable but of a higher cost (from $200 up). Very convenient to use.
- *Electrolytic*—as above, but for very low temperature ranges. Therefore unsuitable for HVAC test work.
- *Gravemeter*—not suitable.

IDENTIFYING STEAM
AND UTILITY COSTS

Steam and utility costs are significant for most plants. It is important to quantify usage, fuel costs as a function of production. Figure 3-5 illustrates a typical Steam and Utility Cost Report. This report enables the plant manager to evaluate the total Btu's of fuel consumed, the total fuel cost, and the total steam generation cost as a function of production. This report is issued monthly.

Since each plant has the same report, plant to plant comparisons are made and the effectiveness of the energy use is measured.

STEAM AND UTILITY COST REPORT

Preparatory Production [   ] Lbs C & F   Month [   ] 19 [ ]

Tire Production _____   Operating Days [ ]

A. STEAM	Quantities	B. OPERATING CONDITIONS AND DATA IN BOILER PLANT

A. STEAM   Quantities

1. Actual Steam Generated [   ] 000 lbs
   a) Press and Temp ___ psi ___ °F ___ 000 lbs
   b) Press and Temp ___ psi ___ °F ___ 000 lbs
   c) Press and Temp ___ psi ___ °F ___ 000 lbs

2. Factor of Evaporation (ASME Standard)
   a) _____
   b) _____
   c) _____

3. Equivalent Steam (No. 1 x No. 2) [   ] 000 lbs

4. Blowdown [   ] lbs
   Heat Loss in BTU from Blowdown _____ BTU
   BTU/Lb-Gal-Cu Ft

	Total $	Total	
5. Coal	[ ]	[ ]	tons
6. Oil	[ ]	[ ]	gals
7. Gas	[ ]	[ ]	MCF
8. Misc	[ ]	[ ]	Units _____

9. Total BTU of Fuel Consumed [   ] 000,000 BTU

10. Boiler Efficiency (970.3 x No. 3) + No. 4 = [   ] %
    No. 9

B. OPERATING CONDITIONS AND DATA IN BOILER PLANT

Quantity of Feedwater Makeup [   ] 000 (Gal.)

Water Treating Chemicals
a) Phosphate _____ lbs
b) Salt _____ lbs
c) _____ _____ lbs
d) _____ _____ lbs
e) _____ _____ lbs
f) _____ _____ lbs
g) _____ _____ lbs

Ave. Feedwater Temp. _____ °F

Max Steam Demand _____ lbs/hr

Equivalent Steam Consumption Per 100 lbs C&F _____ lbs

Units _____ Fuel _____

	Cost	Unit Cost $/1000 Lbs Equiv. Steam
11. Total Fuel Cost (No. 5 + No. 6 + No. 7 + No. 8)	$ [ ]	$ _____
12. Operating Labor (include wage benefits)	$ [ ]	$ _____
13. Operating Supplies (water, chemicals, etc.)	$ _____	
14. Electricity for Power Plant Auxiliaries _____ KWH	$ _____	
15. Maintenance Charges	$ _____	
16. Other Miscellaneous Charges	$ _____	
17. Total Operating Cost (No. 11 thru No. 16)	$ [ ]	$ _____
18. Total Steam Generation Cost (No. 17 + fixed Cost)	$ [ ]	$ _____

Figure 3-5. Steam and Utility Cost Report

CALCULATING THE ENERGY
CONTENT OF THE PROCESS

Knowing the energy content of the plant's process is the first step in understanding how to reduce its cost. Using energy more efficiently reduces the product cost, thus increasing profits. In order to account for the process energy content, all energy that enters and leaves a plant during a given period must be measured. Figure 3-6 illustrates Energy Content of a Process Report. The report applies to any manufacturing operation, whether it is a pulp mill, steel mill, or assembly line. This report enables one to quickly identify energy inefficient operations.

Attention can then be focused on which equipment should be replaced and what maintenance programs should be initiated. This report also focuses attention on the choice of raw materials. By using Btu's per unit of production, measurable goals can be set. This report will also identify opportunities where energy usage can be reduced. The energy content of raw materials can be estimated by using the heating values indicated in Table 3-13.

Example Problem 3-1

Comment on energy content by modifying process No. 1 as follows:

Monthly Usage Rate

	Process No. 1	*Modified Process No. 1*
Ethane	30,000 lb	50,000 lb
Steam	250×10^6 lb	200×10^6 lb
Electricity	0.5×10^6 kWh	0.8×10^6 kWh
Natural gas	350×10^6 ft^3	300×10^6 ft^3

Assuming a Btu content of steam of 1077 Btu/lb, compute the net energy content per process.

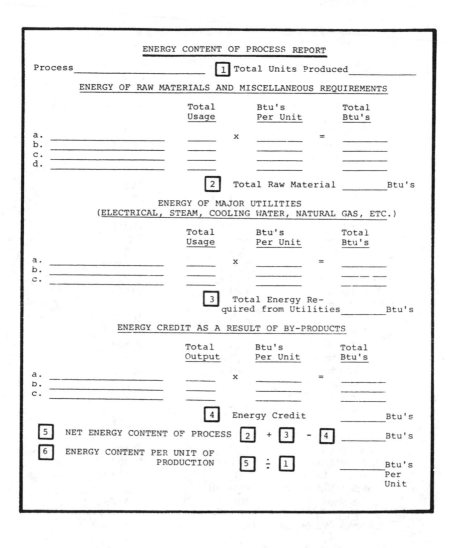

Figure 3-6. Energy Content of Process Report

Table 3-13. Heat of Combustion for Raw Materials

	Formula	Gross Heat of Combustion Btu/lb
Raw Material		
Carbon	C	14,093
Hydrogen	H_2	61,095
Carbon monoxide	CO	4,347
Paraffin Series		
Methane	CH_4	23,875
Ethane	C_2H_4	22,323
Propane	C_3H_8	21,669
n-Butane	C_4H_{10}	21,321
Isobutane	C_4H_{10}	21,271
n-Pentane	C_5H_{12}	21,095
Isopentane	C_5H_{12}	21,047
Neopentane	C_5H_{12}	20,978
n-Hexane	C_6H_{14}	20,966
Olefin Series		
Ethylene	C_2H_4	21,636
Propylene	C_3H_6	21,048
n-Butene	C_4H_8	20,854
Isobutene	C_4H_8	20,737
n-Pentene	C_5H_{10}	20,720
Aromatic Series		
Benzene	C_6H_6	18,184
Toluene	C_7H_8	18,501
Xylene	C_8H_{10}	18,651
Miscellaneous Gases		
Acetylene	C_2H_2	21,502
Naphthalene	$C_{10}H_8$	17,303
Methyl alcohol	CH_3OH	10,258
Ethyl alcohol	C_2H_5OH	13,161
Ammonia	NH_3	9,667

Source: NBS Handbook 115.

Answer

Process No. 1
Energy of Raw Materials

	Total Usage	Btu's Per Unit	Total Btu's
Ethane	30,000 lb	From Table 3-13 22,323 Btu/lb	0.6×10^9

Energy of Major Utilities

Steam	250×10^6 lb	1077 Btu/lb	269.2×10^9
Electricity	0.5×10^6 kWh	From Table 1-2 10,000 Btu/kWh	5×10^9
Natural gas	350×10^6 ft^3	From Table 1-1 1000 Btu/ft^3	350×10^9
		Net energy content for Process No. 1	624×10^9

Modified Process No. 1
Energy of Raw Materials

Ethane	50,000 lb	From Table 3-13 22,323 Btu/lb	1.1×10^9

Energy of Major Utilities

Steam	200×10^6 lb	1077 Btu/lb	215.4×10^9
Electricity	0.8×10^6 lb	From Table 1-2 10,000 Btu/kWh	8×10^9
Natural gas	300×10^6 ft^3	From Table 1-1 1000 Btu/ft^3	300×10^9
		Net energy content for Process No. 1	524×10^9

Modified Process No. 1 saves 100×10^9 Btu's per month.

4

Electrical System Optimization

This chapter is divided into three main sections: *Power Distribution, Lighting Efficiency, and Energy Management Systems.*

By understanding the basic concepts of electrical systems it is possible to reduce energy consumption 25% or more. The first step is to analyze the billing structure. It may be possible to negotiate a better tariff rate with the local utility or modify the facility operation to qualify for a lower rate. In addition, specified charges or discounts for power factor, time of day or demand will determine if certain electrical efficiency measures are economically justified. This chapter reviews the basic parameters required to make sound energy engineering decisions.

POWER DISTRIBUTION

ELECTRICAL RATE TARIFF

The basic electrical rate charges contain the following elements:

Billing Demand—The maximum kilowatt requirement over a 15-, 30-, or 60-minute interval.

Load Factor—The ratio of the average load over a designated period to the peak demand load occurring in that period.

Power Factor—The ratio of resistive power to apparent power. Traditionally electrical rate tariffs have a decreasing kilowatt hour (kWh) charge with usage. This practice is likely to gradually phase out. New tariffs are containing the following elements:

Time of Day—Discounts are allowed for electrical usage during off-peak hours.

Ratchet Rate—The billing demand is based on 80-90% of peak demand for any one month. The billing demand will remain at that ratchet for 12 months even though the actual demand for the succeeding months may be less.

THE POWER TRIANGLE

The total power requirement of a load is made up of two components, namely the resistive part and the reactive part. The resistive portion of a load cannot be added directly to the reactive component since it is essentially 90 degrees out of phase with the other. The pure resistive power is known as the watt, while the reactive power is referred to as the reactive volt amperes. To compute the total volt ampere load, it is necessary to analyze the power triangle indicated below.

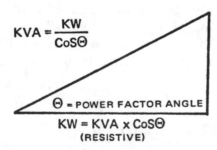

$$KVA = \frac{KW}{CoS\Theta}$$

KVAR = KVA Sin Θ
(REACTIVE)

Θ = POWER FACTOR ANGLE

KW = KVA x CoSΘ
(RESISTIVE)

$$
\begin{aligned}
K &= 1000 \\
W &= \text{Watts} \\
VA &= \text{Volt Amperes} \\
VAR &= \text{Volt Amperes Reactive} \\
\Theta &= \text{Angle Between KVA and KW} \\
CoS\Theta &= \text{Power Factor} \\
Tan\,\Theta &= \frac{KVAR}{KW}
\end{aligned}
$$

The windings of transformers and motors are usually connected in a wye or delta configuration. The relationships for line and phase voltages and currents are illustrated by Figure 4-1.

WYE CONNECTIONS

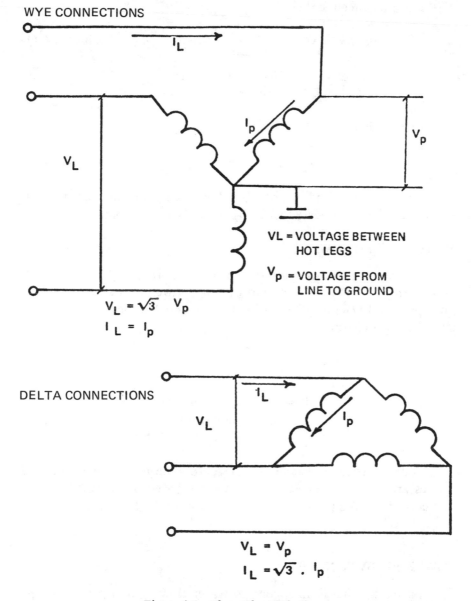

$$V_L = \sqrt{3} \; V_p$$
$$I_L = I_p$$

VL = VOLTAGE BETWEEN
      HOT LEGS

V_p = VOLTAGE FROM
      LINE TO GROUND

DELTA CONNECTIONS

$$V_L = V_p$$
$$I_L = \sqrt{3} \cdot I_p$$

Figure 4-1. Three-Phase Windings

For a balanced 3-phase load

$$\text{Power} = \underbrace{\sqrt{3} \quad V_L \quad I_L}_{} \quad \overbrace{\text{CoS}\Theta} \qquad \textit{Formula (4-1)}$$

Watts =   Volt   Power
                Amperes   Factor

For a balanced 1-phase load

$$P = V_L \ I_L \ \text{CoS}\Theta \qquad\qquad \textit{Formula (4-2)}$$

The primary windings of 13.8 Kv — 480-volt unit substations are usually delta-connected with the secondary wye-connected.

MOTOR HORSEPOWER

The standard power rating of a motor is referred to as a horsepower. In order to relate the motor horsepower to a kilowatt (kW), multiply the horsepower by .746 (Conversion Factor) and divide by the motor efficiency.

$$\text{KVA} = \frac{\text{HP} \times .746}{\eta \times \text{P.F.}} \qquad\qquad \textit{Formula (4-3)}$$

HP = Motor Horsepower
η = Efficiency of Motor
P.F. = Power Factor of Motor

Motor efficiencies and power factors vary with load. Typical values are shown in Table 4-1. Values are based on totally enclosed fan-cooled motors (TEFC) running at 1800 rpm "T" frame.

POWER FLOW CONCEPT

Power flowing is analogous to water flowing in a pipe. To supply several small water users, a large pipe services the plant at a high pressure. Several branches from the main pipe service various loads. Pressure reducing stations lower the main pressure to meet the requirements of each user. Similarly, a large feeder at a high voltage

Table 4-1.

HP RANGE	3-30	40-100
η% at		
½ Load	83.3	89.2
¾ Load	85.8	90.7
Full Load	86.2	90.9
P.F. at		
½ Load	70.1	79.2
¾ Load	79.2	85.4
Full Load	83.5	87.4

services a plant. Through switchgear breakers, the main feeder is distributed into smaller feeders. The switchgear breakers serve as a protector for each of the smaller feeders. Transformers are used to lower the voltage to the nominal value needed by the user.

ELECTRICAL EQUIPMENT

Electrical equipment commonly specified is as follows:

• *Switchgear-Breakers*—used to distribute power.

• *Unit Substation*—used to step down voltage. Consists of a high voltage disconnect switch, transformer and low-voltage breakers. Typical 480-volt transformer sizes are 300 KVA, 500 KVA, 750 KVA, 1000 KVA, 2500 KVA and 3000 KVA.

• *Motor Control Center (M.C.C.)*—a structure which houses starters and circuit breakers or fuses for motor control. It consists of the following:

(1) Thermal overload relays which guard against motor overloads,

(2) Fuse disconnect switches or breakers which protect the cable and motor and can be used as a disconnecting means,

(3) Contactors (relays) whose contacts are capable of opening and closing the power source to the motor.

MOTORS

• *Squirrel Cage Induction Motors* are commonly used. These
motors require three power loads. For two-speed applications several
different types of motors are available. Depending on the process requirements such as constant horsepower or constant torque, the
windings of the motor are connected differently. The theory of two-
speed operation is based on Formula 4-4.

$$\text{Frequency} = \frac{\text{No. of poles} \times \text{speed}}{120} \qquad Formula\ (4\text{-}4)$$

Thus, if the frequency is fixed, the effective number of motor
poles should be changed to change the speed. This can be accomplished by the manner in which the windings are connected. Two-
speed motors require six power leads.

• *D. C. Motors* are used where speed control is essential. The
speed of a D.C. Motor is changed by varying the field voltage through
a rheostat. A D.C. Motor requires two power wires to the armature
and two smaller cables for the field.

• *Synchronous Motors* are used when constant speed operation is essential. Synchronous motors are sometimes cheaper in the
large horsepower categories when slow speed operation is required.
Synchronous motors also are considered for power factor correction.
A .8 P.F. synchronous motor will supply corrective KVARs to the
system. A synchronous motor requires A.C. for power and D.C. for
the field. Since many synchronous motors are self-excited, only the
power cables are required to the motor.

IMPORTANCE OF POWER FACTOR

Transformer size is based on KVA. The closer Θ equals $0°$ or
power factor approaches unity, the smaller the KVA. Many times
utility companies have a power factor clause in their contract with
the customer. The statement usually causes the customer to pay an
additional power rate if the power factor of the plant deviates substantially from unity. The utility company wishes to maximize the
efficiency of its transformers and associated equipment.

POWER FACTOR CORRECTION

One problem facing the energy engineer is to estimate the power factor of a new plant and to install equipment such as capacitor banks or synchronous motors so that the overall power factor will meet the utility company's objectives.

Capacitor banks lower the total reactive KVAR by the value of the capacitors installed.

A second problem is to retrofit an existing plant so that the overall power factor desired is obtained.

Example Problem 4-1

A total motor horsepower load of 854 is made up of motors ranging from 40–100 horsepower. Calculate the connected KVA.

It is desired to operate the plant at a power factor of .95. What approximate capacitor bank is required?

Answer

$$KVA = \frac{HP \times .746}{Motor\ Eff. \times Motor\ P.F.}$$

From Table 4-1 at full load

Eff. = .909 and P.F. = .87

$$KVA = \frac{854 \times .746}{.909 \times .87} = 806$$

The plant is operating at a power factor of .87. The power factor of .87 corresponds to an angle of 29°

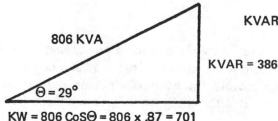

806 KVA

KVAR = 806 Sin 29°
= 806 x .48 = 386

KVAR = 386

Θ = 29°

KW = 806 CoSΘ = 806 x .87 = 701

A power factor of .95 is required.

$$CoS\Theta = .95$$
$$\Theta = 18°$$

The KVAR of 386 needs to be reduced by adding capacitors.

Remember kW does not change with different power factors, but KVA does.

Thus, the desired power triangle would look as follows:

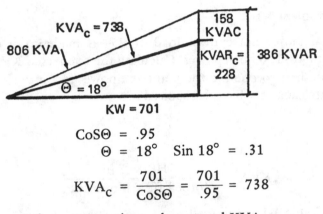

$$CoS\Theta = .95$$
$$\Theta = 18° \quad Sin\ 18° = .31$$

$$KVA_c = \frac{701}{CoS\Theta} = \frac{701}{.95} = 738$$

Note: Power factor correction reduces total KVA.

$$KVAR_c = 738\ Sin\ 18° = 738 \times .31 = 228$$

$$Capacitance\ Bank = 386-228$$
$$158\ KVAC$$

Example Problem 4-2

The client wishes to know the expected power factor for a new plant. The lighting load is 40 kW. The plant is composed of two identical modules (2 motors for each equipment number listed). Remember that KVAs at different power factors cannot be added directly.

Motor List — Module 1

Motor No.	Description	HP	Voltage	Phase
AG-1	Agitator Motor	60	460	3
CF-3	Centrifuge Motor	100	460	3
FP-4	Feed Pump Motor	30	460	3
TP-5	Transfer Pump Motor	10	460	3
CTP-6	Cooling Tower	25	460	3
CT-9	Cooling Tower Motor	20	460	3
HF-10	H&V Supply Fan Motor	40	460	3
HF-11	H&V Exhaust Fan Motor	20	460	3
BC-13	Brine Compressor Motor	50	460	3
C-16	Conveyor Motor	20	460	3
H-17	Hoist Motor	5	460	3

Answer

Based on the motor list, the plant power factor is estimated as follows:

Module 1

Lighting kW_3 = 40 Total

Motors 3-30

30
10
25
20
20
20
5
130

Motors 40-100

60
100
40
50
250

At Full Load:

P.F. = 83.5
η = 86.2

P.F. = 87.4
η = 90.9

$$KVA_1 = \frac{130 \times .746}{.83 \times .86} = 135$$

$$KVA_2 = \frac{250 \times .746}{.90 \times .87} = 238$$

$$kW_1 = KVA \, Cos\Theta$$
$$= KVA \, .83$$
$$= 112$$

$$kW_2 = KVA_2 \, Cos\Theta$$
$$= KVA \, .87$$
$$= 207$$

$$KVAR_1 = KVA_1 \, Sin\Theta$$

$$KVAR_2 = KVA_2 \, Sin\Theta$$

$$\Theta = 33°$$

$$\Theta = 29°$$

$$KVAR_1 = KVA_1 \times .54$$
$$= 135 \times .54$$
$$= 73$$

$$KVAR_2 = KVA_2 \times .48$$
$$KVAR_2 = 115$$

$$kW_{total} = \underbrace{kW_1 + kW_2}_{\text{Module 1}} + \underbrace{kW_1 + kW_2 + kW_3}_{\text{Module 2}} = 678 \, KW$$

$$KVAR_{total} = \underbrace{KVAR_1 + KVAR_2}_{\text{Module 1}} + \underbrace{KVAR_1 + KVAR_2}_{\text{Module 2}}$$
$$= 73 + 115 + 73 + 115 = 376$$

$$KVA_{total} = \sqrt{(678)^2 + (376)^2} = 775 \, KVA$$

$$Cos\Theta = \frac{kW_{total}}{KVA_{total}} = \frac{678}{775} = .87$$

Shortcut Methods

A handy shortcut table which can be used to find the value of the capacitor required to improve the plant power factor is illustrated by Table 4-2.

Table 4-2. Shortcut Method — Power Factor Correction.

KW MULTIPLIERS FOR DETERMINING CAPACITOR KILOVARS

DESIRED POWER-FACTOR IN PERCENTAGE

ORIG. PF	80	81	82	83	84	85	86	87	88	89	90	91	92	93	94	95	96	97	98	99	100
50	.982	1.008	1.034	1.060	1.086	1.112	1.139	1.165	1.192	1.220	1.248	1.276	1.303	1.337	1.369	1.403	1.441	1.481	1.529	1.590	1.732
51	.936	.962	.988	1.014	1.040	1.066	1.093	1.119	1.146	1.174	1.202	1.230	1.257	1.291	1.323	1.357	1.395	1.435	1.483	1.544	1.688
52	.894	.920	.946	.972	.998	1.024	1.051	1.077	1.104	1.132	1.160	1.188	1.215	1.249	1.281	1.315	1.353	1.393	1.441	1.502	1.644
53	.850	.876	.902	.928	.954	.980	1.007	1.033	1.060	1.088	1.116	1.144	1.171	1.205	1.237	1.271	1.309	1.349	1.397	1.458	1.600
54	.809	.835	.861	.887	.913	.939	.966	.992	1.019	1.047	1.075	1.103	1.130	1.164	1.196	1.230	1.268	1.308	1.356	1.417	1.559
55	.769	.795	.821	.847	.873	.899	.926	.952	.979	1.007	1.035	1.063	1.090	1.124	1.156	1.190	1.228	1.268	1.316	1.377	1.519
56	.730	.756	.782	.808	.834	.860	.887	.913	.940	.968	.996	1.024	1.051	1.085	1.117	1.151	1.189	1.229	1.277	1.338	1.480
57	.692	.718	.744	.770	.796	.822	.849	.875	.902	.930	.958	.986	1.013	1.047	1.079	1.113	1.151	1.191	1.239	1.300	1.442
58	.655	.681	.707	.733	.759	.785	.812	.838	.865	.893	.921	.949	.976	1.010	1.042	1.076	1.114	1.154	1.202	1.263	1.405
59	.618	.644	.670	.696	.722	.748	.775	.801	.828	.856	.884	.912	.939	.973	1.005	1.039	1.077	1.117	1.165	1.226	1.368
60	.584	.610	.636	.662	.688	.714	.741	.767	.794	.822	.849	.878	.905	.939	.971	1.005	1.043	1.083	1.131	1.192	1.334
61	.549	.575	.601	.627	.653	.679	.706	.732	.759	.787	.815	.843	.870	.904	.936	.970	1.008	1.048	1.096	1.157	1.299
62	.515	.541	.567	.593	.619	.645	.672	.698	.725	.753	.781	.809	.836	.870	.902	.936	.974	1.014	1.062	1.123	1.265
63	.483	.509	.535	.561	.587	.613	.640	.666	.693	.721	.749	.777	.804	.838	.870	.904	.942	.982	1.030	1.091	1.233
64	.450	.476	.502	.528	.554	.580	.607	.633	.660	.688	.716	.744	.771	.805	.837	.871	.909	.949	.997	1.058	1.200
65	.419	.445	.471	.497	.523	.549	.576	.602	.629	.657	.685	.713	.740	.774	.806	.840	.878	.918	.966	1.027	1.169
66	.388	.414	.440	.466	.492	.518	.545	.571	.598	.626	.654	.682	.709	.743	.775	.809	.847	.887	.935	.996	1.138
67	.358	.384	.410	.436	.462	.488	.515	.541	.568	.596	.624	.652	.679	.713	.745	.779	.817	.857	.905	.966	1.108
68	.329	.355	.381	.407	.433	.459	.486	.512	.539	.567	.595	.623	.650	.684	.716	.750	.788	.828	.876	.937	1.079
69	.299	.325	.351	.377	.403	.429	.456	.482	.509	.537	.565	.593	.620	.654	.686	.720	.758	.798	.840	.907	1.049
70	.270	.296	.322	.348	.374	.400	.427	.453	.480	.508	.536	.564	.591	.625	.657	.691	.729	.769	.811	.878	1.020
71	.242	.268	.294	.320	.346	.372	.399	.425	.452	.480	.508	.536	.563	.597	.629	.663	.701	.741	.783	.850	.992
72	.213	.239	.265	.291	.317	.343	.370	.396	.423	.451	.479	.507	.534	.568	.600	.634	.672	.712	.754	.821	.963
73	.186	.212	.238	.264	.290	.316	.343	.369	.396	.424	.452	.480	.507	.541	.573	.607	.645	.685	.727	.794	.936
74	.159	.185	.211	.237	.263	.289	.316	.342	.369	.397	.425	.453	.480	.514	.546	.580	.618	.658	.700	.767	.909
75	.132	.158	.184	.210	.236	.262	.289	.315	.342	.370	.398	.426	.453	.487	.519	.553	.591	.631	.673	.740	.882
76	.105	.131	.157	.183	.209	.235	.262	.288	.315	.343	.371	.399	.426	.460	.492	.526	.564	.604	.652	.713	.855
77	.079	.105	.131	.157	.183	.209	.236	.262	.289	.317	.345	.373	.400	.434	.466	.500	.538	.578	.620	.687	.829
78	.053	.079	.105	.131	.157	.183	.210	.236	.263	.291	.319	.347	.374	.408	.440	.474	.512	.552	.594	.661	.803
79	.026	.052	.078	.104	.130	.156	.183	.209	.236	.264	.292	.320	.347	.381	.413	.447	.485	.525	.567	.634	.776
80	.000	.026	.052	.078	.104	.130	.157	.183	.210	.238	.266	.294	.321	.355	.387	.421	.458	.499	.541	.608	.750
81	—	.000	.026	.052	.078	.104	.131	.157	.184	.212	.240	.268	.295	.329	.361	.395	.433	.473	.515	.582	.724
82	—	—	.000	.026	.052	.078	.105	.131	.158	.186	.214	.242	.269	.303	.335	.369	.407	.447	.489	.556	.698
83	—	—	—	.000	.026	.052	.079	.105	.132	.160	.188	.216	.243	.277	.309	.343	.381	.421	.463	.530	.672
84	—	—	—	—	.000	.026	.053	.079	.106	.134	.162	.190	.217	.251	.283	.317	.355	.395	.437	.504	.645
85	—	—	—	—	—	.000	.027	.053	.080	.108	.136	.164	.191	.225	.257	.291	.329	.369	.417	.478	.620

ORIGINAL POWER FACTOR IN PERCENTAGE

Example: Total kw input of load from wattmeter reading 100 kw at a power factor of 60%. The leading reactive kvar necessary to raise the power factor to 90% is found by multiplying the 100 kw by the factor found in the table, which is .849. Then 100 kw × 0.849 = 84.9 kvar. Use 85 kvar.

Reprinted by permission of Federal Pacific Electric Company.

WHERE TO LOCATE CAPACITORS

As indicated, the primary purpose of capacitors is to reduce the power consumption. Additional benefits are derived by capacitor location. Figure 4-2 indicates typical capacitor locations. Maximum benefit of capacitors is derived by locating them as close as possible to the load. At this location, its kilovars are confined to the smallest possible segment, decreasing the load current. This, in turn, will reduce power losses of the system substantially. Power losses are proportional to the square of the current. When power losses are reduced, voltage at the motor increases; thus, motor performance also increases.

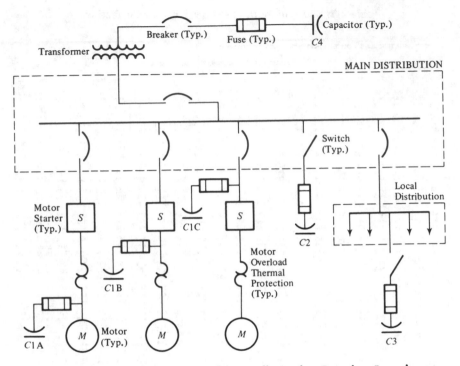

Figure 4-2. Power Distribution Diagram Illustrating Capacitor Locations

Locations $C1A$, $C1B$ and $C1C$ of Figure 4-2 indicate three different arrangements at the load. Note that in all three locations extra switches are not required, since the capacitor is either switched with the motor starter or the breaker before the starter. Case $C1A$ is recommended for new installation, since the maximum benefit is derived and the size of the motor thermal protector is reduced. In Case $C1B$, as in Case $C1A$, the capacitor is energized only when the motor is in operation. Case $C1B$ is recommended in cases where the installation is existing and the thermal protector does not need to be re-sized. In position $C1C$, the capacitor is permanently connected to the circuit but does not require a separate switch, since it can be disconnected by the breaker before the starter.

It should be noted that the rating of the capacitor should *not* be greater than the no-load magnetizing KVAR of the motor. If this condition exists, damaging overvoltage or transient torques can occur. This is why most motor manufacturers specify maximum capacitor ratings to be applied to specific motors.

The next preference for capacitor locations as illustrated by Figure 4-2 is at locations $C2$ and $C3$. In these locations, a breaker or switch will be required. Location $C4$ requires a high voltage breaker. The advantage of locating capacitors at power centers or feeders is that they can be grouped together. When several motors are running intermittently, the capacitors are permitted to be on-line all the time, reducing the total power regardless of load.

ENERGY EFFICIENT MOTORS

Energy efficient motors are now available. These motors are approximately 30% more expensive than their standard counterpart. Based on the energy cost, it can be determined if the added investment is justified. With the emphasis on energy conservation, new lines of energy efficient motors are being introduced. Figures 4-3 and 4-4 illustrate a typical comparison between energy efficient and standard motors.

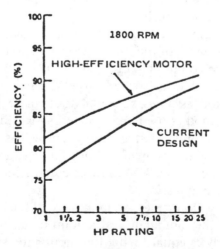

Figure 4-3. Efficiency vs Horsepower Rating (Dripproof Motors)

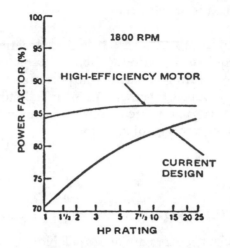

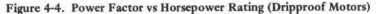

Figure 4-4. **Power Factor vs Horsepower Rating (Dripproof Motors)**

LIGHTING EFFICIENCY

LIGHTING BASICS

About 20 percent of all electricity generated in the United States today is used for lighting.

By understanding the basics of lighting design, several ways to improve the efficiency of lighting systems will become apparent.

There are two common lighting methods used. One is called the "Lumen" method, while the other is the "Point by Point" method. The Lumen method assumes an equal footcandle level throughout the area. This method is used frequently by lighting designers since it is simplest; however, it wastes energy, since it is the light "at the task" which must be maintained and not the light in the surrounding areas. The "Point by Point" method calculates the lighting requirements for the task in question.

The "Point by Point" method makes use of the inverse-square law, which states that the illuminance at a point on a surface perpendicular to the light ray is equal to the luminous intensity of the source at that point divided by the square of the distance between the source and the point of calculation, as illustrated in Formula 4-5.

$$E = \frac{I}{D^2}$$ *Formula (4-5)*

Where

   E = Illuminance in footcandles
   I = Luminous intensity in candles
   D = Distance in feet between the source and the point of calculation.

If the surface is not perpendicular to the light ray, the appropriate trigonometrical functions must be applied to account for the deviation.

Lumen Method

A footcandle is the illuminance on a surface of one square foot in area having a uniformly distributed flux of one lumen. From this definition, the "Lumen Method" is developed and illustrated by Formula 4-6.

$$N = \frac{F_1 \times A}{Lu \times L_1 \times L_2 \times Cu}$$ *Formula (4-6)*

Where

   N is the number of lamps required.
   F_1 is the required footcandle level at the task. A footcandle is a measure of illumination; one standard candle power measured one foot away.
   A is the area of the room in square feet.
   Lu is the Lumen output per lamp. A Lumen is a measure of lamp intensity: its value is found in the manufacturer's catalogue.
   Cu is the coefficient of utilization. It represents the ratio of the Lumens reaching the working plane to the total Lumens generated by the lamp. The coefficient of utilization makes allowances for light absorbed or reflected by walls, ceilings, and the fixture itself. Its values are found in the manufacturer's catalogue.
   L_1 is the lamp depreciation factor. It takes into account that the lamp Lumen depreciates with time. Its value is found in the manufacturer's catalogue.

L_2 is the luminaire (fixture) dirt depreciation factor. It takes into account the effect of dirt on a luminaire and varies with type of luminaire and the atmosphere in which it is operated.

The Lumen method formula illustrates several ways lighting efficiency can be improved.

Faced with the desire to reduce their energy use,[1] lighting consumers have four options: i) reduce light levels, ii) purchase more efficient equipment, iii) provide light when needed at the task at the required level, and iv) add control and reduce lighting loads automatically. The multitude of equipment options to meet one or more of the above needs permits the consumer and the lighting designer-engineer to consider the trade-offs between the initial and operating costs based upon product performance (life, efficacy, color, glare, and color rendering).

Some definitions and terms used in the field of lighting will be presented to help consumers evaluate and select lighting products best suiting their needs. Then, some state-of-the-art advances will be characterized so that their benefits and limitations are explicit.

Lighting Terminology

Efficacy — Is the amount of visible light (lumens) produced for the amount of power (watts) expended. It is a measure of the efficiency of a process but is a term used in place of efficiency when the input (W) has different units than the output (lm) and expressed in lm/W.

Color Temperature — A measure of the color of a light source relative to a black body at a particular temperature expressed in degrees Kelvin (°K). Incandescents have a low color temperature (~2800°K) and have a red-yellowish tone; daylight has a high color temperature (~6000°K) and appears bluish. Today, the phosphors used in fluorescent lamps can be blended to provide any desired color temperature in the range from 2800°K to 6000°K.

---

[1] **Source:** *Lighting Systems Research*, R.R. Verderber

Color Rendering — A parameter that describes how a light source renders a set of colored surfaces with respect to a black body light source at the same color temperature. The color rendering index (CRI) runs from 0 to 100. It depends upon the specific wavelengths of which the light is composed. A black body has a continuous spectrum and contains all of the colors in the visible spectrum. Fluorescent lamps and high intensity discharge lamps (HID) have a spectrum rich in certain colors and devoid in others. For example, a light source that is rich in blues and low in reds could appear white, but when reflected from a substance, it would make red materials appear faded. The same material would appear different when viewed with an incandescent lamp, which has a spectrum that is rich in red.

LIGHT SOURCES[2]

Figure 4-5 indicates the general lamp efficiency ranges for the generic families of lamps most commonly used for both general and supplementary lighting systems. Each of these sources is discussed briefly here. It is important to realize that in the case of fluorescent and high intensity discharge lamps, the figures quoted for "lamp efficacy" are for the lamp only and do not include the associated ballast losses. To obtain the total system efficiency, ballast input watts must be used rather than lamp watts to obtain an overall system lumen per watt figure. This will be discussed in more detail in a later section.

Incandescent lamps have the lowest lamp efficacies of the commonly used lamps. This would lead to the accepted conclusion that incandescent lamps should generally not be used for large-area, general lighting systems where a more efficient source could serve satisfactorily. However, this does not mean that incandescent lamps should never be used. There are many applications where the size, convenience, easy control, color rendering, and relatively low cost of incandescent lamps are suitable for a specific application.

General service incandescent lamps do not have good lumen maintenance throughout their lifetime. This is the result of the tungsten's evaporation off the filament during heating as it deposits on the bulb wall, thus darkening the bulb and reducing the lamp lumen output.

---

[2] **Source:** *Selection Criteria for Lighting Energy Management*, Roger L. Knott

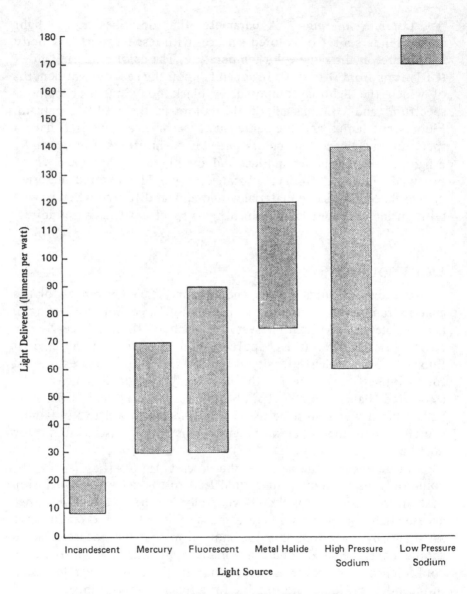

Figure 4-5. Efficiency of Various Light Sources

Efficient Types of Incandescents
for Limited Use

Attempts to increase the efficiency of incandescent lighting while maintaining good color rendition have led to the manufacture of a number of energy-saving incandescent lamps for limited residential use.

Tungsten Halogen—These lamps vary from the standard incandescent by the addition of halogen gases to the bulb. Halogen gases keep the glass bulb from darkening by preventing the filament's evaporation, thereby increasing lifetime up to four times that of a standard bulb. The lumen-per-watt rating is approximately the same for both types of incandescents, but tungsten halogen lamps average 94% efficiency throughout their extended lifetime, offering significant energy and operating cost savings. However, tungsten halogen lamps require special fixtures, and during operation the surface of the bulb reaches very high temperatures, so they are not commonly used in the home.

Reflector or R-Lamps—Reflector lamps are incandescents with an interior coating of aluminum that directs the light to the front of the bulb. Certain incandescent light fixtures, such as recessed or directional fixtures, trap light inside. Reflector lamps project a cone of light out of the fixture and into the room, so that more light is delivered where it is needed. In these fixtures, a 50-watt reflector bulb will provide better lighting and use less energy when substituted for a 100-watt standard incandescent bulb.

Reflector lamps are an appropriate choice for task lighting (because they directly illuminate a work area) and for accent lighting. Reflector lamps are available in 25, 30, 50, 75, and 150 watts. While they have a lower initial efficiency (lumens per watt) than regular incandescents, they direct light more effectively, so that more light is actually delivered than with regular incandescents. (See Figure 4-6.)

PAR Lamps— Parabolic aluminized reflector (PAR) lamps are reflector lamps with a lens of heavy, durable glass, which makes them an appropriate choice for outdoor flood and spot lighting. They are available in 75, 150, and 250 watts. They have longer lifetimes with less depreciation than standard incandescents.

ER Lamps—Ellipsoidal reflector (ER) lamps are ideally suited for recessed fixtures, because the beam of light produced is focused two inches ahead of the lamp to reduce the amount of light trapped in the fixture. In a directional fixture, a 75-watt ellipsoidal reflector lamp delivers more light than a 150-watt R-lamp. (See Figure 4-6.)

Standard Incandescent   **R-Lamp**

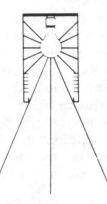

A high percentage   An aluminum
of light output   coating directs light
is trapped in fixture   out of the fixture

ER Lamp

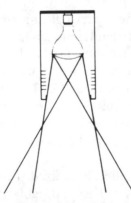

The beam is focused 2 inches
ahead of the lamp, so that very
little light is trapped in the fixture

Figure 4-6. Comparison of Incandescent Lamps

Mercury vapor lamps find limited use in today's lighting systems because fluorescent and other high intensity discharge (HID) sources have surpassed them in both lamp efficacy and system efficiency. Typical ratings for mercury vapor lamps range from about 30 to 70 lumens per watt. The primary advantages of mercury lamps are a good range of color, availability in sizes as low as 30 watts, long life and relatively low cost. However, fluorescent systems are available today which can do many of the jobs mercury used to do and they do it more efficiently. There are still places for mercury vapor lamps in lighting system design, but they are becoming fewer as technology advances in fluorescent and higher efficacy HID sources.

Fluorescent lamps have made dramatic advances in the last 10 years. From the introduction of reduced wattage lamps in the mid-1970s, to the marketing of several styles of low wattage, compact lamps recently, there has been a steady parade of new products. Lamp efficacy now ranges from about 30 lumens per watt to near 90 lumens per watt. The range of colors is more complete than mercury vapor, and lamp manufacturers have recently made significant progress in developing fluorescent and metal halide lamps which have much more consistent color rendering properties allowing greater flexibility in mixing these two sources without creating disturbing color mismatches. The recent compact fluorescent lamps open up a whole new market for fluorescent sources. These lamps permit design of much smaller luminaries which can compete with incandescent and mercury vapor in the low cost, square or round fixture market which the incandescent and mercury sources have dominated for so long. While generally good, lumen maintenance throughout the lamp lifetime is a problem for some fluorescent lamp types.

Energy Efficient "Plus" Fluorescents[3]

The energy efficient "plus" fluorescents represent the second generation of improved fluorescent lighting. These bulbs are available for replacement of standard 4-foot, 40-watt bulbs and require only 32 watts of electricity to produce essentially the same light levels. To the authors' knowledge, they are not available for 8-foot fluores-

---

[3] Source: *Fluorescent Lighting — An Expanding Technology*, R.E. Webb, M.G. Lewis, W.C. Turner.

cent bulb retrofit. The energy efficient plus fluorescents require a ballast change. The light output is similar to the energy efficient bulbs, and the two types may be mixed in the same area if desired.

Examples of energy efficient plus tubes include the Super-Saver Plus by Sylvania and General Electric's Watt Mizer Plus.

Energy Efficient Fluorescents System Change

The third generation of energy efficient fluorescents requires both a ballast and a fixture replacement. The standard 2-foot by 4-foot fluorescent fixture, containing four bulbs and two ballasts, requires approximately 180 watts (40 watts per tube and 20 watts per ballast). The new-generation fluorescent manufacturers claim the following:

- General Electric — "Optimizer" requires only 116 watts with a slight reduction in light output.

- Sylvania — "Octron" requires only 132 watts with little reduction in light level.

- General Electric — "Maximizer" requires 169 watts but supplies 22 percent more light output.

The fixtures and ballasts designed for the third-generation fluorescents are not interchangeable with earlier generations.

Metal halide lamps fall into a lamp efficacy range of approximately 75-125 lumens per watt. This makes them more energy efficient than mercury vapor but somewhat less so than high pressure sodium. Metal halide lamps generally have fairly good color rendering qualities. While this lamp displays some very desirable qualities, it also has some distinct drawbacks including relatively short life for an HID lamp, long restrike time to restart after the lamp has been shut off (about 15-20 minutes at 70°F) and a pronounced tendency to shift colors as the lamp ages. In spite of the drawbacks, this source deserves serious consideration and is used very successfully in many applications.

High pressure sodium lamps introduced a new era of extremely high efficacy (60-140 lumens/watt) in a lamp which operates in fixtures having construction very similar to those used for mercury vapor and metal halide. When first introduced, this lamp suffered from ballast problems. These have now been resolved and luminaries employing high quality lamps and ballasts provide very satisfactory service. The 24,000-hour lamp life, good lumen maintenance and high efficacy of these lamps make them ideal sources for industrial and outdoor applications where discrimination of a range of colors is not critical.

The lamp's primary drawback is the rendering of some colors. The lamp produces a high percentage of light in the yellow range of the spectrum. This tends to accentuate colors in the yellow region. Rendering of reds and greens shows a pronounced color shift. This can be compensated for in the selection of the finishes for the surrounding areas, and, if properly done, the results can be very pleasing. In areas where color selection, matching and discrimination are necessary, high pressure sodium should not be used as the only source of light. It is possible to gain quite satisfactory color rendering by mixing high pressure sodium and metal halide in the proper proportions. Since both sources have relatively high efficacies, there is not a significant loss in energy efficiency by making this compromise.

High pressure sodium has been used quite extensively in outdoor applications for roadway, parking and façade or security lighting. This source will yield a high efficiency system; however, it should be used only with the knowledge that foliage and landscaping colors will be severely distorted where high pressure sodium is the only, or predominant, illuminant. Used as a parking lot source, there may be some difficulty in identification of vehicle colors in the lot. It is necessary for the designer or owner to determine the extent of this problem and what steps might be taken to alleviate it.

Recently lamp manufacturers have introduced high pressure sodium lamps with improved color rendering qualities. However, the improvement in color rendering was not gained without cost—the efficacy of the color-improved lamps is somewhat lower approximately 90 lumens per watt.

Low pressure sodium lamps provide the highest efficacy of any of the sources for general lighting with values ranging up to 180 lumens per watt. Low pressure sodium produces an almost pure yellow light with very high efficacy, and renders all colors gray except yellow or near yellow. This effect results in no color discrimination under low pressure sodium lighting; it is suitable for use in a very limited number of applications. It is an acceptable source for warehouse lighting where it is only necessary to read labels but not to choose items by color. This source has application for either indoor or outdoor safety or security lighting as long as color rendering is not important.

In addition to these primary sources, there are a number of retrofit lamps which allow use of higher efficacy sources in the sockets of existing fixtures. Therefore, metal halide or high pressure sodium lamps can be retrofitted into mercury vapor fixtures, or self-ballasted mercury lamps can replace incandescent lamps. These lamps all make some compromises in operating characteristics, life and/or efficacy.

Figure 4-7 presents data on the efficacy of each of the major lamp types in relation to the wattage rating of the lamps. Without exception, the efficacy of the lamp increases as the lamp wattage rating increases.

The lamp efficacies discussed here have been based on the lumen output of a new lamp after 100 hours of operation or the "initial lumens." Not all lamps age in the same way. Some lamp

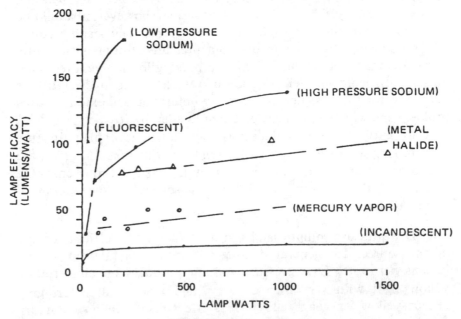

Figure 4-7. Lamp Efficacy
(Does Not Include Ballast Losses)

types, such as lightly loaded fluorescent and high pressure sodium, hold up well and maintain their lumen output at a relatively high level until they are into, or past, middle age. Others, as represented by heavily loaded fluorescent, mercury vapor and metal halide, decay rapidly during their early years and then coast along at a relatively lower lumen output throughout most of their useful life. These factors must be considered when evaluating the various sources for overall energy efficiency.

Incandescent Replacement

The most efficacious lamps that can be used in incandescent sockets are the compact fluorescent lamps. The most popular systems are the twin tubes and double twin tubes. These are closer to the size and weight of the incandescent lamp than the earlier type of fluorescent (circline) replacements.

Twin tubes with lamp wattages from 5 to 13 watts provide amounts of light ranging from 240 to 850 lumens. Table 4-3 lists the characteristics of various types of incandescent and compact fluorescent lamps that can be used in the same type sockets.

Table 4-3. Lamp Characteristics

Lamp Type (Total Input Power)*	Lamp Power (W)	Light Output (lumens)	Lamp Life (hour)	Efficacy (lm/W)
100 W (Incandescent)	100	1750	750	18
75 W (Incandescent)	75	1200	750	16
60 W (Incandescent)	60	890	1000	15
40 W (Incandescent)	40	480	1500	12
25 W (Incandescent)	25	238	2500	10
22 W (Fl. Circline)	18	870	9000	40
44 W (Fl. Circline)	36	1750	9000	40
7 W (Twin)	5	240	10000	34
10 W (Twin)	7	370	10000	38
13 W (Twin)	9	560	10000	43
19 W (Twin)	13	850	10000	45
18 W (Solid-State)**	—	1100	7500	61

*Includes ballast losses.
**Operated at high frequency.

The advantages of the compact fluorescent lamps are larger and increased efficacy, longer life and reduced total cost. The cost per 10^6 lumen hours of operating the 75-watt incandescent and the 18-watt fluorescent is $5.47 and $3.29, respectively. This is based upon an energy cost of $0.075 per kWh and lamp costs of $0.70 and $17 and the 75-watt incandescent and 18-watt fluorescent lamps, respectively. The circline lamps were much larger and heavier than the incandescents and would fit in a limited number of fixtures. The twin tubes are only slightly heavier and larger than the equivalent incandescent lamp. However, there are some fixtures that are too small for them to be employed.

The narrow tube diameter compact fluorescent lamps are now possible because of the recently developed rare earth phosphors. These phosphors have an improved lumen depreciation at high lamp power loadings. The second important characteristic of these narrow band phosphors is their high efficiency in converting the ultraviolet light generated in the plasma into visible light. By proper mixing of these phosphors, the color characteristics (color temperature and color rendering) are similar to the incandescent lamp.

There are two types of compact fluorescent lamps. In one type of lamp system, the ballast and lamp are integrated into a single package; in the second type, the lamp and ballast are separate, and when a lamp burns out it can be replaced. In the integrated system, both the lamp and the ballast are discarded when the lamp burns out.

It is important to recognize when purchasing these compact fluorescent lamps that they provide the equivalent light output of the lamps being replaced. The initial lumen output for the various lamps is shown in Table 4-3.

Lighting Efficiency Options

Several lighting efficiency options are illustrated below: (Refer to Formula 4-6.)

Footcandle Level—The footcandle level required is that level at the task. Footcandle levels can be lowered to one third of the levels for surrounding areas such as aisles. (A minimum 20-footcandle level should be maintained.)

The placement of the lamp is also important. If the luminaire can be lowered or placed at a better location, the lamp wattage may be reduced.

Coefficient of Utilization (Cu)—The color of the walls, ceiling, and floors, the type of luminaire, and the characteristics of the room determine the *Cu*. This value is determined based on manufacturer's literature. The *Cu* can be improved by analyzing components such as lighter colored walls and more efficient luminaires for the space.

Lamp Depreciation Factor and *Dirt Depreciation Factor*— These two factors are involved in the maintenance program. Choosing a luminaire which resists dirt build-up, group relamping and cleaning the luminaire will keep the system in optimum performance. Taking these factors into account can reduce the number of lamps initially required.

The light loss factor *(LLF)* takes into account that the lamp lumen depreciates with time (L_1), that the lumen output depreciates due to dirt build-up (L_2), and that lamps burn out (L_3). Formula 4-7 illustrates the relationship of these factors.

$$LLF = L_1 \times L_2 \times L_3 \qquad \text{Formula (4-7)}$$

To reduce the number of lamps required which in turn reduces energy consumption, it is necessary to increase the overall light loss factor. This is accomplished in several ways. One is to choose the luminaire which minimizes dust build-up. The second is to improve the maintenance program to replace lamps prior to burn-out. Thus if it is known that a group relamping program will be used at a given percentage of rated life, the appropriate lumen depreciation factor can be found from manufacturer's data. It may be decided to use a shorter relamping period in order to increase (L_1) even further. If a group relamping program is used, (L_3) is assumed to be unity.

Figure 4-6 illustrates the effect of dirt build-up on (L_2) for a dustproof luminaire. Every luminaire has a tendency for dirt build-up. Manufacturer's data should be consulted when estimating (L_2) for the luminaire in question.

Solid-State Ballasts

After more than 10 years of development and 5 years of manufacturing experience, operating fluorescent lamps at high frequency (20 to 30 kHz) with solid-state ballasts has achieved credibility. The fact that all of the major ballast manufacturers offer solid-state

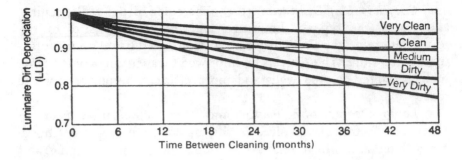

**Figure 4-6. Effect of Dirt Build-Up on Dustproof Luminaires for
Various Atmospheric Conditions**

ballasts and the major lamp companies have designed new lamps to be operated at high frequency is evidence that the solid-state high frequency ballast is now state-of-the-art.

It has been shown that fluorescent lamps operated at high frequency are 10 to 15 percent more efficacious than 60 Hz operation. In addition, the solid-state ballast is more efficient than conventional ballasts in conditioning the input power for the lamps such that the total system efficacy increase is between 20 and 25 percent. That is, for a standard two-lamp, 40-watt F40 T-12 rapid-start system, overall efficacy is increased from 63 lm/W to over 80 lm/W.

In the past few years, continued development of the product has improved reliability and reduced cost. Today, solid-state ballasts can be purchased for less than $30, and, in sufficiently large quantities, some bids have been less than $20. The industry's growth is evidenced by the availability of ballasts for the 8-foot fluorescent lamp, both slimline and high power, as well as the more common F40 (4-ft.) size. In order to be more competitive with initial costs, there are three- and four-lamp ballasts for the F40-type lamps. These multi-lamp ballasts reduce the initial cost per lamp, as well as the installation cost, and are even more efficient than the one- and two-lamp ballast system.

The American National Standards Institute (ANSI) ballast committee has been developing standards for solid-state ballasts for the past few years. The ballast factor is one parameter that will be specified by ANSI. However, solid-state ballasts are available with different ballast factors. The ballast factor is the light output provided by the ballast-lamp system compared to the light output of the lamp specified by the lamp manufacturer. The ANSI ballast factor standard for 40-watt F40 fluorescent lamps is 95 ± 2.5 percent. Because most solid-state ballasts were initially sold in the retrofit market, their ballasts were designed to have a lower ballast factor. Thus, energy was saved not only by the increased efficacy but also by reducing the light output. The thrust was to reduce illumination levels in overlit spaces.

Today, there are solid-state ballasts with a ballast factor exceeding 100 percent. These ballasts are most effectively used in new installations. In these layouts, more light from each luminaire will reduce the number of luminaires, ballasts and lamps, hence reducing both initial and operating costs. It is essential that the lighting designer-engineer and consumer know the ballast factor for the lamp-ballast system. The ballast factor for a ballast also depends upon the lamp. For example, a core-coil ballast will have a ballast factor of 95 ± 2 percent when operating a 40-watt F40 argon-filled lamp and less when operating an "energy saving" 34-watt F40 Krypton-filled lamp. The ballast factor instead will be about 87 ± 2.5 percent with the 34-watt energy saving lamp. Because of this problem, the ANSI standard for the ballast factor for the 34-watt lamps has recently been reduced to 85 percent. Table 4-4 provides some data for several types of solid-state ballasts operating 40-watt and 34-watt F40 lamps and lists some parameters of concern for the consumer.

The table compares several types of solid-state ballasts with a standard core-coil ballast that meets the ANSI standard with the two-lamp, 40-watt F40 lamp. Notice that the system efficacy of any ballast system is about the same operating a 40-watt or a 34-watt F40 lamp. Although the 34-watt "lite white" lamp is about 6 percent more efficient than the 40-watt "cool white" lamp, the ballast losses are greater with the 34-watt lamp due to an increased lamp current. The lite white phosphor is more efficient than the cool white phosphor but has poorer color rendering characteristics.

Note that the percent flicker is drastically reduced when the lamps are operated at high frequency with solid-state ballasts. A recent scientific field study of office workers in the U.K. has shown that complaints of headaches and eye-strain are 50 percent less under high frequency lighting when compared to lamps operating at 50 cycles, the line frequency in the U.K.

Table 4-4. Performance of F40 Fluorescent Lamp Systems

Characteristic	Core-Coil 2 Lamps, T-12 40W	34W	— Solid-State Ballasts — 2 Lamps, T-12 40W	34W	4 Lamps, T-12 40W	34W	2 Lamps, T-8 32W
Power (W)	96	79	72	63	136	111	65
Power Factor (%)	98	92	95	93	94	94	89
Filament Voltage (V)	3.5	3.6	3.1	3.1	2.0	1.6	0
Light Output (lm)	6050	5060	5870	5060	11,110	9250	5820
Ballast Factor	.968	.880	.932	.865	.882	.791	1.003
Flicker (%)	30	21	15	9	1	0	1
System Efficacy (lm/W)	63	64	81	81	82	83	90

Each of the above ballasts has different factors, which are lower when operating the 34-watt Krypton-filled lamp. Table 4-4 also lists the highest system efficacy of 90 lumens per watt for the solid-state ballast and T-8, 32-watt lamp.

All of the above solid-state ballasts can be used in place of core-coil ballasts specified to operate the same lamps. To determine the illumination levels, or the change in illumination levels, the manufacturer must supply the ballast factor for the lamp type employed. The varied light output from the various systems allows the lighting designer-engineer to precisely tailor the lighting level.

CONTROL EQUIPMENT

Table 4-5 lists various types of equipment that can be components of a lighting control system, with a description of the predominant characteristic of each type of equipment. Static equipment can alter light levels semipermanently. Dynamic equipment can alter light levels automatically over short intervals to correspond to the activities in a space. Different sets of components can be used to

Table 4-5. Lighting Control Equipment

System	Remarks
STATIC:	
Delamping	Method for reducing light level 50%.
Impedance Monitors	Method for reducing light level 30, 50%.
DYNAMIC:	
Light Controllers	
Switches/Relays	Method for on-off switching of large banks of lamps.
Voltage/Phase Control	Method for controlling light level continuously 100 to 50%.
Solid-State Dimming Ballasts	Ballasts that operate fluorescent lamps efficiently and can dim them continuously (100 to 10%) with low voltage.
SENSORS:	
Clocks	System to regulate the illumination distribution as a function of time.
Personnel	Sensor that detects whether a space is occupied by sensing the motion of an occupant.
Photocell	Sensor that measures the illumination level of a designated area.
COMMUNICATION:	
Computer/Microprocessor	Method for automatically communicating instructions and/or input from sensors to commands to the light controllers.
Power-Line Carrier	Method for carrying information over existing power lines rather than dedicated hard-wired communication lines.

form various lighting control systems in order to accomplish different combinations of control strategies.

FLUORESCENT LIGHTING CONTROL SYSTEMS

The control of fluorescent lighting systems is receiving increased attention. Two major categories of lighting control are available—personnel sensors and lighting compensators.

Personnel Sensors

There are three classifications of personnel sensors—ultrasonic, infrared and audio.

Ultrasonic sensors generate sound waves outside the human hearing range and monitor the return signals. Ultrasonic sensor systems are generally made up of a main sensor unit with a network of satellite sensors providing coverage throughout the lighted area. Coverage per sensor is dependent upon the sensor type and ranges between 500 and 2,000 square feet. Sensors may be mounted above the ceiling, suspended below the ceiling or mounted on the wall. Energy savings are dependent upon the room size and occupancy. Advertised savings range from 20 to 40 percent.

Several companies manufacture ultrasonic sensors including Novita and Unenco.

Infrared sensor systems consist of a sensor and control unit. Coverage is limited to approximately 130 square feet per sensor. Sensors are mounted on the ceiling and usually directed towards specific work stations. They can be tied into the HVAC control and limit its operation also. Advertised savings range between 30 and 50 percent. (See Figure 4-9.)

Audio sensors monitor sound within a working area. The coverage of the sensor is dependent upon the room shape and the mounting height. Some models advertise coverage of up to 1,600 square feet. The first cost of the audio sensors is approximately one half that of the ultrasonic sensors. Advertised energy savings are approximately the same as the ultrasonic sensors. Advertised energy savings are approximately the same as the ultrasonic sensors. Several restrictions apply to the use of the audio sensors. First, normal background noise must be less than 60 dB. Second, the building should be at least 100 feet from the street and may not have a metal roof.

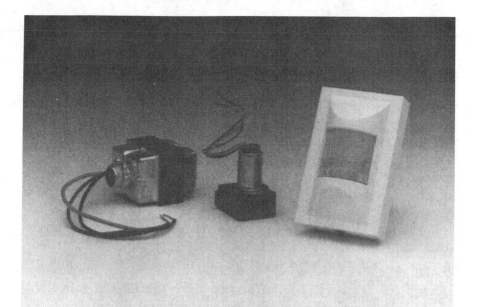

**Figure 4-9. Transformer, relay and wide view infrared sensor
to control lights.** *(Photograph courtesy of Sensorswitch)*

Lighting Compensators

Lighting compensators are divided into two major groups—switched and sensored.

Switched compensators control the light level using a manually operated wall switch. These particular systems are used frequently in residential settings and are commonly known as "dimmer switches." Based on discussions with manufacturers, the switched controls are available for the 40-watt standard fluorescent bulbs only. The estimated savings are difficult to determine, as usually switched control systems are used to control room mood. The only restriction to their use is that the luminaire must have a dimming ballast.

Sensored compensators are available in three types. They may be very simple or very complex. They may be integrated with the building's energy management system or installed as a stand-alone system. The first type of system is the Excess Light Turn-Off (ELTO) system. This sytem senses daylight levels and automatically turns off lights as the sensed light level approaches a programmed upper limit. Advertised paybacks for these types of systems range from 1.8 to 3.8 years.

The second type of system is the Daylight Compensator (DAC) system. This system senses daylight levels and automatically dims lights to achieve a programmed room light level. Advertised savings range from 40 to 50 percent. The primary advantage of this system is it maintains a uniform light level across the controlled system area. The third system type is the Daylight Compensator + Excess Light Turn-Off system. As implied by the name, this system is a combination of the first two systems. It automatically dims light outputs to achieve a designated light level and, as necessary, automatically turns off lights to maintain the desired room conditions.

Specular reflectors: Fluorescent fixtures can be made more efficient by the insertion of a suitably shaped specular reflector. The specular reflector material types are aluminum, silver and multiple dielectric film mirrors. The latter two have the highest reflectivity while the aluminum reflectors are less expensive.

Measurements show the fixture efficiency with higher reflectance specular reflectors (silver or dielectric films) is improved by 15 percent compared to a new fixture with standard diffuse reflectors.

Specular reflectors tend to concentrate more light downward with reduced light at high exit angles. This increases the light modulation in the space, which is the reason several light readings at different sites around the fixture are required for determining the average illuminance. The increased downward component of candle power may increase the potential for reflected glare from horizontal surfaces.

When considering reflectors, information should be obtained on the new candle power characteristics. With this information a lighting designer or engineer can estimate the potential changes in modulation and reflected glare.

ENERGY MANAGEMENT

The availability of computers at moderate costs and the concern for reducing energy consumption have resulted in the application of computer-based controllers to more than just industrial process applications. These controllers, commonly called Energy Management Systems (EMS), can be used to control virtually all non-process energy using pieces of equipment in buildings and industrial plants. Equipment controlled can include fans, pumps, boilers, chillers and lights. This section will investigate the various types of Energy Management Systems which are available and illustrate some of the methods used to reduce energy consumption.

THE TIMECLOCK

One of the simplest and most effective methods of conserving energy in a building is to operate equipment only when it is needed. If, due to time, occupancy, temperature or other means, it can be determined that a piece of equipment does not need to operate, energy savings can be achieved without affecting occupant comfort by turning off the equipment.

One of the simplest devices to schedule equipment operation is the mechanical timeclock. The timeclock consists of a rotating disk which is divided into segments corresponding to the hour of the day and the day of the week. This disk makes one complete revolution in, depending on the type, a 24-hour or a 7-day period. (See Figure 4-10.)

On and off "lugs" are attached to the disk at appropriate positions corresponding to the schedule for the piece of equipment. As the disk rotates, the lugs cause a switch contact to open and close, thereby controlling equipment operation.

A common application of timeclocks is scheduling office building HVAC equipment to operate during business hours Monday through Friday and to be off all other times. As is shown in the following problem, significant savings can be achieved through the correct application of timeclocks.

Example Problem 4-3

An office building utilizes two 50 hp supply fans and two 15 hp return fans which operate continuously to condition the building. What are the annual savings that result from installing a timeclock to operate these fans from 7:00 a.m. to 5:00 p.m., Monday through Friday? Assume an electrical rate of $0.08/kWh.

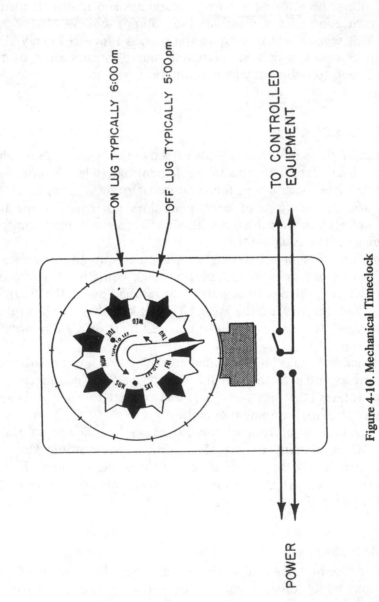

Figure 4-10. Mechanical Timeclock

Answer

> Annual Operation Before Timeclock =
> 52 weeks X 7 days/week X 24 hours/day = 8736 hours
>
> Annual Operation After Timeclock =
> 52 X (5 days/week X 10 hours/day) = 2600 hours
>
> Savings = 130 hp X 0.746 kW/hp X (8736–2600) hours X
> $0.08/kWh = $47,600

Although most buildings today utilize some version of a time-clock, the magnitude of the savings value in this example illustrates the importance of correct timeclock operation and the potential for additional costs if this device should malfunction or be adjusted inaccurately. Note that the above example also ignores heating and cooling savings which would result from the installation of a time-clock.

PROBLEMS WITH MECHANICAL TIMECLOCKS

Although the use of mechanical timeclocks in the past has resulted in significant energy savings, they are being replaced by Energy Management Systems because of problems that include the following:

- The on/off lugs sometimes loosen or fall off.
- Holidays, when the building is unoccupied, cannot easily be taken into account.
- Power failures require the timeclock to be reset or it is not synchronized with the building schedule.
- Inaccuracies in the mechanical movement of the timeclock prevent scheduling any closer than ±15 minutes of the desired times.
- There are a limited number of on and off cycles possible each day.
- It is a time-consuming process to change schedules on multiple timeclocks.

Energy Management Systems, or sometimes called electronic timeclocks, are designed to overcome these problems plus provide increased control of building operations.

ENERGY MANAGEMENT SYSTEMS

Recent advances in digital technology, dramatic decreases in the cost of this technology and increased energy awareness have resulted in the increased application of computer-based controllers (i.e., Energy Management Systems) in commercial buildings and industrial plants. These devices can control anywhere from one to a virtually unlimited number of items of equipment.

By concentrating the control of many items of equipment at a single point, the EMS allows the building operator to tailor building operation to precisely satisfy occupant needs. This ability to maximize energy conservation, while preserving occupant comfort, is the ultimate goal of an energy engineer.

Microprocessor Based

Energy Management Systems can be placed in one of two broad, and sometimes overlapping, categories referred to as microprocessor-based and mini-computer based.

Microprocessor-based systems can control from 1 to 40 input/outpoints and can be linked together for additional loads. Programming is accomplished by a keyboard or hand-held console, and an LED display is used to monitor/review operation of the unit. A battery maintains the programming in the event of power failure. (See Figures 4-11 and 4-12.)

Capabilities of this type of EMS are generally pre-programmed so that operation is relatively straightforward. Programming simply involves entering the appropriate parameters (e.g., the point number and the on and off times) for the desired function. Microprocessor-based EMS can have any or all of the following capabilities:

- Scheduling
- Duty Cycling
- Demand Limiting

- Optimal Start
- Monitoring
- Direct Digital Control

Scheduling

Scheduling with an EMS is very much the same as it is with a timeclock. Equipment is started and stopped based on the time cf day and the day of week. Unlike a timeclock, however, multiple start/stops can be accomplished very easily and accurately (e.g.,

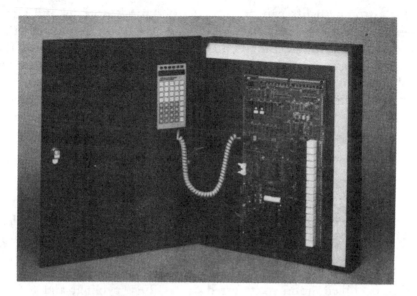

Figure 4-11. Microprocessor-Based Programmable EMS
(Photograph Courtesy Control Systems International)

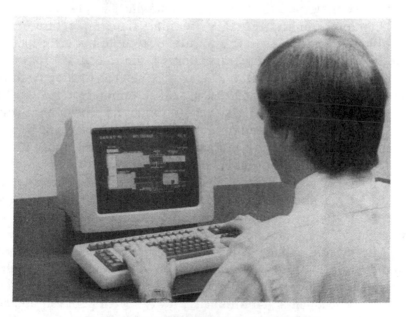

Figure 4-12. Keyboard Used to Access EMS
(Photo Courtesy Control Systems International)

in a classroom, lights can be turned off during morning and after-noon break periods and during lunch). It should be noted that this single function, if accurately programmed and depending on the type of facility served, can account for the largest energy savings attributable to an EMS.

Additionally, holiday dates can be entered into the EMS a year in advance. When the holiday occurs, regular programming is over-ridden and equipment can be kept off.

Duty Cycling

Most HVAC fan systems are designed for peak load conditions, and consequently these fans are usually moving much more air than is needed. Therefore, they can sometimes be shut down for short periods each hour, typically 15 minutes, without affecting occupant comfort. Turning equipment off for pre-determined periods of time during occupied hours is referred to as duty cycling, and can be accomplished very easily with an EMS. Duty cycling saves fan and pump energy but does not reduce the energy required for space heating or cooling since the thermal demand must still be met.

The more sophisticated EMS's monitor the temperature of the conditioned area and use this information to automatically modify the duty cycle length when temperatures begin to drift. If, for example, the desired temperature in an area is 70° and at this temperature equipment is cycled 50 minutes on and 10 minutes off, a possible temperature-compensated EMS may respond as shown in Figure 4-13. As the space temperature increases above (or below if so pro-grammed) the setpoint, the equipment off time is reduced until, at 80° in this example, the equipment operates continuously.

Duty cycling is best applied in large, open-space offices which are served by a number of fans. Each fan could be programmed so that the off times do not coincide, thereby assuring adequate air flow to the offices at all times.

Duty cycling of fans which provide the only air flow to an area should be approached carefully to insure that ventilation require-ments are maintained and that varying equipment noise does not annoy the occupants. Additionally, duty cycling of equipment imposes extra stress on motors and associated equipment. Care should be taken, particularly with motors over 20 hp, to prevent

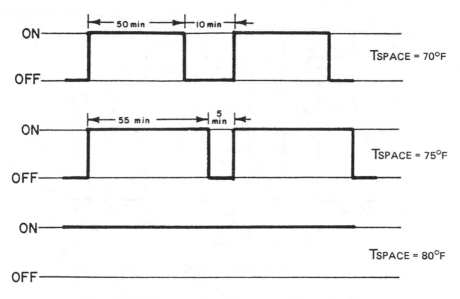

Figure 4-13. Temperature Compensated Duty Cycling

starting and stopping of equipment in excess of what is recommended by the manufacturer.

Demand Charges

Electrical utilities charge commercial customers based not only on the amount of energy used (kWh) but also on the peak demand (kW) for each month. Peak demand is very important to the utility so that they may properly size the required electrical service and insure that sufficient peak generating capacity is available to that given facility.

In order to determine the peak demand during the billing period, the utility establishes short periods of time called the demand interval (typically 15, 30, or 60 minutes). The billing demand is defined as the highest average demand recorded during any one demand interval within the billing period. (See Figure 4-14.) Many utilities now utilize "ratchet" rate charges. A "ratchet" rate means that the billed demand for the month is based on the highest demand in the previous 12 months, or an average of the current month's peak demand and the previous highest demand in the past year.

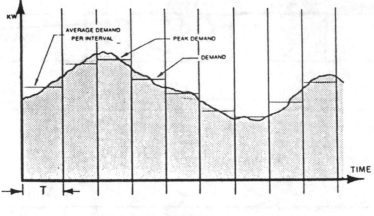

T - DEMAND INTERVAL

Figure 4-14. Peak Demand

Depending on the facility, the demand charge can be a significant portion, as much as 20%, of the utility bill. The user will get the most electrical energy per dollar if the load is kept constant, thereby minimizing the demand charge. The objective of demand control is to even out the peaks and valleys of consumption by deferring or rescheduling the use of energy during peak demand periods.

A measure of the electrical efficiency of a facility can be found by calculating the load factor. The load factor is defined as the ratio of energy usage (kWh) per month to the peak demand (kW) X the facility operating hours.

Example Problem 4-4

What is the load factor of a continuously operating facility that consumed 800,000 kWh of energy during a 30-day billing period and established a peak demand of 2000 kW?

Answer

$$\text{Load Factor} = \frac{800,000 \text{ kWh}}{2000 \text{ kW X 30 days X 24 hours/day}} = 0.55$$

The ideal load factor is 1.0, at which demand is constant; therefore, the difference between the calculated load factor and 1.0 gives an indication of the potential for reducing peak demand (and demand charges) at a facility.

Demand Limiting

Energy Management Systems with demand limiting capabilities utilize either pulses from the utility meter or current transformers to predict the facility demand during any demand interval. If the facility demand is predicted to exceed the user-entered setpoint, equipment is "shed" to control demand. Figure 4-15 illustrates a typical demand chart before and after the actions of a demand limiter.

Electrical load in a facility consists of two major categories: essential loads which include most lighting, elevators, escalators, and most production machinery; and non-essential ("sheddable") loads such as electric heaters, air conditioners, exhaust fans, pumps, snow melters, compressors and water heaters. Sheddable loads will not,

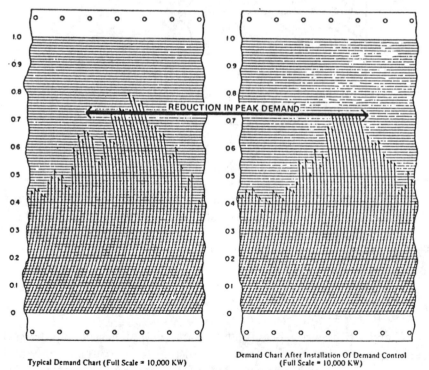

Typical Demand Chart (Full Scale = 10,000 KW)

Demand Chart After Installation Of Demand Control
(Full Scale = 10,000 KW)

Figure 4-15. Demand Limiting Comparison

when turned off for short periods of time to control demand, affect productivity or comfort.

To prevent excessive cycling of equipment, most Energy Management Systems have a deadband that demand must drop below before equipment operation is restored (See figure 4-16). Additionally, minimum on and maximum off times and shed priorities can be entered for each load to protect equipment and insure that comfort is maintained.

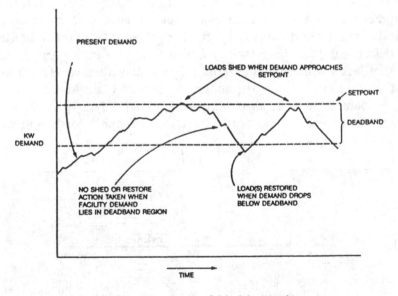

Figure 4-16. Demand Limiting Actions

It should be noted that demand shedding of HVAC equipment in commercial office buildings should be applied with caution. Since times of peak demand often occur during times of peak air conditioning loads, excessive demand limiting can result in occupant discomfort.

Time of Day Billing

Many utilities are beginning to charge their larger commercial users based on the time of day that consumption occurs. Energy and demand during peak usage periods (i.e., summer weekday afternoons and winter weekday evenings) are billed at much higher rates than consumption during other times. This is necessary because utilities must augment the power production of their large power plants during periods of peak demand with small generators which are expensive to operate. Some of the more sophisticated Energy Management Systems can now account for these peak billing periods with different demand setpoints based on the time of day and day of week.

Optimal Start

External building temperatures have a major influence on the amount of time it takes to bring the building temperature up to occupied levels in the morning. Buildings with mechanical time clocks usually start HVAC equipment operation at an early enough time in the morning (as much as 3 hours before occupancy time) to bring the building up to temperature on the coldest day of the year. During other times of the year when temperatures are not as extreme, building temperatures can be up to occupied levels several hours before it is necessary, and consequently unnecessary energy is used. (See Figure 4-17.)

Energy Management Systems with optimal start capabilities, however, utilize indoor and outdoor temperature information, along with learned building characteristics, to vary start time of HVAC equipment so that building temperatures reach desired values just as occupancy occurs. Consequently, if a building is scheduled to be occupied at 8:00 a.m., on the coldest day of the year, the HVAC equipment may start at 5:00 a.m. On milder days, however, equipment may not be started until 7:00 a.m. or even later, thereby saving significant amounts of energy.

Most Energy Management Systems have a "self-tuning" capability to allow them to learn the building characteristics. If the building is heated too quickly or too slowly on one day, the start time is adjusted the next day to compensate.

Monitoring

Microprocessor-based EMS can usually accomplish a limited amount of monitoring of building conditions including the following:

- Outside air temperature

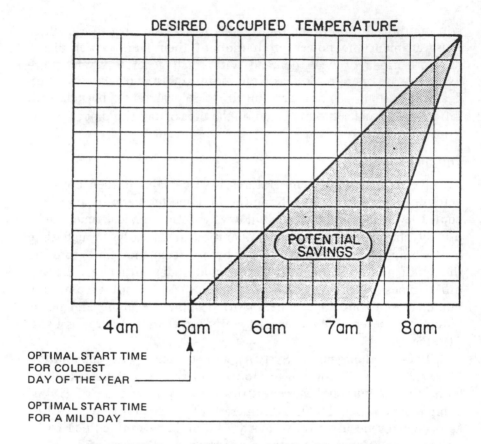

Figure 4-17. Typical Variation in Building Warm-Up Times

- Several indoor temperature sensors
- Facility electrical energy consumption and demand
- Several status input points

The EMS can store this information to provide a history of the facility. Careful study of these trends can reveal information about facility operation that can lead to energy conservation strategies that might not otherwise be apparent.

Direct Digital Control

The most sophisticated of the microprocessor-based EMSs provide a function referred to as direct digital control (DDC). This capability allows the EMS to provide not only sophisticated energy management but also basic temperature control of the building's HVAC systems.

Direct digital control has taken over the majority of all process control application and is now becoming an important part of the HVAC industry. Traditionally, pneumatic controls were used in most commercial facilities for environmental control.

The control function in a traditional facility is performed by a pneumatic controller which receives its input from pneumatic sensors (i.e., temperature, humidity) and sends control signals to pneumatic actuators (valves, dampers, etc.). Pneumatic controllers typically perform a single, fixed function which cannot be altered unless the controller itself is changed or other hardware is added. (See Figure 4-18 for a typical pneumatic control configuration.)

With direct digital control, the microprocessor functions as the primary controller. Electronic sensors are used to measure variables such as temperature, humidity and pressure. This information is used, along with the appropriate application program, by the microprocessor to determine the correct control signal, which is then sent directly to the controlled device (valve or damper actuator). (See Figure 4-18 for a typical DDC configuration.)

Direct digital control (DDC) has the following advantage over pneumatic controls:

- Reduces overshoot and offset errors, thereby saving energy
- Flexibility to easily and inexpensively accomplish changes of control strategies
- Calibration is maintained more accurately, thereby saving energy and providing better performance.

To program the DDC functions, a user programming language is utilized. This programming language uses simple commands in English to establish parameters and control strategies.

Mini-Computer Based

Mini-computer based EMS can provide all the functions of the microprocessor based EMS, as well as the following:

- Extensive graphics

- Special reports and studies
- Fire and security monitoring and detection
- Custom programs

These devices can control and monitor from 50 to an unlimited number of points and form the heart of a building's (or complex's) operations.

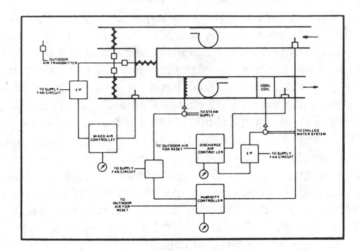

CONVENTIONAL PNEUMATIC CONTROL SYSTEM

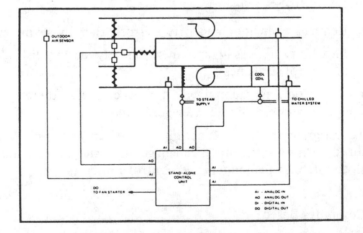

DIRECT DIGITAL CONTROL SYSTEM

Figure 4-18. Comparison of Pneumatic and DDC Controls

Figure 4-19 shows a typical configuration for this type of system.

The "central processing unit" (CPU) is the heart of the EMS. It is a mini-computer with memory for the operating system and applications software. The CPU performs arithmetic and logical decisions necessary to perform central monitoring and control.

Data and programs are stored or retrieved from the memory or mass storage devices (generally a disk storage system). The CPU has programmed I/O ports for specific equipment, such as printers and cathode ray tube (CRT) consoles. During normal operation, it coordinates operation of all other EMS components.

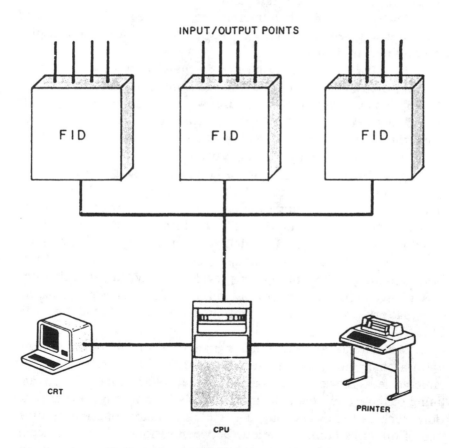

Figure 4-19. Mini-Computer Based EMS

A cathode ray tube console (CRT), either color and/or black and white, with a keyboard is used for operator interaction with the EMS. It accepts operator commands, displays data and graphically displays systems controlled or monitored by the EMS. A "printer," (or printers) provides a permanent copy of system operations and historical data.

A "field interface device" (FID) provides an interface to the points which are monitored and controlled, performs engineering conversions to or from a digital format, performs calculations and logical operations, accepts and processes CPU commands and is capable, in some versions, of stand-alone operations in the event of CPU or communications link failure.

The FID is essentially a microprocessor based EMS as described in the previous section. It may or may not have a keyboard/display unit on the front panel.

The FIDs are generally located in the vicinity of the points to be monitored and/or controlled and are linked together and to the CPU by a single twisted pair of wires which carries multiplexed data (i.e., data from a number of sources combined on a single channel) from the FID to the CPU and back. In some versions, the FIDs can communicate directly with each other.

Early versions of mini-computer based EMS used the CPU to perform all of the processing with the FID used merely for input and output. A major disadvantage of this type of "centralized" system is that the loss of the CPU disables the entire control system. The development of "intelligent" FIDs in a configuration known as "distributed processing" helped to solve this problem. This system, which is becoming prevalent today, utilizes microprocessor-based FIDs to function as remote CPUs. Each panel has its own battery pack to insure continued operation should the main CPU fail.

Each intelligent panel sends signals back to the main CPU only upon a change of status rather than continuously transmitting the same value as previous "centralized" systems have done. This streamlining of data flow to the main CPU frees it to perform other functions such as trend reporting. The CPU's primary function becomes one of directing communications between various FID panels, generating reports and graphics and providing operator interface for programming and monitoring.

Features

The primary difference in operating functions of the mini-computer based EMS is its increased capability to monitor building operations. For this reason, these systems are sometimes referred to as Energy Monitoring and Control Systems (EMCS). Analog inputs such as temperature and humidity can be monitored, as well as digital inputs such as pump or valve status.

The mini-computer based EMS is also designed to make operator interaction very easy. Its operation can be described as "user friendly" in that the operator, working through the keyboard, enters information in English in a question and response format. In addition, custom programming languages are available so that powerful programs can be created specifically for the building through the use of simplified English commands.

The graphics display CRT can be used to create HVAC schematics, building layouts, bar charts, etc. to better understand building systems operation. These graphics can be "dynamic" so that values and statuses are continuously updated.

Many mini-computer based EMS can also easily incorporate fire and security monitoring functions. Such a configuration is sometimes referred to as a Building Automation System (BAS). By combining these functions with energy management, savings in initial equipment costs can be achieved. Reduced operating costs can be achieved as well by having a single operator for these systems.

The color graphics display can be particularly effective in pinpointing alarms as they occur within a building and guiding quick and appropriate response to that location. In addition, management of fan systems to control smoke in a building during a fire is facilitated with a system that combines energy management and fire monitoring functions.

Note, however, that the incorporation of fire, security and energy management functions into a single system increases the complexity of that system. This can result in longer start-up time for the initial installation and more complicated troubleshooting if problems occur. Since the function of fire monitoring is critical to building operation, these disadvantages must be weighed against the previously mentioned advantages to determine if a combined BAS is desired.

DATA TRANSMISSION METHODS

A number of different transmission systems can be used in an EMS for communications between the CPU and FID panels. These transmission systems include telephone lines, coaxial cables, electrical power lines, radio frequency, fiber optics and microwave. Table 4-6 compares the various transmission methods.

Table 4-6. Transmission Method Comparisons

Method	First Cost	Scan Rates	Reliability	Maint. Effort	Expandibility	Compatibility with Future Requirements
Coaxial	high	fast	excellent	min.	unlimited	unlimited
Twisted pair	low	med.	very good	min.	unlimited	unlimited
RF	med.	fast but limited	low	high	very limited	very limited
Microwave	very high	very fast	excellent	high	unlimited	unlimited
Telephone	very low	slow	low to high	min.	limited	limited
Fiber optics	high	very fast	excellent	min.	unlimited	unlimited
Power Line Carrier	med.	med.	med.	high	limited	limited

Twisted Pair

One of the most common data transmission methods for an EMS is a twisted pair of wires. A twisted pair consists of two insulated conductors twisted together to minimize interference from unwanted signals.

Twisted pairs are permanently hardwired lines between the equipment sending and receiving data that can carry information over a wide range of speeds, depending on line characteristics. To maintain a particular data communication rate, the line bandwidth, time delay or the signal-to-noise ratio may require adjustment by conditioning the line.

Data transmission in twisted pairs, in most cases, is limited to 1200 bps (bits per second) or less. By using signal conditioning, operating speeds up to 9600 bps may be obtained.

Voice Grade Telephone Lines

Voice grade lines used for data transmission are twisted pair circuits defined as type 3002 by the Bell Telephone Company. The local telephone company charges a small connection fee for this service, plus a monthly equipment lease fee. Maintenance is included in the monthly lease fee, with a certain level of service guaranteed.

Two of the major problems involve the quality of telephone pairs provided to the installer and the transmission rate. The minimum quality of each line intended for use in the trunk wiring system must be clearly defined.

The most common voice grade line used for data communication is the unconditioned type 3002 allowing transmission rates up to 1200 bps. The 3002 type line may be used for data transmission up to 9600 bps with the proper line conditioning. Voice grade lines must be used with the same constraints and guidelines for twisted pairs.

Coaxial Cable

Coaxial cable consists of a center conductor surrounded by a shield. The center conductor is separated from the shield by a dielectric. The shield protects against electromagnetic interference. Coaxial cables can operate at data transmission rates in the megabits per second range, but attenuation becomes greater as the data transmission rate increases.

The transmission rate is limited by the data transmission equipment and not by the cable. Regenerative repeaters are required at specific intervals depending on the data rate, nominally every 2000 feet to maintain the signal at usable levels.

Power Line Carrier

Data can be transmitted to remote locations over electric power lines using carrier current transmission that superimposes a low power RF (radio frequency) signal, typically 100 kHz, onto the 60 Hz power distribution system. Since the RF carrier signal cannot operate across transformers, all communicating devices must be connected to the same power circuit (same transformer secondary and phase) unless RF couplers are installed across transformers permitting the transmitters and receivers to be connected over a wider area of the power system. Transmission can be either one-way or two-way.

Note that power line carrier technology is sometimes used in microprocessor based EMS retrofit applications to control single loads in a facility where hardwiring would be difficult and expensive (e.g., wiring between two buildings). Figure 4-20 shows a basic power line carrier system configuration.

Radio Frequency

Modulated RF signals, usually VHF or FM radio, can be used as a data transmission method with the installation of radio receivers and transmitters. RF systems can be effectively used for two-way communication between CPU and FID panels where other data transmission methods are not available or suitable for the application. One-way RF systems can be effectively used to control loads at remote locations such as warehouses and unitary heaters and for family housing projects.

The use of RF at a facility, however, must be considered carefully to avoid conflict with other existing or planned facility RF systems. Additionally, there may be a difficulty in finding a frequency on which to transmit, since there are a limited number available.

The kinds of signals sent over an FM radio system are also limited as are the distances over which the signals can be transmitted. The greater the distance, the greater the likelihood that erroneous signals will be received.

Fiber Optics

Fiber optics uses the wideband properties of infrared light traveling through transparent fibers. Fiber optics is a reliable communications media which is rapidly becoming cost competitive when compared to other high-speed data transmission methods.

The bandwidth of this media is virtually unlimited, and extremely high transmission rates can be obtained. The signal attenuation of high-quality fiber optic cable is lower than the best coaxial cables.

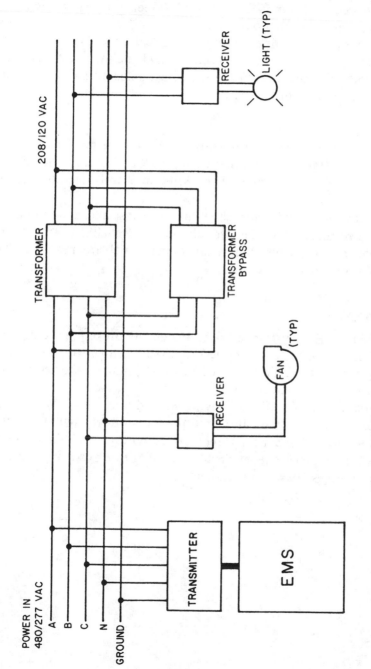

Figure 4-20. Power Line Carrier System Configuration

Repeaters required nominally every 2000 feet, for coaxial cable, are 3 to 6 miles apart in fiber optics systems.

Fiber optics terminal equipment selection is limited to date, and there is a lack of skilled installers and maintenance personnel familiar with this media. Fiber optics must be carefully installed and cannot be bent at right angles.

Microwave Transmission

For long distance transmission, a microwave link can be used. The primary drawback of microwave links is first cost. Receivers/transmitters are needed at each building in a multi-facility arrangement.

Microwave transmission rates are very fast and are compatible with present and future data requirements. Reliability is excellent, too, but knowledgeable maintenance personnel are required. The only limit on expansion is cost.

SUMMARY

The term Energy Management System denotes equipment whose functions can range from simple timeclock control to sophisticated building automation. Two broad and overlapping categories of these systems are microprocessor and mini-computer based.

Capabilities of EMS can include scheduling, duty cycling, demand limiting, optimal start, monitoring, direct digital control, fire detection and security. Direct digital control capability enables the EMS to replace the environmental control system so that it directly manages HVAC operations.

5

Waste Heat Recovery

INTRODUCTION

Waste heat is heat which is generated in a process but then "dumped" to the environment even though it could still be reused for some useful and economic purpose.

The essential quality of heat is not the amount but rather its "value."

The strategy of how to recover this heat depends in part on the temperature of the waste heat gases and the economics involved.

This chapter will present the various methods involved in traditionally recovering waste heat.

Portions of material used in Chapters 5 and 6 are based upon the *Waste Heat Management Guidebooks* published by the U.S. Department of Commerce/National Bureau of Standards. The authors express appreciation to Kenneth G. Kreider; Michael B. McNeil; W. M. Rohrer, Jr.; R. Ruegg; B. Leidy and W. Owens who have contributed extensively to this publication.

SOURCES OF WASTE HEAT

Sources of waste energy can be divided according to temperature into three temperature ranges. The high temperature range refers to temperatures above 1200°F. The medium temperature range is between 450°F and 1200°F, and the low temperature range is below 450°F.

High and medium temperature waste heat can be used to produce process steam. If high temperature waste heat exists, instead of producing steam directly, consider the possibility of using the high temperature energy to do useful work before the waste heat is extracted. Both gas and steam turbines are useful and fully developed heat engines.

In the low temperature range, waste energy which would be otherwise useless can sometimes be made useful by application of mechanical work through a device called the heat pump.

HIGH TEMPERATURE HEAT RECOVERY

The combustion of hydrocarbon fuels produces product gases in the high temperature range. The maximum theoretical temperature possible in atmospheric combustors is somewhat under 3500°F, while measured flame temperatures in practical combustors are just under 3000°F. Secondary air or some other dilutant is often admitted to the combustor to lower the temperature of the products to the required process temperature (for example, to protect equipment) thus lowering the practical waste heat temperature.

Table 5-1 gives temperatures of waste gases from industrial process equipment in the high temperature range. All of these result from direct fuel fired processes.

Table 5-1

Type of Device	Temperature F
Nickel refining furnace	2500–3000
Aluminum refining furnace	1200–1400
Zinc refining furnace	1400–2000
Copper refining furnace	1400–1500
Steel heating furnaces	1700–1900
Copper reverberatory furnace	1650–2000
Open hearth furnace	1200–1300
Cement kiln (Dry process)	1150–1350
Glass melting furnace	1800–2800
Hydrogen plants	1200–1800
Solid waste incinerators	1200–1800
Fume incinerators	1200–2600

MEDIUM TEMPERATURE HEAT RECOVERY

Table 5-2 gives the temperatures of waste gases from process equipment in the medium temperature range. Most of the waste heat in this temperature range comes from the exhausts of directly fired process units. Medium temperature waste heat is still hot enough to allow consideration of the extraction of mechanical work from the waste heat, by a steam or gas turbine. Gas turbines can be economically utilized in some cases at inlet pressures in the range of 15 to 30 lb/in² g. Steam can be generated at almost any desired pressure and steam turbines used when economical.

Table 5-2

Type of Device	Temperature F
Steam boiler exhausts	450–900
Gas turbine exhausts	700–1000
Reciprocating engine exhausts	600–1100
Reciprocating engine exhausts (turbocharged)	450–700
Heat treating furnaces	800–1200
Drying and baking ovens	450–1100
Catalytic crackers	800–1200
Annealing furnace cooling systems	800–1200

LOW TEMPERATURE HEAT RECOVERY

Table 5-3 lists some heat sources in the low temperature range. In this range it is usually not practical to extract work from the source, though steam production may not be completely excluded if there is a need for low pressure steam. Low temperature waste heat may be useful in a supplementary way for preheating purposes. Taking a common example, it is possible to use economically the energy from an air-conditioning condenser operating at around 90°F to heat the domestic water supply. Since the hot water must be heated to about 160°F, obviously the air-conditioner waste heat is not hot enough. However, since the cold water enters the domestic water system at about 50°F, energy interchange can take place to raise the water to something less than 90°F. Depending upon the relative air-conditioning load and hot water requirements, any excess condenser heat can be rejected, and the additional energy required by the hot water can be provided by the usual electrical or fired heater.

Table 5-3

Source	Temperature F
Process steam condensate	130–190
Cooling water from:	
Furnace doors	90–130
Bearings	90–190
Welding machines	90–190
Injection molding machines	90–190
Annealing furnaces	150–450
Forming dies	80–190
Air compressors	80–120
Pumps	80–190
Internal combustion engines	150–250
Air conditioning and refrigeration condensers	90–110
Liquid still condensers	90–190
Drying, baking and curing ovens	200–450
Hot processed liquids	90–450
Hot processed solids	200–450

WASTE HEAT RECOVERY APPLICATIONS

To use waste heat from sources such as those above, one often wishes to transfer the heat in one fluid stream to another (e.g., from flue gas to feedwater or combustion air). The device which accomplishes the transfer is called a heat exchanger. In the discussion immediately below is a listing of common uses for waste heat energy and, in some cases, the name of the heat exchanger that would normally be applied in each particular case.

The equipment that is used to recover waste heat can range from something as simple as a pipe or duct to something as complex as a waste heat boiler.

Some applications of waste heat are as follows:

• Medium to high temperature exhaust gases can be used to preheat the combustion air for:

Boilers using air preheaters

Furnaces using recuperators

Ovens using recuperators

Gas turbines using regenerators

• Low to medium temperature exhaust gases can be used to preheat boiler feedwater or boiler makeup water using *economizers*, which are simply gas-to-liquid water heating devices.

• Exhaust gases and cooling water from condensers can be used to preheat liquid and/or solid feedstocks in industrial processes. Finned tubes and tube-in-shell *heat exchangers* are used.

• Exhaust gases can be used to generate steam in *waste heat boilers* to produce electrical power, mechanical power, process steam, and any combination of above.

• Waste heat may be transferred to liquid or gaseous process units directly through pipes and ducts or indirectly through a secondary fluid such as steam or oil.

• Waste heat may be transferred to an intermediate fluid by heat exchangers or waste heat boilers, or it may be used by circulating the hot exit gas through pipes or ducts. Waste heat can be used to operate an absorption cooling unit for air conditioning or refrigeration.

THE WASTE HEAT RECOVERY SURVEY

In order to identify sources of waste heat, a survey is usually made. Figure 5-1 illustrates a survey form which can be used for the Waste Heat Audit. It is important to record flow and temperature of waste gases.

Composition data is required for heat recovery and system design calculations. Be sure to note contaminants since this factor could limit the type of heat recovery equipment to apply. Contaminants can foul or plug heat exchangers.

Operation schedule affects the economics and type of equipment to be specified. For example, an incinerator that is only used one shift per day may require a different method of recovering discharges than if it were used three shifts a day. A heat exchanger used for waste heat recovery in this service would soon deteriorate due to metal fatigue. A different type of heat recovery incinerator utilizing heat storage materials such as rock or ceramic would be more suitable.

SURVEY FORM FOR INDUSTRIAL PROCESS UNITS

NAME OF PROCESS UNIT _____ INVENTORY NUMBER _____

LOCATION OF PROCESS UNIT, PLANT NAME _____ BUILDING _____

MANUFACTURER _____ MODEL _____ SERIAL NUMBER _____

| | | FIRING RATE | HHV | TEMPERATURE OF | | | FLUE GAS COMPOSITION % VOLUME | | | | |
	NAME			COMB. AIR	FUEL	STACK	CO_2	O_2	CO	CH	N_2
PRIMARY FUEL											
FIRST ALTERNAT.											
SECOND ALTERNAT.											

	FLOW PATH 1	FLOW PATH 2	FLOW PATH 3	FLOW PATH 4
FLUID COMPOSITION				
FLOW RATE				
INLET TEMPERATURE				
OUTLET TEMPERATURE				
DESCRIPTION				

ANNUAL HOURS OPERATION _____ ANNUAL CAPACITY FACTOR, % _____

ANNUAL FUEL CONSUMPTION: PRIMARY FUEL _____; FIRST ALTERN. _____ SEC. ALTERN. _____

PRESENT FUEL COST: PRIMARY FUEL _____; FIRST ALTERN. _____ SEC. ALTERN. _____

ANNUAL ELECTRICAL ENERGY CONSUMPTION, KWHR. _____

PRESENT ELECTRICAL ENERGY RATE _____

CONTAMINANTS _____

Figure 5-1. Waste Heat Survey

WASTE HEAT RECOVERY CALCULATIONS

From the heat balance (Chapter 6), the heat recovered from the source is determined by Formula 5-1.

$$q = m \, c_p \, \Delta T \qquad\qquad \textit{Formula (5-1)}$$

Where q = heat recovered, Btu/hr
m = mass flow rate lb/hr
c_p = specific heat of fluid, Btu/lb°F
ΔT = temperature change of gas or liquid during heat recovery °F

If the flow is air, then Formula 5-1 can be expressed as

$$q = 1.08 \text{ cfm } \Delta T \qquad\qquad \textit{Formula (5-2)}$$

Where   cfm = volume flow rate in standard cubic feet per minute

If the flow is water, then Formula 5-1 can be expressed as

$$q = 500 \text{ gpm } \Delta T \qquad\qquad \textit{Formula (5-3)}$$

Where gpm = volume flow rate in gallons per minute

Example Problem 5-1

A Waste Heat Audit survey indicates 10,000 lb/hr of water at 190°F is discharged to the sewer. How much heat can be saved by utilizing this fluid as makeup to the boiler instead of the 70°F feedwater supply? Fuel cost is $6 per million Btu, boiler efficiency .8, and hours of operation 4000.

Analysis

$q = mc_p \ \Delta T = 10,000 \times (190{-}70) = 1.2 \times 10^6$ Btu/hr
Savings $= 1.2 \times 10^6 \times 4000 \times \$6/10^6/.8 = \$36,000$

Heat Transfer by Convection

Convection is the transfer of heat to or from a fluid, gas, or liquid. Formula 5-4 is indicative of the basic form of convective heat transfer. U_0, in this case, represents the convection film conductance, Btu/ft^2 · hr · °F.

Heat transferred for heat exchanger applications is predominantly a combination of conduction and convection expressed as

$$q = U_0 \, A \Delta T_m \qquad \text{Formula (5-4)}$$

Where q = rate of heat flow by convection, Btu/hr

U_0 = is the overall heat transfer coefficient Btu/ft^2 · hr · °F

A = is the area of the tubes in square feet

ΔT_m = is the logarithmic mean temperature difference and represents the situation where the temperature of two fluids change as they transverse the surface.

$$\Delta T_m = \frac{\Delta T_1 - \Delta T_2}{\mathrm{Log}_e \; [\Delta T_1 / \Delta T_2]} \qquad \text{Formula (5-5)}$$

To understand the different logarithmic mean temperature relationships, Figure 5-2 should be used. Referring to Figure 5-2, the ΔT_m for the counterflow heat exchanger is

$$\Delta T_m = \frac{(t_1 - t_2') - (t_2 - t_1')}{\mathrm{Log}_e \; [t_1 - t_2'/t_2 - t_1']} \qquad \text{Formula (5-6)}$$

The ΔT_m for the parallel flow heat exchanger is

$$\Delta T_m = \frac{(t_1 - t_1') - (t_2 - t_2')}{\mathrm{Log}_e \; [t_1 - t_1'/t_2 - t_2']} \qquad \text{Formula (5-7)}$$

HEAT TRANSFER BY RADIATION

Radiation is the transfer of heat energy by electromagnetic means between two materials whose surfaces "see" each other. The governing equation is known as the Stefan-Boltzmann equation, and is written

$$q = \sigma F_e F_a A \; (T^4{}_{aba_1} - T^4{}_{aba_2}) \qquad \text{Formula (5-8)}$$

Again, q and A are as defined for conduction and convection, and T_{aba_1} and T_{aba_2} are the absolute temperatures of the two surfaces involved. The factor F_e is a function of the condition of the radiation surfaces and in some cases the areas of the surfaces. The factor F_a is

A. COUNTERFLOW

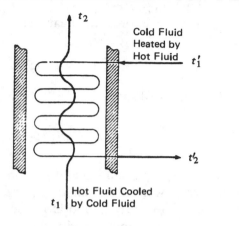

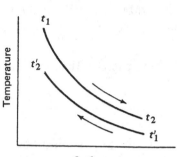

B. PARALLEL FLOW

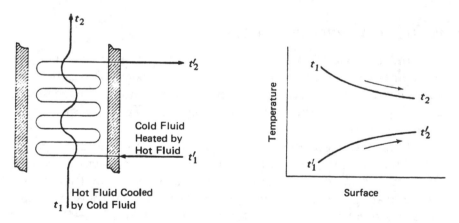

Figure 5-2. Temperature Relationships for Heat Exchangers

the configuration factor and is a function of the areas and their positions. Both F_e and F_a are dimensionless. The Stefan-Boltzmann constant σ is equal to 0.1714 But/h·ft^2·R^4. Although radiation is usually associated with solid surfaces, certain gases can emit and absorb radiation. These include the so-called nonpolar molecular gases such as H_2O, CO_2, CO, SO_2, NH_3 and the hydrocarbons. Some of these gases are present in every combustion process.

It is convenient in certain calculations to express the heat transferred by radiation in the form of

$$q = h_r A (T_1 - T_2) \qquad \text{Formula (5-9)}$$

where the coefficient of radiation, h_r, is defined as

$$h_r = \frac{\sigma F_e F_a (T^4 aba_1 - T^4 aba_2)}{(T_1 - T_1)} \qquad \text{Formula (5-10)}$$

Note that h_r is still dependent on the factors F_e and F_a and also on the absolute temperatures to the fourth power. The main advantage is that in Formula 5-10 the rate of heat flow by radiation is a function of the temperature difference and can be combined with the coefficient of convection to determine the total heat flow to or from a surface.

WASTE HEAT RECOVERY EQUIPMENT

Industrial heat exchangers have many pseudonyms. They are sometimes called recuperators, regenerators, waste heat steam generators, condensers, heat wheels, temperature and moisture exchangers, etc. Whatever name they may have, they all perform one basic function: the transfer of heat.

Heat exchangers are characterized as single or multipass gas to gas, liquid to gas, liquid to liquid, evaporator, condenser, parallel flow, counterflow, or crossflow. The terms single or multipass refer to the heating or cooling media passing over the heat transfer surface once or a number of times. Multipass flow involves the use of internal baffles. The next three terms refer to the two fluids between which heat is transferred in the heat exchanger and imply that no phase changes occur in those fluids. Here the term "fluid" is used in the most general sense. Thus, we can say that these terms apply to non-evaporator and noncondensing heat exchangers. The term evaporator applies to a heat exchanger in which heat is transferred to an evaporating (boiling) liquid, while a condenser is a heat exchanger in which heat is removed from a condensing vapor. A parallel flow heat exchanger is one in which both fluids flow in approximately the same direction, whereas in counterflow the two fluids move in opposite directions. When the two fluids move at right angles to each other, the heat exchanger is considered to be of the crossflow type.

The principal methods of reclaiming waste heat in industrial plants make use of heat exchangers. The heat exchanger is a system which separates the stream containing waste heat and the medium which is to absorb it but which allows the flow of heat across the separation boundaries. The reasons for separating the two streams may be any of the following:

(1) A pressure difference may exist between the two streams of fluid. The rigid boundaries of the heat exchanger can be designed to withstand the pressure difference.

(2) In many, if not most, cases the one stream would contaminate the other, if they were permitted to mix. The heat exchanger prevents mixing.

(3) Heat exchangers permit the use of an intermediate fluid better suited than either of the principal exchange media for transporting waste heat through long distances. The secondary fluid is often steam, but another substance may be selected for special properties.

(4) Certain types of heat exchangers, specifically the heat wheel, are capable of transferring liquids as well as heat. Vapors being cooled in the gases are condensed in the wheel and later re-evaporated into the gas being heated. This can result in improved humidity and/or process control, abatement of atmospheric air pollution and conservation of valuable resources.

The various names or designations applied to heat exchangers are partly an attempt to describe their function and partly the result of tradition within certain industries. For example, a recuperator is a heat exchanger which recovers waste heat from the exhaust gases of a furnace to heat the incoming air for combustion. This is the name used in both the steel and the glass making industries. The heat exchanger performing the same function in the steam generator of an electric power plant is termed an air preheater, and in the case of a gas turbine plant, a regenerator.

However, in the glass and steel industries the word regenerator refers to two chambers of brick checkerwork which alternately absorb heat from the exhaust gases and then give up part of that heat to the incoming air. The flows of flue gas and of air are periodically reversed by valves so that one chamber of the regenerator is being heated by the products of combustion while the other is being cooled by the incoming air. Regenerators are often more expensive to buy and more expensive to maintain than are recuperators, and their application is primarily in glass melt tanks and in open hearth steel furnaces.

It must be pointed out, however, that although their functions are similar, the three heat exchangers mentioned above may be structurally quite different as well as different in their principal modes of heat transfer. A more complete description of the various industrial heat exchangers follows later in this chapter, and details of their differences will be clarified.

The specification of an industrial heat exchanger must include the heat exchange capacity, the temperatures of the fluids, the allowable pressure drop in each fluid path, and the properties and volumetric flow of the fluids entering the exchanger. These specifications will determine construction parameters and thus the cost of the heat exchanger. The final design will be a compromise between pressure drop, heat exchanger effectiveness, and cost. Decisions leading to that final design will balance out the cost of maintenance and operation of the overall system against the fixed costs in such a way as to minimize the total. Advice on selection and design of heat exchangers is available from vendors.

The essential parameters which should be known in order to make an optimum choice of waste heat recovery devices are

- Temperature of waste heat fluid
- Flow rate of waste heat fluid
- Chemical composition of waste heat fluid
- Minimum allowable temperature of waste heat fluid
- Temperature of heated fluid
- Chemical composition of heated fluid
- Maximum allowable temperature of heated fluid
- Control temperature, if control required

In the rest of this chapter, some common types of waste heat recovery devices are discussed in some detail.

GAS TO GAS HEAT EXCHANGERS

Recuperators

The simplest configuration for a heat exchanger is the metallic radiation recuperator which consists of two concentric lengths of metal tubing as shown in Figure 5-3.

The inner tube carries the hot exhaust gases while the external annulus carries the combustion air from the atmosphere to the air inlets of the furnace burners. The hot gases are cooled by the incoming combustion air which now carries additional energy into the combustion chamber. This is energy which does not have to be supplied by the fuel; consequently, less fuel is burned for a given furnace loading. The saving in fuel also means a decrease in combustion air and therefore stack losses are decreased not only by lowering the stack gas temperatures but also by discharging smaller quantities of exhaust gas. This particular recuperator gets its name from the fact that a substantial portion of the heat transfer from the hot gases to

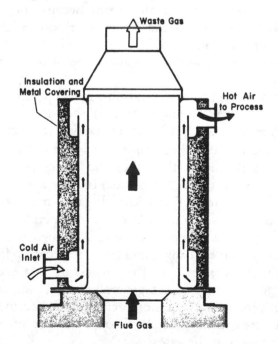

Figure 5-3. Diagram of Metallic Radiation Recuperator

the surface of the inner tube takes place by radiative heat transfer. The cold air in the annulus, however, is almost transparent to infrared radiation so that only convection heat transfer takes place to the incoming air. As shown in the diagram, the two gas flows are usually parallel, although the configuration would be simpler and the heat transfer more efficient if the flows were opposed in direction (or counterflow). The reason for the use of parallel flow is that recuperators frequently serve the additional function of cooling the duct carrying away the exhaust gases and consequently extending its service life.

The inner tube is often fabricated from high temperature materials such as stainless steels of high nickel content. The large temperature differential at the inlet causes differential expansion, since the outer shell is usually of a different and less expensive material. The mechanical design must take this effect into account. More elaborate designs of radiation recuperators incorporate two sections: the bottom operating in parallel flow and the upper section using the more efficient counterflow arrangement. Because of the large axial expansions experienced and the stress conditions at the bottom of the recuperator, the unit is often supported at the top by a free-standing support frame with an expansion joint between the furnace and recuperator.

A second common configuration for recuperators is called the tube type or convective recuperator. As seen in the schematic diagram of Figure 5-4, the hot gases are carried through a number of parallel small diameter tubes, while the incoming air to be heated enters a shell surrounding the tubes and passes over the hot tubes one or more times in a direction normal to their axes.

If the tubes are baffled to allow the gas to pass over them twice, the heat exchanger is termed a two-pass recuperator; if two baffles are used, a three-pass recuperator, etc. Although baffling increases both the cost of the exchanger and the pressure drop in the combustion air path, it increases the effectiveness of heat exchange. Shell- and tube-type recuperators are generally more compact and have a higher effectiveness than radiation recuperators, because of the larger heat transfer area made possible through the use of multiple tubes and multiple passes of the gases.

The principal limitation on the heat recovery of metal recuperators is the reduced life of the liner at inlet temperatures exceeding 2000F. At this temperature, it is necessary to use the less efficient arrangement of parallel flows of exhaust gas and coolant in order to maintain sufficient cooling of the inner shell. In addition, when furnace combustion air flow is dropped back because of reduced load, the heat transfer rate from hot waste gases to preheat combustion air becomes excessive, causing rapid surface deterioration. Then, it is usually necessary to provide an ambient air bypass to cool the exhaust gases.

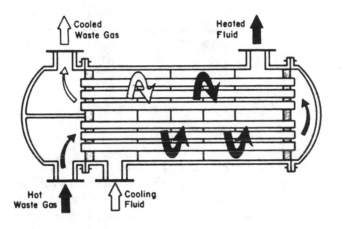

Figure 5-4. Diagram of Convective-Type Recuperator

In order to overcome the temperature limitations of metal recuperators, ceramic tube recuperators have been developed whose materials allow operation on the gas side to 2800°F and on the preheated air side to 2200°F on an experimental basis and to 1500°F on a more or less practical basis. Early ceramic recuperators were built of tile and joined with furnace cement, and thermal cycling caused cracking of joints and rapid deterioration of the tubes. Later developments introduced various kinds of short silicon carbide tubes which can be joined by flexible seals located in the air headers. This kind of patented design illustrated in Figure\5-5 maintains the seals at comparatively low temperatures and has reduced the seal leakage rates to a few percent.

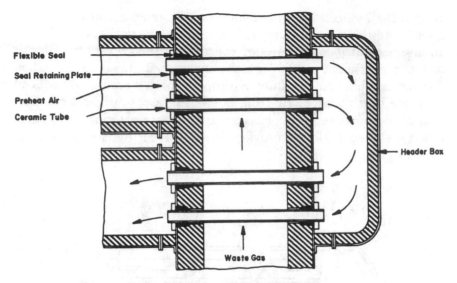

Figure 5-5. Ceramic Recuperator

Earlier designs had experienced leakage rates from 8 to 60 percent. The new designs are reported to last two years with air preheat temperatures as high as 1300°F, with much lower leakage rates.

An alternative arrangement for the convective type recuperator, in which the cold combustion air is heated in a bank of parallel vertical tubes that extend into the flue gas stream, is shown schematically in Figure 5-6. The advantage claimed for this arrangement is the ease of replacing individual tubes, which can be done during full capacity furnace operation. This minimizes the cost, the inconvenience and possible furnace damage due to a shutdown forced by recuperator failure.

For maximum effectiveness of heat transfer, combinations of radiation type and convective type recuperators are used, with the convective type always following the high temperature radiation recuperator. A schematic diagram of this arrangement is seen in Figure 5-7.

Although the use of recuperators conserves fuel in industrial furnaces and although their original cost is relatively modest, the purchase of the unit is often just the beginning of a somewhat more

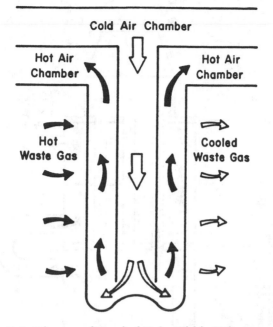

Figure 5-6. Diagram of Vertical Tube-Within-Tube Recuperator

extensive capital improvement program. The use of a recuperator, which raises the temperature of the incoming combustion air, may require purchase of high temperature burners, larger diameter air lines with flexible fittings to allow for expansion, cold air lines for cooling the burners, modified combustion controls to maintain the required air/fuel ratio despite variable recuperator heating, stack dampers, cold air bleeds, controls to protect the recuperator during blower failure or power failures and larger fans to overcome the additional pressure drop in the recuperator. It is vitally important to protect the recuperator against damage due to excessive temperatures, since the cost of rebuilding a damaged recuperator may be as high as 90 percent of the initial cost of manufacture, and the drop in efficiency of a damaged recuperator may easily increase fuel costs by 10 to 15 percent.

Figure 5-8 shows a schematic diagram of one radiant tube burner fitted with a radiation recuperator. With such a short stack, it is necessary to use two annuli for the incoming air to achieve reasonable heat exchange efficiencies.

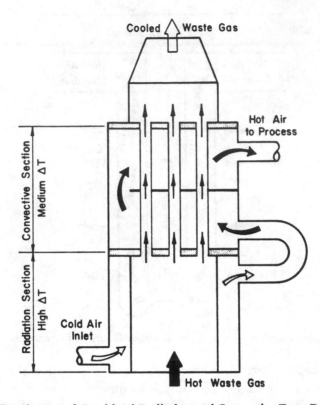

Figure 5-7. Diagram of Combined Radiation and Convective Type Recuperator

Recuperators are used for recovering heat from exhaust gases to heat other gases in the medium to high temperature range. Some typical applications are in soaking ovens, annealing ovens, melting furnaces, afterburners and gas incinerators, radiant-tube burners, reheat furnaces, and other gas to gas waste heat recovery applications in the medium to high temperature range.

Heat Wheels

A rotary regenerator (also called an air preheater or a heat wheel) is finding increasing applications in low to medium temperature waste heat recovery. Figure 5-9 is a sketch illustrating the application of a heat wheel. It is a sizable porous disk, fabricated from some material having a fairly high heat capacity, which rotates be-

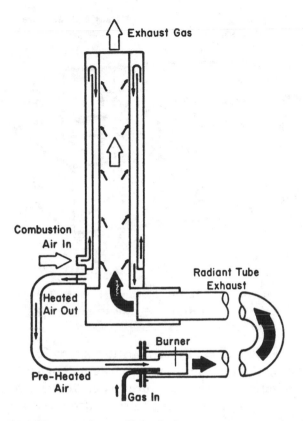

Figure 5-8. Diagram of a Small Radiation-Type Recuperator Fitted to a Radiant Tube Burner

tween two side-by-side ducts: one a cold gas duct, the other a hot gas duct. The axis of the disk is located parallel to, and on the partition between, the two ducts. As the disk slowly rotates, sensible heat (and, in some cases, moisture that contains latent heat) is transferred to the disk by the hot air and, as the disk rotates, from the disk to the cold air. The overall efficiency of sensible heat transfer for this kind of regenerator can be as high as 85 percent. Heat wheels have been built as large as 70 feet in diameter with air capacities up to 40,000 ft³/min. Multiple units can be used in parallel. This may help to prevent a mismatch between capacity requirements and the limited number of sizes available in packaged units. In very large installations such as

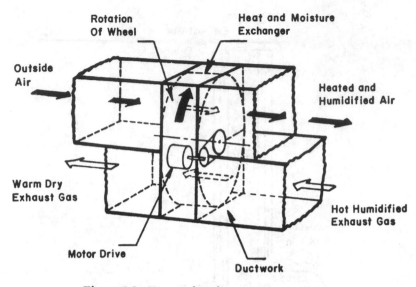

Figure 5-9. Heat and Moisture Recovery Using a
Heat Wheel Type Regenerator

those required for preheating combustion air in fixed station electrical generating stations, the units are custom designed.

The limitation on temperature range for the heat wheel is primarily due to mechanical difficulties introduced by uneven expansion of the rotating wheel when the temperature differences mean large differential expansion, causing excessive deformations of the wheel and thus difficulties in maintaining adequate air seals between duct and wheel.

Heat wheels are available in four types. The first consists of a metal frame packed with a core of knitted mesh stainless steel or aluminum wire, resembling that found in the common metallic kitchen pot scraper; the second, called a laminar wheel, is fabricated from corrugated metal and is composed of many parallel flow passages; the third variety is also a laminar wheel but is constructed from a ceramic matrix of honeycomb configuration. This type is used for higher temperature applications with a present-day limit of about 1600°F. The fourth variety is of laminar construction, but the flow passages are coated with a hygroscopic material so that latent heat may be recovered. The packing material of the hygroscopic wheel may be any of a number of materials. The hygroscopic material is often termed a desiccant.

Most industrial stack gases contain water vapor (since water vapor is a product of the combustion of all hydrocarbon fuels and since water is introduced into many industrial processes) and part of the process water evaporates as it is exposed to the hot gas stream. Each pound of water requires approximately 1000 Btu for its evaporation at atmospheric pressure; thus each pound of water vapor leaving in the exit stream will carry 1000 Btu of energy with it. This latent heat may be a substantial fraction of the sensible energy in the exit gas stream. A hygroscopic material is one such as lithium chloride (LiCl) which readily absorbs water vapor. Lithium chloride is a solid which absorbs water to form a hydrate, $LiCl \cdot H_2O$, in which one molecule of lithium chloride combines with one molecule of water. Thus, the ratio of water to lithium chloride in $LiCl \cdot H_2O$ is 3/7 by weight. In a hygroscopic heat wheel, the hot gas stream gives up part of its water vapor to the coating; the cool gases which enter the wheel to be heated are drier than those in the inlet duct, and part of the absorbed water is given up to the incoming gas stream. The latent heat of the water adds directly to the total quantity of recovered waste heat. The efficiency of recovery of water vapor can be as high as 50 percent.

Since the pores of heat wheels carry a small amount of gas from the exhaust to the intake duct, cross contamination can result. If this contamination is undesirable, the carryover of exhaust gas can be partially eliminated by the addition of a purge section where a small amount of clean air is blown through the wheel and then exhausted to the atmosphere, thereby clearing the passages of exhaust gas. Figure 5-10 illustrates the features of an installation using a purge section. Note that additional seals are required to separate the purge ducts. Common practice is to use about six air changes of clean air for purging. This limits gas contamination to as little as 0.04 percent and particle contamination to less than 0.2 percent in laminar wheels, and cross contamination to less than 1 percent in packed wheels. If inlet gas temperature is to be held constant, regardless of heating loads and exhaust gas temperatures, then the heat wheel must be driven at variable speed. This requires a variable speed drive and a speed control system using an inlet air temperature sensor as the control element. This feature, however, adds considerably

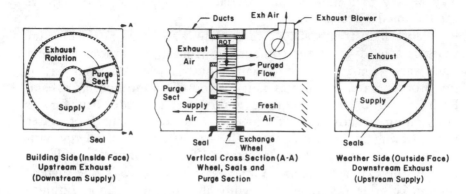

Figure 5-10. Heat Wheel Equipped with Purge Section to
Clear Contaminants from the Heat Transfer Surface

to the cost and complexity of the system. When operating with outside air in periods of high humidity and sub-zero temperatures, heat
wheels may require preheat systems to prevent frost formation.
When handling gases which contain water-soluble, greasy or adhesive
contaminants or large concentrations of process dust, air filters may
be required in the exhaust system upstream from the heat wheel.

One application of heat wheels is in space heating situations
where unusually large quantities of ventilation air are required for
health or safety reasons. As many as 20 or 30 air changes per hour
may be required to remove toxic gases or to prevent the accumulation of explosive mixtures. Comfort heating for that quantity of ventilation air is frequently expensive enough to make the use of heat
wheels economical. In the summer season the heat wheel can be
used to cool the incoming air from the cold exhaust air, reducing the
air-conditioning load by as much as 50 percent. It should be pointed
out that in many circumstances where large ventilating requirements
are mandatory, a better solution than the installation of heat wheels
may be the use of local ventilation systems to reduce the hazards
and/or the use of infrared comfort heating at principal work areas.

Heat wheels are finding increasing use for process heat recovery
in low and moderate temperature environments. Typical applications
would be curing or drying ovens and air preheaters in all sizes for
industrial and utility boilers.

Air Preheaters

Passive gas to gas regenerators, sometimes called air preheaters, are available for applications which cannot tolerate any cross contamination. They are constructed of alternate channels (see Figure 5-11) which put the flows of the heating and the heated gases in close contact with each other, separated only by a thin wall of conductive metal. They occupy more volume and are more expensive to construct than are heat wheels, since a much greater heat transfer surface area is required for the same efficiency. An advantage, besides the absence of cross contamination, is the decreased mechanical complexity since no drive mechanism is required. However, it becomes more difficult to achieve temperature control with the passive regeneration, and, if this is a requirement, some of the advantages of its basic simplicity are lost.

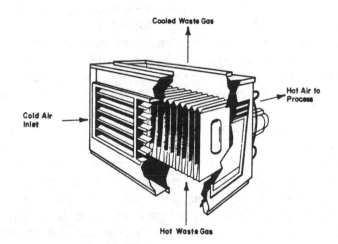

Figure 5-11. A Passive Gas to Gas Regenerator

Gas to gas regenerators are used for recovering heat from exhaust gases to heat other gases in the low to medium temperature range. A list of typical applications follows:

- Heat and moisture recovery from building heating and ventilation systems
- Heat and moisture recovery from moist rooms and swimming pools

- Reduction of building air-conditioner loads
- Recovery of heat and water from wet industrial processes
- Heat recovery from steam boiler exhaust gases
- Heat recovery from gas and vapor incinerators
- Heat recovery from baking, drying, and curing ovens
- Heat recovery from gas turbine exhausts
- Heat recovery from other gas to gas applications in the low through high temperature range.

Heat-Pipe Exchangers

The heat pipe is a heat transfer element that has only recently become commercial, but it shows promise as an industrial waste heat recovery option because of its high efficiency and compact size. In use, it operates as a passive gas to gas finned-tube regenerator. As can be seen in Figure 5-12, the elements form a bundle of heat pipes which extend through the exhaust and inlet ducts in a pattern that resembles the structured finned coil heat exchangers. Each pipe, however, is a separate sealed element consisting of an annular wick on the inside of the full length of the tube, in which an appropriate heat transfer fluid is entrained.

Figure 5-13 shows how the heat absorbed from hot exhaust gases evaporates the entrained fluid, causing the vapor to collect in the center core. The latent heat of vaporization is carried in the vapor to the cold end of the heat pipe located in the cold gas duct. Here the vapor condenses, giving up its latent heat. The condensed liquid is then carried by capillary (and/or gravity) action back to the hot end where it is recycled. The heat pipe is compact and efficient because: (1) the finned-tube bundle is inherently a good configuration for convective heat transfer in both gas ducts, and (2) the evaporative-condensing cycle within the heat tubes is a highly efficient way of transferring the heat internally. It is also free from cross contamination. Possible applications include

- Drying, curing and baking ovens
- Waste steam reclamation
- Air preheaters in steam boilers
- Air dryers

- Brick kilns (secondary recovery)
- Reverberatory furnaces (secondary recovery)
- Heating, ventilating and air-conditioning systems

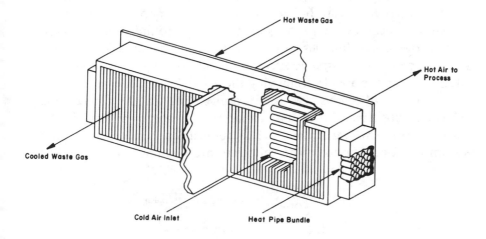

Figure 5-12. Heat Pipe Bundle Incorporated in Gas to Gas Regenerator

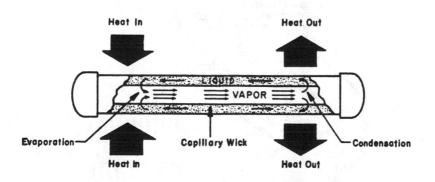

Figure 5-13. Heat Pipe Schematic

GAS OR LIQUID TO LIQUID REGENERATORS

Finned-Tube Heat Exchangers

When waste heat in exhaust gases is recovered for heating liquids for purposes such as providing domestic hot water, heating the feedwater for steam boilers, or for hot water space heating, the finned-tube heat exchanger is generally used. Round tubes are connected together in bundles to contain the heated liquid, and fins are welded or otherwise attached to the outside of the tubes to provide additional surface area for removing the waste heat in the gases.

Figure 5-14 shows the usual arrangement for the finned-tube exchanger positioned in a duct and details of a typical finned-tube construction. This particular type of application is more commonly known as an economizer. The tubes are often connected all in series

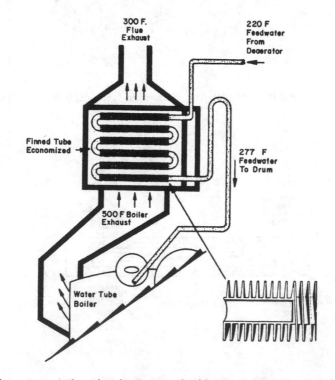

Figure 5-14. Finned-Tube Gas to Liquid Regenerator (Economizer)

but can also be arranged in series-parallel bundles to control the liquid side pressure drop. The air side pressure drop is controlled by the spacing of the tubes and the number of rows of tubes within the duct.

Finned-tube exchangers are available prepackaged in modular sizes or can be made up to custom specifications very rapidly from standard components. Temperature control of the heated liquid is usually provided by a bypass duct arrangement which varies the flow rate of hot gases over the heat exchanger. Materials for the tubes and the fins can be selected to withstand corrosive liquids and/or corrosive exhaust gases.

Finned-tube heat exchangers are used to recover waste heat in the low to medium temperature range from exhaust gases for heating liquids. Typical applications are domestic hot water heating, heating boiler feedwater, hot water space heating, absorption-type refrigeration or air conditioning and heating process liquids.

Shell and Tube Heat Exchanger

When the medium containing waste heat is a liquid or a vapor which heats another liquid, then the shell and tube heat exchanger must be used since both paths must be sealed to contain the pressures of their respective fluids. The shell contains the tube bundle, and usually internal baffles, to direct the fluid in the shell over the tubes in multiple passes. The shell is inherently weaker than the tubes so that the higher pressure fluid is circulated in the tubes while the lower pressure fluid flows through the shell. When a vapor contains the waste heat, it usually condenses, giving up its latent heat to the liquid being heated. In this application, the vapor is almost invariably contained within the shell. If the reverse is attempted, the condensation of vapors within small diameter parallel tubes causes flow instabilities. Tube and shell heat exchangers are available in a wide range of standard sizes with many combinations of materials for the tubes and shells.

Typical applications of shell and tube heat exchangers include heating liquids with the heat contained by condensates from refrigeration and air-conditioning systems; condensate from process steam; coolants from furnace doors, grates, and pipe supports; coolants from engines, air compressors, bearings, and lubricants; and the condensates from distillation processes.

Waste Heat Boilers

Waste heat boilers are ordinarily water tube boilers in which the hot exhaust gases from gas turbines, incinerators, etc., pass over a number of parallel tubes containing water. The water is vaporized in the tubes and collected in a steam drum from which it is drawn off for use as heating or processing steam.

Figure 5-15 indicates one arrangement that is used, where the exhaust gases pass over the water tubes twice before they are exhausted to the air. Because the exhaust gases are usually in the medium temperature range and in order to conserve space, a more compact boiler can be produced if the water tubes are finned in order to increase the effective heat transfer area on the gas side. The diagram shows a mud drum, a set of tubes over which the hot gases make a double pass, and a steam drum which collects the steam generated above the water surface. The pressure at which the steam is generated and the rate of steam production depend on the temperature of the hot gases entering the boiler, the flow rate of the hot gases, and the efficiency of the boiler. The pressure of a pure vapor in the presence of its liquid is a function of the temperature of the liquid from which it is evaporated. The steam tables tabulate this relationship between saturation pressure and temperature. Should the waste heat in the exhaust gases be insufficient for generating the required amount of process steam, it is sometimes possible to add auxiliary burners which burn fuel in the waste heat boiler or to add an afterburner to the exhaust gas duct just ahead of the boiler. Waste heat boilers are built in capacities from less than a thousand to almost a million ft^3/min. of exhaust gas.

Typical applications of waste heat boilers are to recover energy from the exhausts of gas turbines, reciprocating engines, incinerators, and furnaces.

Gas and Vapor Expanders

Industrial steam and gas turbines are in an advanced state of development and readily available on a commercial basis. Recently,

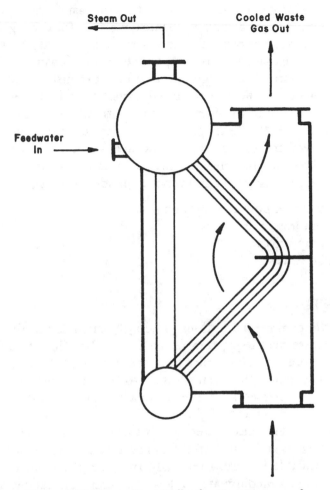

**Figure 5-15. Waste Heat Boiler for Heat Recovery from
Gas Turbines or Incinerators**

special gas turbine designs for low pressure waste gases have become
available; for example, a turbine is available for operation from the
top gases of a blast furnace. In this case, as much as 20 MW of power
could be generated, representing a recovery of 20 to 30 percent of
the available energy of the furnace exhaust gas stream. Maximum top
pressures are of the order of 40 lb/in² g.

Perhaps of greater applicability than the last example are steam turbines used for producing mechanical work or for driving electrical generators. After removing the necessary energy for doing work, the steam turbine exhausts partially spent steam at a lower pressure than the inlet pressure. The energy in the turbine exhaust stream can then be used for process heat in the usual ways. Steam turbines are classified as back-pressure turbines, available with allowable exit pressure operation above 400 lb/in^{2}g, or condensing turbines which operate below atmospheric exit pressures. The steam used for driving the turbines can be generated in direct fired or waste heat boilers. A list of typical applications for gas and vapor expanders follows:

- Electrical power generation
- Compressor drives
- Pump drives
- Fan drives

Heat Pumps

In the commercial options previously discussed in this chapter, we find waste heat being transferred from a hot fluid to a fluid at a lower temperature. Heat must flow spontaneously "downhill", that is, from a system at high temperature to one at a lower temperature. This can be expressed scientifically in a number of ways—all the variations of the statement of the second law of thermodynamics. The practical impact of these statements is that energy, as it is transformed again and again and transferred from system to system, becomes less and less available for use. Eventually that energy has such low intensity (resides in a medium at such low temperature) that it is no longer available at all to perform a useful function. It has been taken as a general rule of thumb in industrial operations that fluids with temperatures less than 250°F are of little value for waste heat extraction; flue gases should not be cooled below 250°F (or, better, 300°F to provide a safe margin), because of the risk of condensation of corrosive liquids. However, as fuel costs continue to rise, such waste heat can be used economically for space heating and other low temperature applications. It is possible to reverse the direction of spontaneous energy flow by the use of a thermodynamic system known as a heat pump.

This device consists of two heat exchangers, a compressor and an expansion device. A liquid or a mixture of liquid and vapor of a pure chemical species flows through an evaporator, where it absorbs heat at low temperature and, in doing so, is completely vaporized. The low temperature vapor is compressed by a compressor which requires external work. The work done on the vapor raises its pressure and temperature to a level where its energy becomes available for use. The vapor flows through a condenser where it gives up its energy as it condenses to a liquid. The liquid is then expanded through a device back to the evaporator where the cycle repeats. The heat pump was developed as a space heating system where low temperature energy from the ambient air, water, or earth is raised to heating system temperatures by doing compression work with an electric motor-driven compressor. The performance of the heat pump is ordinarily described in terms of the coefficient of performance or COP, which is defined as

$$COP = \frac{\text{Heat transferred in condenser}}{\text{Compressor work}} \qquad \textit{Formula (5-11)}$$

which in an ideal heat pump is found as

$$COP = \frac{T_H}{T_H - T_L} \qquad \textit{Formula (5-12)}$$

where T_L is the temperature at which waste heat is extracted from the low temperature medium and T_H is the high temperature at which heat is given up by the pump as useful energy. The coefficient of performance expresses the economy of heat transfer.

In the past, the heat pump has not been applied generally to industrial applications. However, several manufacturers are now redeveloping their domestic heat pump systems as well as new equipment for industrial use. The best applications for the device in this new context are not yet clear, but it may well make possible the use of large quantities of low-grade waste heat with relatively small expenditures of work.

Summary

Table 5-4 presents the collation of a number of significant attributes of the most common types of industrial heat exchangers in matrix form. This matrix allows rapid comparisons to be made in selecting competing types of heat exchangers. The characteristics given in the table for each type of heat exchanger are allowable temperature range, ability to transfer moisture, ability to withstand large temperature differentials, availability as packaged units, suitability for retrofitting, and compactness and the allowable combinations of heat transfer fluids.

Table 5-4. Operation and Application Characteristics of Industrial Heat Exchangers

SPECIFICATIONS FOR WASTE RECOVERY UNIT / COMMERCIAL HEAT TRANSFER EQUIPMENT	Low Temperature Sub-Zero – 250°F	Intermediate Temp. 250°F – 1200°F	High Temperature 1200°F – 2000°F	Recovers Moisture	Large Temperature Differentials Permitted	Packaged Units Available	Can Be Retrofit	No Cross-Contamination	Compact Size	Gas-to-Gas Heat Exchange	Gas-to-Liquid Heat Exchanger	Liquid-to-Liquid Heat Exchanger	Corrosive Gases Permitted with Special Construction
Radiation Recuperator			•		•	1	•	•		•			•
Convection Recuperator		•	•		•	•	•	•		•			•
Metallic Heat Wheel	•	•		2		•	•	3	•	•			•
Hygroscopic Heat Wheel	•			•		•	•	3	•	•			
Ceramic Heat Wheel		•	•			•	•	•	•	•			
Passive Regenerator	•	•			•	•	•	•	•	•			
Finned-Tube Heat Exchanger	•	•			•	•	•	•	•		•		4
Tube Shell-and-Tube Exchanger	•	•			•	•	•	•	•		•	•	
Waste Heat Boilers	•	•				•	•	•				•	4
Heat Pipes	•	•			5	•	•	•	•	•			•

1. Off-the-shelf items available in small capacities only.
2. Controversial subject. Some authorities claim moisture recovery. Do not advise depending on it.
3. With a purge section added, cross-contamination can be limited to less than 1% by mass.
4. Can be constructed of corrosion-resistant materials, but consider possible extensive damage to equipment caused by leaks or tube ruptures.
5. Allowable temperatures and temperature differential limited by the phase equilibrium properties of the internal fluid.

6

Utility System Optimization

BASIS OF THERMODYNAMICS

Thermodynamics deals with the relationships between heat and work. It is based on two basic laws of nature: the first and second laws of thermodynamics. The principles are used in the design of equipment such as steam engines, turbines, pumps, and refrigerators, and in practically every process involving a flow of heat or a chemical equilibrium.

First Law: The first law states that energy can neither be created nor destroyed, thus, it is referred to as the law of conservation of energy. Formula 6-1 expresses the first law for the steady state condition.

$$E_2 - E_1 = Q - W \qquad \text{Formula (6-1)}$$

Where

$E_2 - E_1$ is the change in stored energy at the boundary states 1 and 2 of the system

Q is the heat added to the system

W is the work done by the system

Figure 6-1 illustrates a thermodynamic process where mass enters and leaves the system. The potential energy (Z) and the kinetic energy $(V^2/64.2)$ plus the enthalpy represent the stored energy of the mass. Note, Z is the elevation above the reference point in feet, and V is the velocity of the mass in ft/sec. In the case of the steam turbine, the change in Z, V, and Q are small in comparison to the change in enthalpy. Thus, the energy equation reduces to

$$W/778 = h_1 - h_2 \qquad \text{Formula (6-2)}$$

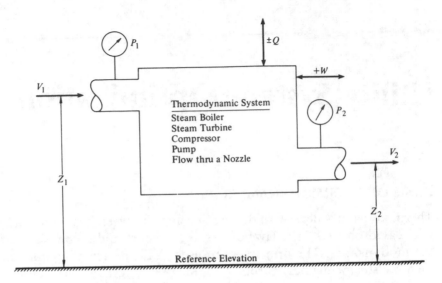

Figure 6-1. System Illustrating Conservation of Energy

Where

W is the work done in ft · lb/lb
h_1 is the enthalpy of the entering steam, Btu/lb
h_2 is the enthalpy of the exhaust steam, Btu/lb
And 1 Btu equals 778 ft · lb

Second Law: The second law qualifies the first law by discussing the conversion between heat and work. All forms of energy, including work, can be converted to heat, but the converse is not generally true. The Kelvin-Planck statement of the second law of thermodynamics says essentially the following: Only a portion of the heat from a heat work cycle, such as a steam power plant, can be converted to work. The remaining heat must be rejected as heat to a sink of lower temperature (to the atmosphere, for instance).

The Clausius statement, which also deals with the second law, states that heat, in the absence of some form of external assistance, can only flow from a hotter to a colder body.

THE CARNOT CYCLE

The Carnot cycle is of interest because it is used as a comparison of the efficiency of equipment performance. The Carnot cycle offers the maximum thermal efficiency attainable between any given temperatures of heat source and sink. A thermodynamic cycle is a series of processes forming a closed curve on any system of thermodynamic coordinates. The Carnot cycle is illustrated on a temperature-entropy diagram, Figure 6-2A, and on the Mollier Diagram for superheated steam, Figure 6-2B.

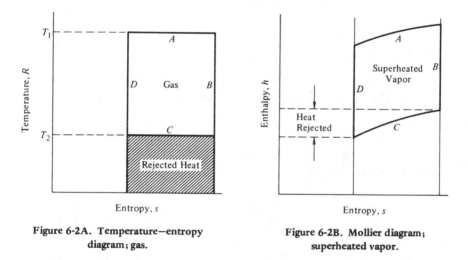

Figure 6-2A. Temperature—entropy   Figure 6-2B. Mollier diagram;
diagram; gas.   superheated vapor.

Figure 6-2. Carnot Cycles

The cycle consists of the following:

1. Heat addition at constant temperature, resulting in expansion work and changes in enthalpy.
2. Adiabatic isentropic expansion (change in entropy is zero) with expansion work and an equivalent decrease in enthalpy.
3. Constant temperature heat rejection to the surroundings, equal to the compression work and any changes in enthalpy.
4. Adiabatic isentropic compression returning to the starting temperature with compression work and an equivalent increase in enthalpy.

The Carnot cycle is an example of a reversible process and has no counterpart in practice. Nevertheless, this cycle illustrates the principles of thermodynamics. The thermal efficiency for the Carnot cycle is illustrated by Formula (6-3).

$$\text{Thermal efficiency} = \frac{T_1 - T_2}{T_1} \qquad Formula\ (6\text{-}3)$$

Where

T_1 = Absolute temperature of heat source, $°R$ (Rankine)
T_2 = Absolute temperature of heat sink, $°R$

and absolute temperature is given by Formula 6-4.

Absolute temperature = 460 + temperature in Fahrenheit.
Formula (6-4)

T_2 is usually based on atmospheric temperature, which is taken as $500°R$.

PROPERTIES OF STEAM PRESSURE AND TEMPERATURE

Water boils at $212°F$ when it is in an open vessel under atmospheric pressure equal to 14.7 psia (pounds per square inch, absolute). Absolute pressure is the amount of pressure exerted by a system on its boundaries and is used to differentiate it from gage pressure. A pressure gage indicates the difference between the pressure of the system and atmospheric pressure.

psia = psig + atmospheric pressure in psia   *Formula (6-5)*

Changing the pressure of water changes the boiling temperature. Thus, water can be vaporized at $170°F$, at $300°F$, or any other temperature, as long as the applied pressure corresponds to that boiling point.

SOLID, LIQUID AND VAPOR STATES OF A LIQUID

Water, as well as other liquids, can exist in three states: solid, liquid, and vapor. In order to change the state from ice to water or from water to steam, heat must be added. The heat required to change a solid to liquid is called the *latent heat of fusion.* The heat required to change a liquid to a vapor is called the *latent heat of vaporization.*

In condensing steam, heat must be removed. The quantity is exactly equal to the latent heat that went into the water to change it to steam.

Heat supplied to a fluid, during the change of state to a vapor, will not cause the temperature to rise; thus, it is referred to as the latent heat of vaporization. Heat given off by a substance when it condenses from steam to a liquid is called *sensible* heat. Physical properties of water, such as the latent heat of vaporization, also change with variations in pressure.

USE OF THE STEAM TABLES

Steam properties are illustrated in Chapter 15 by Tables 15-13, 15-14 and 15-15 (ASME Steam Tables). Table 15-13 is referred to as the steam table for saturated steam. Steam properties are shown and correlated to temperature. Table 15-14 is another form of the steam table, where steam properties are shown and correlated to pressure. (Pressure must be converted to psia.) The properties of superheated steam are indicated by Table 15-15.

The term h_{fg} is the latent heat or enthalpy of vaporization. Thus, from Table 15-14, at atmospheric pressure the latent heat of vaporization is 970.3 Btu/lb. At 200 psia the latent heat of vaporization is 842.8 Btu/lb.

The enthalpy h_f represents the amount of heat required to raise one pound of water from 32°F to a liquid state at another temperature. As an example, from Table 15-13, to raise water from 32°F to 170°F will require 137.9 Btu/lb.

Example Problem 6-1

How much heat is required to raise 100 pounds of water at 126°F to 170°F?

Answer

From Table 15-13:

$$\text{At } 126°F - h_f = 93.9 \text{ Btu/lb}$$
$$\text{At } 170°F - h_f = 137.9 \text{ Btu/lb}$$
$$Q = 100(137.9 - 93.9) = 44 \times 10^2 \text{ Btu.}$$

USE OF THE SPECIFIC HEAT CONCEPT

Another physical property of a material is the *specific heat*. The specific heat is defined as the amount of heat required to raise a unit of mass of a substance one degree. For water, it can be seen from the previous example that one Btu of heat is required to raise one lb water 1°F; thus, the specific heat of water $C_p = 1$. Specific heats for other materials are illustrated in Table 6-1. This leads to two equations.

$$Q = wCp \, \Delta T \qquad \qquad \textit{Formula (6-6)}$$

Table 6-1. Specific Heat of Various Substances

SUBSTANCE	SPECIFIC HEAT	SUBSTANCE	SPECIFIC HEAT
SOLIDS		**LIQUIDS**	
ALUMINUM	0.230	ALCOHOL	0.600
ASBESTOS	0.195	AMMONIA	1.100
BRASS	0.086	BRINE, CALCIUM (20% SOLUTION)	0.730
BRICK	0.220	BRINE, SODIUM (20% SOLUTION)	0.810
BRONZE	0.086	CARBON TETRACHLORIDE	0.200
CHALK	0.215	CHLOROFORM	0.230
CONCRETE	0.270	ETHER	0.530
COPPER	0.093	GASOLINE	0.700
CORK	0.485	GLYCERINE	0.576
GLASS, CROWN	0.161	KEROSENE	0.500
GLASS, FLINT	0.117	MACHINE OIL	0.400
GLASS, THERMOMETER	0.199	MERCURY	0.033
GOLD	0.030	PETROLEUM	0.500
GRANITE	0.192	SULPHURIC ACID	0.336
GYPSUM	0.259	TURPENTINE	0.470
ICE	0.480	WATER	1.000
IRON, CAST	0.130	WATER, SEA	0.940
IRON, WROUGHT	0.114		
LEAD	0.031	**GASES**	
LEATHER	0.360	AIR	0.240
LIMESTONE	0.216	AMMONIA	0.520
MARBLE	0.210	BROMINE	0.056
MONEL METAL	0.128	CARBON DIOXIDE	0.200
PORCELAIN	0.255	CARBON MONOXIDE	0.243
RUBBER	0.481	CHLOROFORM	0.144
SILVER	0.055	ETHER	0.428
STEEL	0.118	HYDROGEN	3.410
TIN	0.045	METHANE	0.593
WOOD	0.330	NITROGEN	0.240
ZINC	0.092	OXYGEN	0.220
		SULPHUR DIOXIDE	0.154
		STEAM (SUPERHEATED, 1 PSI)	0.450

Reprinted by permission of The Trane Company.

Where

    Q = quantity of heat, Btu
    w = weight of substance, lb
    Cp = specific heat of substance, Btu per lb °F
    ΔT = temperature change of substance °F

$$Q = MCp \ \Delta T \hspace{3cm} Formula\ (6\text{-}7)$$

Where

    Q = quantity of heat, Btu/hr (Btu/hr is sometimes abbreviated as Btuh)
    M = flow rate, lbs/hr
    Cp = specific heat, Btu per lb °F
    ΔT = temperature change of substance°F

Example Problem 6-2

    Check answer to Example Problem 6-1 using Formula 6-6.

Answer

    Q = $WCp \ \Delta T$
        = $100 \times 1 \times (170 - 126) = 44 \times 10^2$ Btu.

    The saturated vapor enthalpy h_g represents the amount of heat necessary to change water at 32°F to steam at a specified temperature and pressure.

Example Problem 6-3

    How much heat is required to raise 100 pounds of water at 126°F to 15 psig steam?

Answer

    At 126°F $- h_f$ = 93.9 Btu/lb
    15 psig corresponds to 30 psia at 250°F
    At 30 psia, h_g = 1164.1 Btu/lb
    $Q = 100(1164.1 - 93.9) = 1070.2 \times 10^2$ Btu.

USE OF THE LATENT HEAT CONCEPT

When water is raised from $32°F$ to $212°F$ steam, the total heat required is composed of two components:

(a)   The heat required to raise the temperature of water from $32°F$ to $212°F$; $h_f = 180.17$ Btu/lb from Table 15-13.

(b)   The heat required to evaporate the water at $212°F$; $h_{fg} = 970.3$ Btu/lb from Table 15-13.

Thus, $h_g = h_f + h_{fg} = 180.17 + 970.3 = 1150.5$ Btu/lb, which agrees with the value of h_g found in Steam Table 15-13.

Example Problem 6-4

30 psig steam is used for a heat exchanger and returns to the system as 30 psig condensate. What amount of heat is given off to the process fluid?

Answer

30 psig = 45 psia. This corresponds to an h_{fg} of approximately 928 Btu/lb.

THE USE OF THE SPECIFIC VOLUME CONCEPT

Another property of water is the specific volume v_f of water and the specific volume v_g of steam. *Specific volume* is defined as the space occupied by one pound of a material. For example, at 50 psia, water occupies 0.01727 ft^3 per lb and steam occupies 8.515 ft^3 per lb.

The *specific weight* is simply the weight of one cubic foot of a material and is the reciprocal of the specific volume.

THE MOLLIER DIAGRAM

A visual tool for understanding and using the properties of steam is illustrated by the Mollier Diagram, Figure 6-3. The Mollier Diagram enables one to find the relationship between temperature, pressure, enthalpy, and entropy for steam. Constant temperature and pressure curves illustrate the effect of various processes on steam.

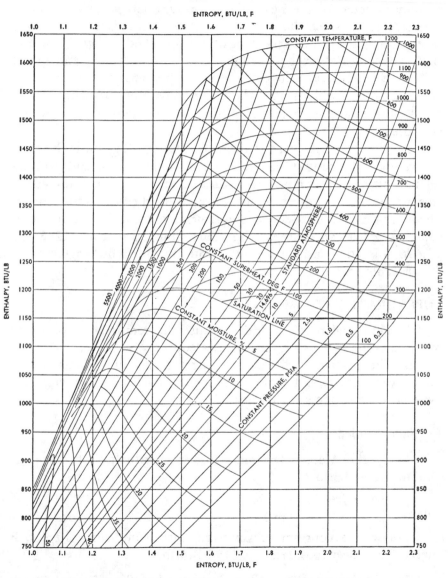

Figure 6-3. Mollier Diagram *(Courtesy Babcock & Wilcox Company and ASME)*

For a constant temperature process (isothermal), the change in entropy is equal to the heat added (or subtracted) divided by the temperature at which the process is carried out. This is a simple way of explaining the physical meaning of entropy. The *change* in entropy is of interest. Referring to Table 15-13, the value of entropy at 32°F is zero. Increases in entropy are a measure of the portion of heat in a process which is unavailable for conversion to work. Entropy has a close relationship to the second law of thermodynamics, discussed earlier in this chapter.

Another item of interest from the Mollier Diagram is the saturation line. The saturation line indicates temperature and pressure relationships corresponding to saturated steam. Below this curve, steam contains a % moisture, as indicated by the % moisture curves. Steam at temperatures above the saturation curve is referred to as superheated. As an example, this chart indicates that at 212°F and 14.696 psia, the enthalpy h_g is 1150 Btu/lb, which agrees with Steam Table 15-14.

SUPERHEATED STEAM

If additional heat is added to raise the temperature of steam over the point at which it was evaporated, the steam is termed superheated. Thus, steam at the same temperature as boiling water is saturated steam. Steam at temperatures higher than boiling water, at the same pressure, is superheated steam.

Example Problem 6-5

Using the Mollier Diagram, find the enthalpy of steam at 14.696 psia and 300°F.

Answer

Follow the constant pressure line until it intersects the 300°F curve. The answer to the right is 1192 Btu/lb.

In addition to the Mollier Diagram, the Superheated Steam Table 15-15 is helpful.

Steam cannot be superheated in the presence of water because the heat supplied will only evaporate the water. Thus, the water will be evaporated prior to becoming superheated. Superheated steam is condensed by first cooling it down to the boiling point corresponding to its pressure. When the steam has been de-superheated, further removal of heat will result in its condensation.

In the generation of power, superheated steam has many uses.

HEAT BALANCE

A heat balance is an analysis of a process which shows where all the heat comes from and where it goes. This is a vital tool in assessing the profit implications of heat losses and proposed waste heat utilization projects. The heat balance for a steam boiler, process furnace, air conditioner, etc. must be derived from measurements made during actual operating periods. Chapter 3 provides information on the instrumentation available to make these measurements. The measurements that are needed to get a complete heat balance involve energy inputs, energy losses to the environment and energy discharges.

Energy Input

Energy enters most process equipment either as chemical energy in the form of fossil fuels, of sensible enthalpy of fluid streams, of latent heat in vapor streams, or as electrical energy.

For each input it is necessary to meter the quantity of fluid flowing or the electrical current. This means that if accurate results are to be obtained, submetering for each flow is required (unless all other equipment served by a main meter can be shut down so that the main meter can be used to measure the inlet flow to the unit). It is not necessary to continuously submeter every flow since temporary installations can provide sufficient information. In the case of furnaces and boilers that use pressure ratio combustion controls, the control flow meters can be utilized to yield the correct information. It should also be pointed out that for furnaces and boilers only the fuel should be metered. Tests of the exhaust products provide sufficient information to derive the oxidant (usually air) flow if accurate fuel flow data are available. For electrical energy inflows, the current is measured with an ammeter, or a kilowatt hour meter may be installed as a submeter. Ammeters using split core transformers are available for measuring alternating current flow without opening the line. These are particularly convenient for temporary installations.

In addition to measuring the flow for each inlet stream, it is necessary to know the chemical composition of the stream. For air, water, and other pure substances no tests for composition are required, but for fossil fuels the composition must be determined by chemical analysis or secured from the fuel supplier. For vapors one should know the quality—this is the mass fraction of vapor present in the mixture of vapor and droplets. Measurement of quality is made with a vapor calorimeter which requires only a small sample of the vapor stream.

Other measurements that are required are the entering temperatures of the inlet stream of fluid and the voltages of the electrical energy entering (unless kilowatt-hour meters are used).

The testing routines discussed above involve a good deal of time, trouble and expense. However, they are necessary for accurate analyses and may constitute the critical element in the engineering and economic analyses required to support decisions to expend capital on waste heat recovery equipment.

Energy Losses

Energy loss from process equipment to the ambient environment is usually by radiative and convective heat transfer. Radiant heat transfer, that is, heat transfer by light or other electromagnetic radiation, is discussed in the section of Chapter 3 dealing with infrared thermography. Convective heat transfer, which takes place by hot gas at the surface of the hot material being displaced by cooler gas, may be analyzed using Newton's law of cooling.

$$Q = UA (T_s - T_o) \hspace{2cm} \textit{Formula (6-8)}$$

Where

Q = rate of heat loss in energy units Btu/h
U = heat transfer coefficient in Btu/h·ft²·F
A = area of surface losing heat in ft²
T_s = surface temperature
T_o = ambient temperature

Although heat flux meters are available, it is usually easier to measure the quantities above and derive the heat loss from the equation. The problems encountered in using the equation involve the measurement of surface temperatures and the finding of accurate values for the heat transfer coefficient.

Unfortunately, the temperature distribution over the surface of a process unit can be very nonuniform so that an estimate of the overall average is quite difficult. New infrared measurement techniques, which are discussed in Chapter 3, make the determination somewhat more accurate. The heat transfer coefficient is not only a strong function of surface and ambient temperatures but also depends on geometric considerations and surface conditions. Thus for given surface and ambient temperatures a flat, vertical plate will have a different h_{cr} value than will a horizontal or inclined plate.

Energy Discharges

The composition, discharge rate and temperature of each outflow from the process unit are required in order to complete the heat balance. For a fuel-fired unit, only the composition of the exhaust products, the flue gas temperature and the fuel input rate to the unit are required to derive

(1) air input rate
(2) exhaust gas flow rate
(3) energy discharge rate from exhaust stack

The composition of the exhaust products can be determined from an Orsat analysis, a chromatographic test or, less accurately, from a determination of the volumetric fraction of oxygen or CO_2. Figure 6-4 can be used for determining the quantity of excess or deficiency of air in the combustible mixture. It is based on the fact that chemical reactions occur with fixed ratios of reactants to form given products.

For example, natural gas with the following composition:

$$
\begin{array}{rl}
CO_2 & - \ 0.7\% \ volume \\
O_2 & - \ 0.0 \\
CH_4 & - \ 92.0 \\
C_2H_6 & - \ \ \ 6.8 \\
N_2 & - \underline{\ \ \ 0.5} \\
& \ \ \ 100.00
\end{array}
$$

is burned to completion with the theoretical amount of air indicated in the volume equation below:

FUEL: $0.92\ CH_4 + 0.068\ C_2H_6$
(1 ft³) $+ 0.007\ CO_2 + 0.005\ N_2$
plus
AIR: $2.078\ O_2 + 7.813\ N_2\ \rightarrow$
(9.891 ft³)
yields
DRY PRODUCTS: $1.063\ CO_2 + 7.818\ N_2$
(10.925 ft³) $+ 2.044\ H_2O$ *Formula (6-9)*

The equation is based upon the laws of conservation of mass and elemental chemical species. The ratio of N_2 and O_2 in the combustible mixture results from the approximate volumetric ratio of N_2 to O_2 in air, i.e.,

$$20.9\%\ O_2, 79.1\%\ N_2\ \text{or}\ \frac{79.1}{20.9} = 3.76 = \frac{\text{Volume}\ N_2}{\text{Volume}\ O_2}.$$

For gases the coefficients of the chemical equation represent relative volumes of each species reacting. Ordinarily, excess air is provided to

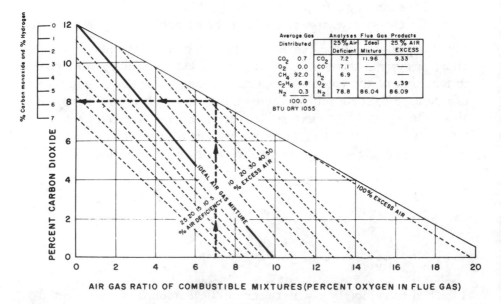

Figure 6-4. Natural Gas Combustion Chart

the fuel so that every fuel molecule will react with the necessary number of oxygen molecules even though the physical mixing process is imperfect. If 10 percent excess air were supplied, this mixture of reactants and products would give a chemical equation appropriately modified as given below.

FUEL: $0.92\ CH_4 + 0.068\ C_2H_6$
(1 ft³) $+ 0.007\ CO_2 + 0.005\ N_2$

plus

AIR: $2.286\ O_2 + 8.594\ N_2 \rightarrow$
(10.880 ft³)

yields

DRY PRODUCTS: $1.063\ CO_2 + 0.208\ O_2 + 8.599\ N_2$
(11.914 ft³) $+ 2.044\ H_2O$ *Formula (6-10)*

where an additional term representing the excess oxygen appears in the products, along with a corresponding increase in the nitrogen.

As an example let us assume that an oxygen meter has indicated a reading of 7% for the products of combustion from the natural gas whose composition was given previously and that the exhaust gas temperature was measured as 700°F. Figure 6-4 is used as indicated to determine that 45% excess air is mixed with fuel. The combustion equation then becomes:

FUEL: $0.92\ CH_4 + 0.068\ C_2H_6$
(1 ft³) $+ 0.007\ CO_2 + 0.005\ N_2$

plus

AIR: $3.013\ O_2 + 11.329\ N_2 \rightarrow$
(14.342 ft³)

yields

DRY PRODUCTS: $1.063\ CO_2 + 0.935\ O_2 + 11.334\ N_2$
(15.376 ft³) $+ 2.044\ H_2O$

For each 1 ft³ of fuel, 14.342 ft³ of air is supplied and 15.376 ft³ of exhaust products (at mixture temperature) is formed. Each cubic foot of fuel contains 1055 Btu of energy, so the fuel energy input is 250,000ft³/h X 1055 = 263,750,000 Btu/h. From Figure 6-5 we compute the exhaust gas losses at 700°F as

CO_2: 250,000 $\times$ 1.063 $\times$ 17.5 =   4,651,000 Btu/h
H_2O: 250,000 $\times$ 2.044 $\times$ 14.0 =   7,154,000
 O_2 : 250,000 $\times$ 0.935 $\times$ 12.3 =   2,875,000
 N_2 : 250,000 $\times$ 11.334 $\times$ 11.5 = 32,585,000
 47,265,000 Btu/h

or 18% of the fuel energy supplied. Some of this could be recovered by suitable waste heat equipment.

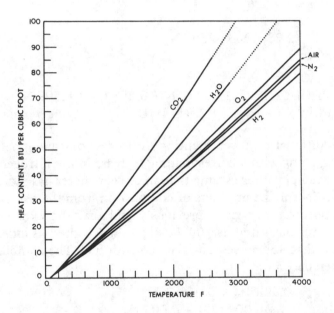

Figure 6-5. Heat Content of Gases Found in Flue Products, Based on Gas Volumes at 60°F (dashed line indicates dissociation)

Heat Balance on a Boiler

Let us consider a further example, a heat balance on a boiler. A process steam boiler has the following specifications:

- Natural gas fuel with HHV = 1001.2 Btu/ft^3
- Gas firing rate = 2126.5 ft^3/min.
- Steam discharge at 150 lb/in^2 g saturated
- Steam capacity of 100,000 lb/h
- Condensate returned at 180°F

The heat balance on the burner is derived from measurements made *after* the burner controls had been adjusted for an optimum air/fuel ratio corresponding to 10% excess air. All values of the heat content of the fluid streams are referred to a base temperature of 60°F. Consequently the computations for each fluid stream entering or leaving the boiler are made by use of the equation below.

$$\dot{H} = \dot{m} \ (h - h_o) \qquad\qquad \text{Formula (6-11)}$$

Where

$\dot{H}$ = the enthalpy rate for entrance or exit fluids

$\dot{m}$ = mass flow rates for entrance or exit fluids

h = specific enthalpy at the fluid temperature of the fluid entering or leaving the entrance or exit

h_o = specific enthalpy of that fluid at the reference temperature $T_o = 60°F$

The first law of thermodynamics for the boiler is expressed as: Sum of all the enthalpy rates of substances entering = Sum of all enthalpy rates of substances leaving + q where q is the rate of heat loss to the surroundings. This can be expressed as

$$\Sigma_{in} H_i = \Sigma_{out} H_i + q \qquad\qquad \text{Formula (6-12)}$$

where Σ is the summation sign, and H_i is the enthalpy rate of substance i.

For gaseous fuels the computation for the heat content of the gases is more conveniently expressed in the form

$$H = J_o C_{pm} \ (T - T_o) \qquad\qquad \text{Formula (6-13)}$$

Where

J_o = the volume rate of the gas stream corrected back to 1.0 atmosphere and 60°F ($T_o = 60°F$)

C_{pm} = the specific heat given on the basis of a standard volume of gas averaged over the temperature range $(T - T_o)$ and the gas mixture components.

$$C_{pm} = \Sigma_i X_i C_{pi} \qquad\qquad \text{Formula (6-14)}$$

Where

X_i = percent by volume of a component in one of the flow paths

C_{pi} = average specific heat over temperature range for each component

From Formula 6-10 we derive the volume fractions of each gas component as follows:

Component	X
CO_2	8.9
H_2O	17.2
N_2	72.2
O_2	1.7
	100.0

The average specific heat is found (using Figure 6-6).

$$
\begin{aligned}
C_{pm} = \quad &.089 \times 0.0275 = 0.00245 \\
&0.172 \times 0.0220 = 0.00378 \\
&0.722 \times 0.0186 = 0.01344 \\
&0.017 \times 0.0195 = \underline{0.00033} \\
&\qquad\qquad\qquad\quad\; 0.02 \text{ Btu/Scf} \cdot \text{F}
\end{aligned}
$$

The combustion equation also tells that when the products are at standard conditions 11.88 ft³ fuel and air generate 11.915 ft³ of products and that for every cubic foot of gas burned, 10.880 ft³ of air is introduced. Thus for a firing rate of 2126.5 ft³/min. the air required is 2126.5 × 10.88 = 23,136 ft³/min. or 23,136 × 60 = 1,388,179 ft³/h. The total fuel and air flow rate is then almost exactly equal to 1,388,179 + 2126.5 × 60 = 1,515,769 ft³/h. This corresponds to a flue gas discharge rate of 1,519,128 ft³/h.

For a flue gas discharge rate of 1,519,128 ft³/h and a temperature of 702°F, the total exhaust heat rate is found as

$$
\dot{H}_{\text{EX GAS}} = 1,519,128 \; \frac{\text{ft}^3}{\text{h}} \times 0.02 \; \frac{\text{Btu}}{\text{ft}^3 \cdot \text{F}}
$$
$$
\times \; (702-60)\text{F} = 19,505,604 \text{ Btu/h}
$$

For the steam leaving the boiler (100,000 lb/h at 150 lb/in² g saturated) the energy flow rate is found using the Steam Tables in Chapter 15 and the following equation:

$$\dot{H}_{\text{Steam}} = \dot{m}\,(h-h_o)$$
$$= 100{,}000 \text{ lb/h } (1195.6-28.08) \text{ Btu/lb}$$
$$= 116{,}752{,}000 \text{ Btu/h}$$

where 1195.6 is the specific enthalpy of saturated steam at 150 lb/in^2g and 28.08 is the specific enthalpy of saturated liquid water at 60° F, since we are using that temperature as our standard reference temperature for the heat balance. We have used 60°F as a reference temperature; this is not universal practice and in the boiler industry 70°F is more common, whereas in other areas 25°C is normal.

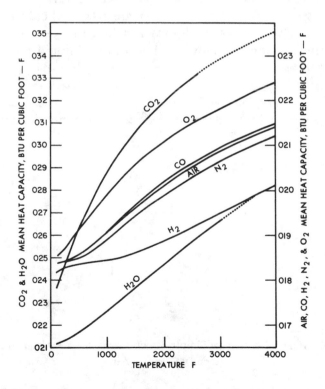

Figure 6-6. Mean Heat Capacity of Gases Found in Flue Products from 60 to T F (dashed line indicates dissociation)

CHEMICAL ENERGY IN FUEL

To determine the heat content of the chemical energy in fuel, find the higher heating value (HHV) for the fuel and multiply it by the volumetric flow rate for a gaseous fuel or the mass flow rate for a liquid or solid fuel. The assumed higher heating value for the natural gas used in the boiler of our example is 1001.1 Btu/ft^3 and the heat content rate is then

$$\dot{H} = 2126.5 \ \frac{ft^3}{min.} \ \times \ 60 \ \frac{min.}{h} \ \times \ 1001.1 \ \frac{Btu}{ft^3}$$
$$= 127{,}743{,}000 \ \frac{Btu}{h}$$

The enthalpy rates for the condensate return and make-up water are derived from data in the steam tables where the specific enthalpy of the compressed liquids is taken to be almost exactly equal to the specific enthalpy of the *saturated* liquid found at the same temperature.

The complete heat balance derived in the manner detailed above is presented in Figure 6-7.

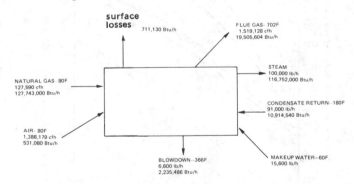

Figure 6-7. Heat Balance for a Simple Steam Generator Burning Natural Gas with 10% Excess Air

Waste heat is available from the combustion products leaving the stack. It amounts to 19,505,604 Btu/h at a temperature of 702°F. Some energy is also available from the condensers.

WASTE HEAT RECOVERY

The energy exhausted to the atmosphere should not be discarded. A portion of it can be recovered by using a heat exchanger. Any requirement for energy at a temperature in excess of 200°F can be satisfied. It is necessary to identify the prospective uses for the waste energy, make an economic analysis of the costs and savings involved in each of the options, and decide among those options on the basis of the economics of each. An important option in every case is that of rejecting all options if none proves economic. For the illustrative example, let us assume that the following uses for the waste heat have been identified:

- Preheating the combustion air
- Preheating the boiler feedwater
- Heating the domestic hot water supply (370 gal/hr from 50° to 170°F)
- A combination of the preceding

The first two are practices which are standard in energy-intensive high technology industries (e.g., electric companies).

Two rapid calculations give the waste heat available in the exhaust gases between 702°F and 220°F and the heat requirements for the domestic hot water supply to be 14,063,000 Btu/h and 369,900 Btu/h respectively. Since the latter constitutes only 2.6% of the available waste energy, we should reject that option. The remaining options are to preheat the combustion air and the feedwater or to divide the waste energy between those 2 options. For a small boiler we probably would not find the purchase of two separate heat exchangers an economic option, so that we shall limit ourselves to one of the other of the first two options. Without going into a detailed analysis, we may note that preheating the combustion air or the feedwater provides a double benefit. Since the preheated air or water requires less fuel to produce the same steam capacity, a direct fuel saving results. But the smaller quantity of fuel means a smaller air requirement, and this in turn means a smaller quantity of exhaust products and thus smaller stack loss at a given stack temperature.

The economic benefits can be estimated as follows: The air preheater is estimated to save 6% of the fuel. With an average boiler loading of 60% for the 8760 h in a year, this amounts to an annual fuel saving of

$$.6 \times 0.06 \times 8760 \times 127{,}590 = 40{,}237 \ \frac{ft^3}{yr}$$

which is worth, at an average rate of $5.50/1000 ft^3, an annual dollar saving of

$$\frac{40{,}237}{1000} \times 5.50 = \$221{,}000/yr$$

If the costs for installing the air preheater are assumed to be

Cost of preheater	$ 52,000
Cost of installation	57,200
New burners, air piping, controls and fan	56,600
	$165,800

For the feedwater heater (or economizer) the fuel savings is 9.2% or a total annual fuel saving of

$$0.092 \times 127{,}600 \times 8760 \times 0.60 = 61{,}696{,}000 \ \frac{ft^3}{yr}$$

and the economic benefit is

$$\frac{61{,}696{,}000 \ \frac{ft^3}{yr}}{1000} \times \$5.50 = \$340{,}000/yr$$

The cost of the economizer installed is estimated to be $134,000. This is clearly the best option for this particular boiler, especially as there is no need in this case for modifications to the boiler and accessories beyond the heat exchanger retrofit. There are several reasons for its superiority. The first is that in the case of air preheating, we are exchanging the waste heat in the gases to the incoming air which has almost the same mass flow rate and almost the same specific heat. Thus we can expect that the final temperature of the preheater air will be almost the arithmetic mean of the ambient air temperature and the exhaust gas temperature entering the economizer. In the case of the feedwater, the mass flow rate times specific heat is over four and one-quarter times that of the combustion air. Thus we can expect to transfer more energy to the water which results in a lower flue gas temperature leaving the stack.

The second reason is that preheating the air quite often (as in this case) affects the boiler accessories which require additional modifications and thus related capital expenditures.

In the preceding example of the process steam boiler, we are analyzing an efficient process unit. The heat available in the product (process steam) constituted a large percentage of the energy introduced in the fuel. The efficiency in percentage terms is computed as

$$\eta = \frac{\text{Useful output}}{\text{Energy input}} = \frac{Q_{steam}}{Q_{fuel}} \times 100$$

$$\eta = \frac{100,000 \text{ lb/h } (1195.6-148) \times 100}{127,590 \text{ ft}^3/\text{h} \times 1001.2 \text{ Btu/ft}^3}$$

$$= \frac{1.0476 \times 10^{10}}{1.2775 \times 10^8} = 0.82, \text{ or } 82\%$$

HEAT RECOVERY IN STEEL TUBE FURNACE

As a second and very different example, note the steel tube furnace illustrated below in Figure 6-8.

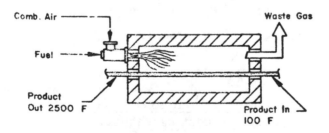

Figure 6-8. Continuous Steel Tube Furnace

The tubing enters the furnace from the right at a temperature of 100°F.

The specifications for the steel tube heating furnace are

Product capacity — 50 ton/h
Product specifications — 0.23% carbon steel
Final product temperature — 2000°F
Air/Fuel inlet temperature — 100°F
Air/Fuel mixture —10% excess air

Fuel — No. 5 fuel oil gravity API° 16
Fuel firing rate 48.71 gpm at 240°F (factory usage)
Utilization factor — 0.62

The useful heat leaves the furnace in the steel at 2000°F. This is called the useful furnace output and equals

$$Q_{prod} = \dot{m}_{prod} C_P (T_{out} - T_{in})$$

$$= 50 \frac{ton}{h} \times 2200 \text{ lb/ton}$$

$$\times .179 \text{ Btu/lb} \times (2000-100) \text{ F}$$

$$= 0.3741 \times 10^8 \frac{Btu}{h}$$

0.179 is used as the average specific heat of steel over the 100°F–2000°F range. The heat input to the furnace is the chemical energy in the fuel oil. The heating value for No. 5 fuel oil is found from Table 6-2.

The heat input is determined by

$$Q_{fuel} = 48.71 \text{ gpm} \times 142,300 \frac{Btu}{gal}$$

$$= 4.159 \times 10^8 \text{ Btu/h}$$

The percent efficiency of the furnace is:

$$\eta = \frac{output \times 100}{input}$$

$$= \frac{\text{Enthalpy added to steel} \times 100}{\text{Enthalpy entering with fuel}}$$

$$= \frac{0.3741 \times 10^8}{4.159 \times 10^8} \times 100 = 9\%$$

This means that of the 415.9 MBtu introduced per hour to the furnace, that 378.4 million are released to the atmosphere. At the present average cost of $2/gal. for No. 5 fuel oil, the heat wasted is

$$Q_{waste} = 48.71 \text{ gpm} \times (1-0.09)\frac{gal \text{ wasted}}{gal \text{ used}} \times .62$$

$$\times 60 \text{ min./h} \times \$2/gal = \$3297/h$$

$$= \$3297/h \times 8760h/yr = \$28,885,000/yr$$

Table 6-2. Physical Properties of Fuel Oil at 60°F

Fuel oil (CS-12-48) Grade No.	Gravity, API	Sp gr	Lb per gal	Btu per lb	Net Btu per gal
6	3	1.0520	8.76	18,190	152,100
6	4	1.0443	8.69	18,240	151,300
6	5	1.0366	8.63	18,290	149,400
6	6	1.0291	8.57	18,340	148,800
6	7	1.0217	8.50	18,390	148,100
6	8	1.0143	8.44	18,440	147,500
6	9	1.0071	8.39	18,490	146,900
6	10	1.0000	8.33	18,540	146,200
6	11	0.9930	8.27	18,590	145,600
6	12	.9861	8.22	18,640	144,900
6, 5	14	.9725	8.10	18,740	143,600
6, 5	16	.9593	7.99	18,840	142,300
5	18	.9465	7.89	18,930	140,900
4, 5	20	.9340	7.78	19,020	139,600
4, 5	22	.9218	7.68	19,110	138,300
4, 5	24	.9100	7.58	19,190	137,100
4, 2	26	.8984	7.49	19,270	135,800
4, 2	28	.8871	7.39	19,350	134,600
2	30	.8762	7.30	19,420	133,300
2	32	.8654	7.21	19,490	132,100
2	34	.8550	7.12	19,560	130,900
1, 2	36	.8448	7.04	19,620	129,700
1, 2	38	.8348	6.96	19,680	128,500
1	40	.8251	6.87	19,750	127,300
1	42	0.8156	6.79	19,810	126,200

The relation between specific gravity and degrees API is expressed by the formula:

$$\frac{141.5}{131.5 + API} = \text{sp gr at 60 F.}$$

For each 10 F above 60 F add 0.7 API.
For each 10 F below 60 F subtract 9.7 AP

In order to construct the combustion equation we must note from Table 6-3 that the carbon-hydrogen ratio is about 7.3. Thus

$$C_{6.08}H_{10} + 9.44\,O_2 + 35.49\,N_2 \rightarrow 6.08\,CO_2$$
$$+ 5\,H_2O + 0.86\,O_2 + 35.49\,N_2$$

Again referring to Table 6-2 the density of the liquid fuel is 7.99 lb/gal. Therefore, the above equation represents the reaction for 100 lb fuel/7.00 lb/gal = 12.51 gal and for 35.49 + 9.44 or 44.93

Table 6-3. Typical Properties of Commercial Petroleum Products Sold in the East and Midwest

| | Premium Fuels | | | | | Fuel Oils | | | | |
| | Gasolines | | | | Reduced crude, heavy gas oil or premium residuum* | | | | | |
	12 psia Reid vapor pressure natural gasoline	Straight run gasoline	No. 1 fuel, light diesel, or stove oil	Diesel or "gas house" gas oil		No. 2	No. 3	Cold No. 5	No. 5	Bunker C or No. 6
Gravity, °API	79	63	38–42	35–38	19–22	32–35	28–32	20–25	16-24	6–14
Viscosity	...	...	34–36 †	34–36 †	15–50 ‡	34–36 †	35–45 †	<20 ‡	20–40 ‡	50–300 ‡
Conradson carbon, wt %	None	None	Trace	Trace	3–7	Trace	<0.15	1–2	2–4	6–15
Pour point, °F	<0	<0	<0	<10	...	<10	<20	<15	15–60	>50
Sulfur, wt %	<0.1	<0.2	<0.1	0.1–0.6	<2.0	0.1–0.6	0.2–1.0	0.5–2.0	0.5–2.0	1–4
Water and sed., wt %	None	None	Trace	Trace	<1.0	<0.05	<0.10	0.1–0.5	0.5–1.0	0.5–2.0
Distillation, °F										
10%	110	150	390–410	400–440	550–650	420–440	450–500	<600	<600	600–700
50%	150	230	440–460	490–510	800–900	490–530	540–560	...	...	850–950
90%	235	335	490–520	580–600	...	580–620	600–670	<700	>700	...
End point	315	370	510–560	630–660	...	630–660	650–700	...	...	...
Flash point, °F	<100	<100	100–140	110–170	>150	130–170	>130	>130	>130	>150
Aniline point, °F	145	125	140–160	150–170	...	120–140	120–140	...	...	...
C to H ratio	5.2	5.6	6.1–6.4	6.2–6.5	6.9–7.3	6.6–7.1	7.0–7.3	7.3–7.7	7.3–7.7	7.7–9.0
Average MBtu recovered/gal of oil:										
Carbureted water gas	101	101	102	101	93	91	86	82	83	74
High-Btu oil gas	90	89	90	90	82	80	76	72	73	64
Tar + carbon, wt % of oil										
Carbureted water gas	...	...	20	22	35	31	36	42	42	52
High-Btu oil gas	...	...	32	34	45	42	46	52	52	60

(Data on gasolines and data below C/H ratio not given in reports.)
* The 6 wt % Conradson carbon oil used in the Hall High Btu Oil Gas Tests (A.G.A. Gas Production Research Committee. Hall High Btu Oil Gas Process, New York, 1949.) and "New England Gas Enriching Oil" are typical examples of this group.
† Saybolt Universal seconds at 100 F.
‡ Saybolt Furol seconds at 122 F.

Remarks:
No. 1 Fuel Oil—A distillate oil intended for vaporizing pot-type burners. A volatile fuel.
No. 2 Fuel Oil—For general purpose domestic heating; for use in burners not requiring No. 1 oil. Moderately volatile.
No. 3 Fuel Oil—Formerly a distillate oil for use in burners requiring low viscosity oil. Now incorporated as a part of No. 2.
No. 4 Fuel Oil—For burner installations not equipped with preheating facilities.
No. 5 Fuel Oil—A residual type oil. Requires preheating to 170–220 F.
No. 6 Fuel Oil—Preheating to 220–260 F suggested. A high viscosity oil.

lb mol of air. 44.93 lb mol of air weighs 1301 lb (using the molecular weight for air found in Table 6-4). The specific volume of air at standard conditions is found from the same table as 378.5 ft^3/lb mol. Therefore, the volume of air required to burn 12.51 gal of fuel is 44.93 × 378.5 = 17,000 ft^3 at standard temperature and pressure. The air input is then

$$m = \frac{17{,}000 \text{ ft}^3}{12.51 \text{ gal}} \cdot 48.71 \text{ gpm} = 66{,}200 \text{ ft}^3/\text{m}$$

or

$$m_{air} = 66{,}200 \times 60 = 3{,}972{,}000 \text{ ft}^3/\text{h}$$
$$Q_{air} = 1.2 \text{ Btu/ft}^3 \times 3{,}972{,}000$$
$$= 4{,}766{,}000 \text{ Btu/h}$$

Table 6-4. Gas Constants and Volume of the Pound-Mol for Certain Gases

	$R' = R/M$, specific gas constant, ft per F	M, mol wt, lb per lb-mol	R, universal gas constant, ft-lb per (lb-mol)(F)	Mv, cu ft per lb-mol *
Hydrogen	767.04	2.016	1546	378.9
Oxygen	48.24	32.000	1544	378.2
Nitrogen	55.13	28.016	1545	378.3
Nitrogen, "atmospheric"†	54.85	28.161	1545	378.6
Air	53.33	28.966	1545	378.5
Water vapor	85.72	18.016	1544	378.6
Carbon dioxide	34.87	44.010	1535	376.2
Carbon monoxide	55.14	28.010	1544	378.3
Hydrogen sulfide	44.79	34.076	1526	374.1
Sulfur dioxide	23.56	64.060	1509	369.6
Ammonia	89.42	17.032	1523	373.5
Methane	96.18	16.042	1543	378.2
Ethane	50.82	30.068	1528	374.5
Propane	34.13	44.094	1505	368.7
n-Butane	25.57	58.120	1486	364.3
iso-Butane	25.79	58.120	1499	367.4
Ethylene	54.70	28.052	1534	376.2
Propylene	36.01	42.078	1515	379.1

* At 60 F, 30 in. Hg, dry.
† Includes other inert gases in trace amounts.

The volume ratio of stack gases at standard conditions to air inlet rate is found from the combustion equation as

$$\frac{6.08 + 5 + 0.86 + 35.49}{9.44 + 35.49} = \frac{47.43}{44.93} = 1.056$$

$$Q_{ex} = 1.056 \times 3,972,000$$
$$= 4,194,000 \text{ ft}^3/\text{h}$$

The volume fractions of the stack gas components are found from the combustion equation as

	χ
CO_2	0.128
H_2O	0.106
O_2	0.018
N_2	0.748
	1.000

The specific enthalpy of the stack gas components may be calculated by use of Figure 6-5 for the exhaust temperature of 2200°F leading to the average specific enthalpy as indicated on next page.

$$
\begin{array}{llllll}
 & & & & & b \\
CO_2 & 0.128 & \times & 69.3 & = & 8.870 \\
H_2O & 0.106 & \times & 54.0 & = & 5.724 \\
O_2 & 0.018 & \times & 45.4 & = & 0.817 \\
N_2 & 0.748 & \times & 43.4 & = & \underline{32.463} \\
 & & & & & 47.874
\end{array}
$$

Therefore

$$Q_{ex} = 4,194,000 \text{ ft}^3/\text{h} \times 47.874 \frac{\text{Btu}}{\text{ft}^3}$$

$$= 200,784,000 \frac{\text{Btu}}{\text{h}}$$

The heat losses from the surface equal the fuel enthalpy input less the exhaust gas enthalpy and the produce enthalpy.

$$Q = 415,880,000 + 4,766,000 - 200,784,000$$

$$- 37,410,000 = 182,451,270 \frac{\text{Btu}}{\text{h}}$$

The complete heat balance diagram is shown in Figure 6-9. Now suppose that we install a metallic recuperator and preheat the combustion air to 750°F. The new heat balance is shown in Figure 6-10.

The economic analysis of this situation is as follows:

The waste heat saved is equal to the fuel savings times the heating value of the fuel.

$$Q_{saved} = (48.71 - 31.66) \text{ gpm} \times 142,300 \frac{\text{Btu}}{\text{gal}}$$

$$= 145,554,000 \text{ or } 35\% \text{ saving}$$

which equals

$$0.35 \times 48.71 \text{ gpm} \times 60 \times \$2/\text{gal} = \$2,045.82/\text{h}$$

In using Figures 6-12 and 6-13 remember that the combustion air volume (in SCF) is reduced directly proportional to the reduction in firing rate. If we maintain the same air/fuel ratio for the furnace (10% excess air), the air rate is reduced from the initial value of 3,972,000 ft³/h at a firing rate of 48.71 gpm to a rate of 31.66 gpm/ 48.71 gpm $\times$ 3,972,000 ft³/h = 2,581,800 ft³/h with the lower firing rate of 31.66 gpm when the recuperator is used.

If, as in our example, the furnace is used 0.62 $\times$ 8760 or 6531 h/ yr, then the annual saving is 5431 $\times$ \$2,045.80 = \$11.10 $\times$ 10⁶/yr.

Since a metallic recuperator can be purchased for a fraction of the savings, the savings will allow a payoff in less than a year.

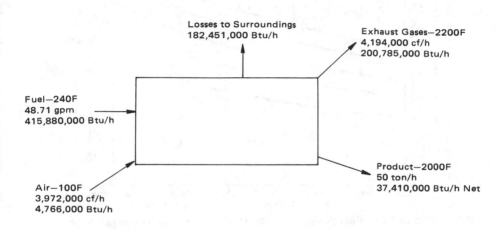

Figure 6-9. Heat Balance for a Simple Continuous Steel Tube Furnace

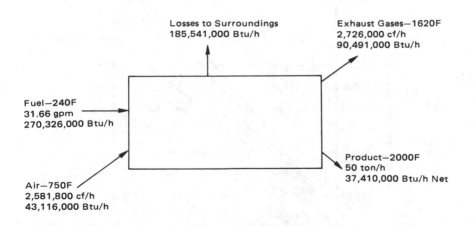

Figure 6-10. Heat Balance for a Continuous Steel Tube Furnace Equipped with a Recuperator Preheating Combustion Air to 750°F

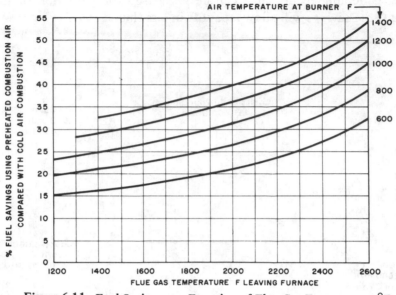

Figure 6-11. Fuel Savings as a Function of Flue Gas Temperature °F
Leaving Furnace

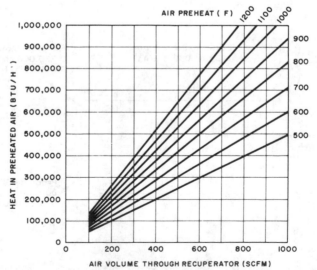

Figure 6-12. Heat Recovered as a Function of Air Volume through Recuperator
(SCFM)

BOILERS

Boiler Configurations and Components

Industrial boiler designs are influenced by fuel characteristics and firing method, steam demand, steam pressures, firing characteristics and the individual manufacturers. Industrial boilers can be classified as either firetube or watertube, indicating the relative position of the hot combustion gases with respect to the fluid being heated.

Firetube Boilers

Firetube units pass the hot products of combustion through tubes submerged in the boiler water. Conventional units generally employ from 2 to 4 passes to increase the surface area exposed to the hot gases and thereby increase efficiency. Multiple passes, however, require greater fan power, increased boiler complexity and larger shell dimensions. (Refer to Figure 6-13.) Maximum capacity of firetube units is currently limited to 25,000 lbs of steam per hour (750 boiler hp) with an operating pressure of 250 psi due to economic factors related to material strength and thickness.

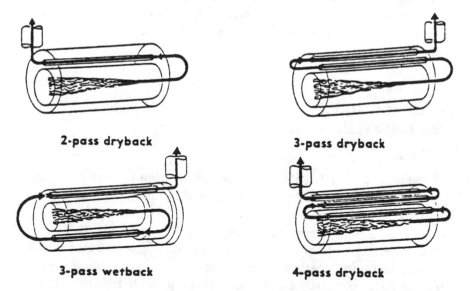

2-pass dryback

3-pass dryback

3-pass wetback

4-pass dryback

Figure 6-13. Typical Firetube Boiler Gas Flow Patterns

Advantages of firetube units include: (1) ability to meet wide and sudden load fluctuations with only slight pressure changes, (2) low initial costs and maintenance and (3) simple foundation and installation procedures.

Watertube Boilers

Watertube units circulate the boiler water inside the tubes and the flue gases outside. Water circulation is generally provided by the density variation between cold feed water and the hot water/steam mixture in the riser as illustrated in Figure 6-14. Watertube boilers may be subclassified into different groups by tube shape, by drum number and location and by capacity. Refer to Figure 6-15.

Another important determination is "field" versus "shop" erected units. Many engineers feel that shop assembled boilers can meet closer tolerance than field assembled units and therefore may be more efficient; however, this has not been fully substantiated. Watertube units range in size from as small as 1000 lbs of steam per hour to the giant utility boilers in the 1000 MW class. The largest industrial boilers are generally taken to be about 500,000 lbs of steam per hour. Important elements of a steam generator include the firing mechanism, the furnace water walls, the superheaters, convective regions, the economizer and air preheater and the associated ash and dust collectors.

FUEL HANDLING AND FIRING SYSTEMS

Gas Fired

Natural gas fuel is the simplest fuel to burn in that it requires little preparation and mixes readily with the combustion air supply. Industrial boilers generally use low-pressure burners operating at a pressure of 1/8 to 4 psi. Gas is generally introduced at the burner through several orifices that generate gaseous jets that mix rapidly with the incoming combustion air supply. There are many designs in use that differ primarily in the orientation of the burner orifices and their locations in the burner housing.

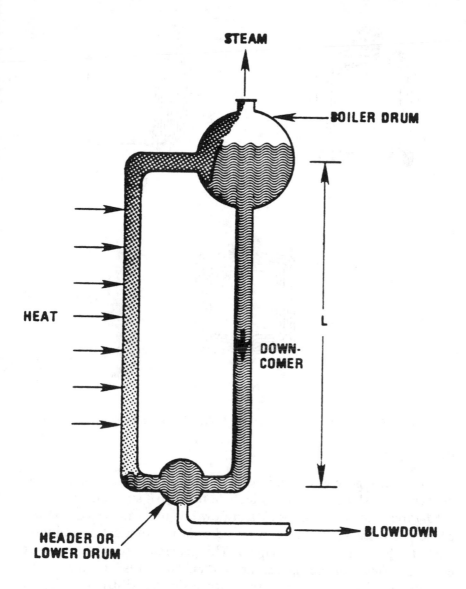

Figure 6-14. Water Circulation Pattern in a Watertube Boiler

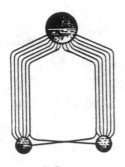

A Type

D Type

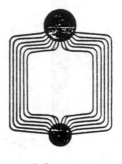

O Type

Figure 6-15. Classification of Watertube Boilers by Basic Tube Arrangement

Oil Fired

Oil fuels generally require some type of pretreatment prior to delivery to the burner. This may include the use of strainers to remove solid foreign material and tank and flow line preheaters to assure the proper viscosity. Oil must be atomized prior to vaporization and mixing with the combustion air supply. This generally requires the use of either air, steam or mechanical atomizers. The oil is introduced into the furnace through a gun fitted with a tip that distributes the oil into a fine spray that allows mixing between the oil droplets and the combustion air supply. Oil cups that spin the oil into a fine mist are also employed on small units. An oil burner may be equipped with diffusers that act as flame holders by inducing strong recirculation patterns near the burner. In some burners, primary air nozzles are employed.

Pulverized Coal Fired

The pulverizer system provides four functions: pulverizing, drying, classifying to the required fineness and transporting the coal to the burner's main air stream. The furnace may be designed for dry ash removal in the hopper bottom or for molten ash removal as in a slag tap furnace. The furnace is dependent on the burning and ash characteristics of the coal as well as the firing system and type of furnace bottom. The primary objectives are to control furnace ash deposits and provide sufficient cooling of the gases leaving the furnace to reduce the buildup of slag in the convective regions. Pulverized coal fired systems are generally considered to be economical for units with capacities in excess of 200,000 lbs of steam per hour.

Stoker Fired

Coal stoker units are characterized by bed combustion on the boiler grate with the bulk of the combustion air supplied through the grate. Several stoker firing methods currently in use on industrial sized boilers include underfed, overfed and spreader. In underfed and overfed stokers, the coal is transferred directly on to the burning bed. In a spreader stoker the coal is hurled into the furnace when it is partially burned in suspension before lighting on the grate. Several grate configurations can be used with overfed and spreader stokers including stationary, chain, traveling, dumping and vibrating grates. Each grate configuration has its own requirements as to coal fineness and ash characteristics for optimum operation. Spreader stoker units have the advantage that they can burn a wide variety of fuels including waste products. Underfed and overfed units have the disadvantage that they are relatively slow to respond to load variations. Stoker units can be designed for a wide range of capacities from 2,000 to 350,000 lbs of steam per hour. Spreader stoker units are generally equipped with overfire air jets to induce turbulence for improved mixing and combustible burnout. Stoker units are also equipped with ash reinjection systems that allow collected ash that contains a significant portion of unburned carbon to be reintroduced into the furnace for burning.

COMBUSTION CONTROL SYSTEMS

Combustion controls have two purposes: (1) maintain constant steam conditions under varying loads by adjusting fuel flow, and (2) maintain an appropriate combustion air-to-fuel flow.

Classification

Combustion control systems can be classified as series, parallel and series/parallel.

In series control, either the fuel or air is monitored and the other is adjusted accordingly. For parallel control systems, changes in steam conditions result in a change in both air and fuel flow. In series/parallel systems, variations in steam pressure affect the rate of fuel input, and simultaneously the combustion air flow is controlled by the steam flow.

Combustion controls can be also classified as positioning and metering controls.

Positioning controls respond to system demands by moving to a present position. In metering systems, the response is controlled by actual measurements of the fuel air flows.

Application

The application and degree of combustion controls vary with the boiler size and is dictated by system costs. The parallel positioning jackshaft system has been extensively applied to industrial boilers based on minimum system costs. The combustion control responds to changes in steam pressure and can be controlled by a manual override. The control linkage and cam positions for the fuel and air flow are generally calibrated on start-up.

Improved control of excess air can be obtained by substituting electric or pneumatic systems for the mechanical linkages. In addition, relative position of fuel control and combustion air dampers can be modified. More advanced systems are pressure ratio control of the fuel and air pressure, direct air and fuel metering and excess air correction systems using flue gas O_2 monitoring. Factors that have limited the application of the most sophisticated control systems to industrial boilers include cost, reliability and maintenance.

SAFETY

It is essential that energy engineers conducting boiler evaluation tests and tune-ups understand and be aware of boiler safety devices. Occasionally operators who don't understand safety have been known to bypass safety features in order to keep a unit operating.

Summary of requirements found in the NATIONAL BOILER SAFETY CODES:

* PREPURGE—4 to 8 air changes to insure no fuel vapors or gases remain in the boiler which could ignite or explode when the pilot light is off.

* PILOT PROVING—10 to 15 second proving period pilot must ignite and be proven before the main fuel valves open.

* MAIN FLAME TRIAL—usually 10 to 15 second trial period for natural gas and oil after main fuel valves open.

* FLAME FAILURE RESPONSE TIME—usually 3 to 4 seconds after the main flame goes out the Fuel Shut Off valves should automatically close to prevent a build up of explosive vapor concentrations within the boiler.

* INTERLOCKS—automatically shut the boiler down if certain safety features are not in the safe condition.

Loss of atomizing means (steam/air)
High/Low gas pressure
Low fire for light off
Low fuel oil temperature
Low fuel oil pressure
Main combustion air
Low water level

* FUEL VALVES—Safety Shut Off Valves (SSOV). Usually two valves; closing time 1 to 5 seconds depending on fuel and size of boiler.

* FLAME DETECTORS—should be positioned and capable of detecting flame only and not sparks from the spark ignitor or hot refractory so as not to give a false flame signal.

OPERATING GUIDES

Consult the plant engineer and the boiler manufacturers technical manuals for complete boiler operating procedures. Typical items to check before conducting efficiency checks are as follows:

Oil Burners

Make sure the atomizer is of the proper design and size and the burner is centered with dimensions according to manufacturer's drawings.

Inspect oil-tip passages and orifices for wear (use proper size drill as a feeler gage) and remove any coke or gum deposits to assure the proper oil-spray pattern.

Verify proper oil pressure and temperature at the burner.

Verify proper atomizing-steam pressure.

Make sure that the burner diffuser (impeller) is not damaged and is properly located to the oil-gun tip.

Check to see that the oil gun is positioned properly within the burner throat and that the throat refractory is in good condition.

Gas Burners

Inspect gas-ingestion orifices and verify that all passages are unobstructed. Also, be sure filters and moisture traps are in place, clean and operating properly, to prevent plugging of gas orifices.

Confirm proper location and orientation of diffusers, spuds, etc. Look for any burned off or missing burner parts.

Combustion Controls

Inspect all fuel valves to verify proper movement; clean valve internal surfaces if necessary.

Eliminate "play" in control linkages on dampers. Any play, no matter how slight, will cause a loss of efficiency since a precise tune-up will be impossible.

Make sure fuel-supply inlet pressure to pressure regulators are high enough to assure constant regulator-outlet pressures for all firing rates.

Correct any control elements that fail to respond smoothly to varying steam demand. Unnecessary hunting caused by improperly adjusted regulators or automatic master controllers can waste fuel.

Check that all gages are functioning and are calibrated.

Furnace

Inspect boiler furnace and gas side surfaces for excessive deposits and fouling. These lead to higher stack temperatures and lower boiler efficiencies.

Inspect furnace refractory and insulation for cracks that may cause leaks and missing refractory.

Clean furnace inspection parts and make sure that burner throat, furnace walls, and leading connection passes are visible through them since flame observation is an essential part of efficient boiler operation and testing.

BOILER EFFICIENCY IMPROVEMENT

The boiler plant should be designed and operated to produce the maximum amount of usable heat from a given amount of fuel.

Combustion is a chemical reaction of fuel and oxygen which produces heat. Oxygen is obtained from the input air which also contains nitrogen. Nitrogen is useless to the combustion process. The carbon in the fuel can combine with air to form either CO or CO_2. Incomplete combustion can be recognized by a low CO_2 and high CO content in the stack. Excess air causes more fuel to be burned than required. Stack losses are increased and more fuel is needed to raise ambient air to stack temperatures. On the other hand, if insufficient air is supplied, incomplete combustion occurs and the flame temperature is lowered.

Boiler Efficiency

Boiler efficiency (E) is defined as

$$\%E = \frac{\text{Heat Out of Boiler}}{\text{Heat Supplied to Boiler}} \times 100 \qquad \textit{Formula (6-15)}$$

For steam-generating boilers

$$\%E = \frac{\text{Evaporation Ratio} \times \text{Heat Content of Steam}}{\text{Calorific Value of Fuel}} \times 100 \qquad \textit{Formula (6-16)}$$

For hot water boilers

$$\%E = \frac{\text{Rate of Flow from Boiler} \times \text{Heat Output of Water}}{\text{Caloric Value of Fuel} \times \text{Fuel Rate}} \times 100$$

$$\textit{Formula (6-17)}$$

The relationship between steam produced and fuel used is called the evaporation ratio.

Boilers are usually designed to operate at the maximum efficiency when running at rated output. Figure 6-16 illustrates boiler efficiency as a function of time on line.

Full boiler capacity for heating occurs only a small amount of the time. On the other hand, part loading of 60% or less occurs approximately 90% of the time.

Where the present boiler plant has deteriorated, consideration should be given to replacement with modular boilers sized to meet the heating load.

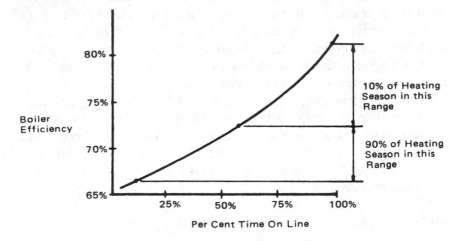

Figure 6-16. Effect of Cycling to Meet Part Loads

The overall thermal efficiency of the boiler and the various losses of efficiency of the system are summarized in Figure 6-17.

To calculate dry flue gas loss, Formula 6-18 is used.

$$\text{Flue gas loss} = \frac{K(T-t)}{CO_2} \qquad \textit{Formula (6-18)}$$

Where

K = constant for type of fuel = 0.39 Coke
                                  = 0.37 Anthracite
                                  = 0.34 Bituminous Coal
                                  = 0.33 Coal Tar Fuel
                                  = 0.31 Fuel Oil

$$T \quad = \quad \text{temperature of flue gases in } °F$$

t = temperature of air supply to furnace in $°F$

CO_2 = percentage CO_2 content of flue gas measured volumetrically.

It should be noted that this formula does not apply to the combustion of any gaseous fuels such as natural gas, propane, butane, etc. Basic combustion formulas or nomograms should be used in the gaseous fuel case.

---

1. Overall thermal efficiency . . . . ____

2. Losses due to flue gases

   (a) Dry Flue Gas . . . . ____

The loss due to heat carried up the stack in dry flue gases can be determined, if the carbon dioxide (CO_2) content of the flue gases and the temperatures of the flue gas and air to the furnace are known.

   (b) Moisture % Hydrogen . . . . ____

   (c) Incomplete combustion . . . .
                                                                    ____

3. Balance of account, including radiation and other unmeasured

   losses . . . . ____

              TOTAL 100%

Figure 6-17. Thermal Efficiency of Boiler

The savings in fuel as related to the change in efficiency is given by Formula 6-19.

$$\text{Savings in Fuel} = \frac{\text{New Efficiency} - \text{Old Efficiency}}{\text{New Efficiency}}$$
$$\times \text{Fuel Consumption} \qquad Formula\ (6\text{-}19)$$

Figure 6-19 can be used to estimate the effect of flue gas composition, excess air, and stack temperature on boiler efficiency.

To estimate losses due to moisture, Figure 6-18 is used.

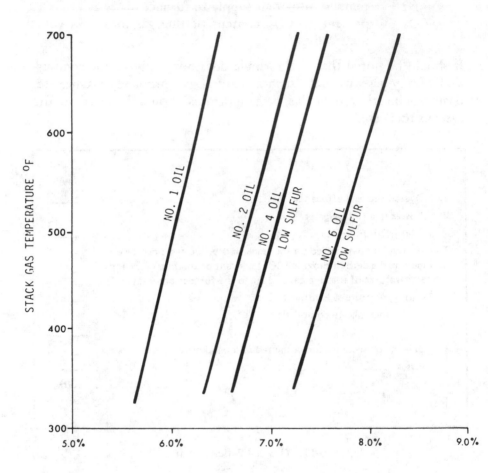

NOTE:
1. The figure gives a simple reference to heat loss in stack gases due to the formation of water in burning the hydrogen in various fuel oils.
2. The graph assumes a boiler room temperature of 80°F.

Figure 6-18. Heat Loss Due to Burning Hydrogen in Fuel
(Source: Instructions For Energy Auditors, Volume 1)

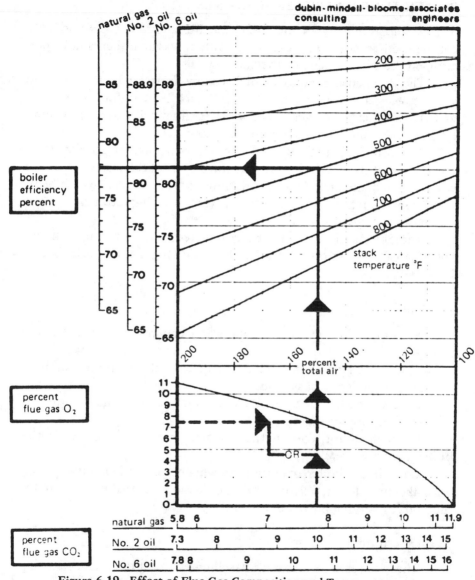

**Figure 6-19. Effect of Flue Gas Composition and Temperature on
Boiler Efficiency**

*(Source: Guidelines for Saving Energy in Existing Buildings—Engineers, Architects and
Operators Manual, ECM-2)*

BOILER TUNE-UP TEST PROCEDURES

As illustrated by Figures 6-20 and 6-21, either % CO_2 or % O_2 can be used to determine excess air as long as the boiler is not operating on the fuel rise side of the curve.

The detailed procedures which follow illustrate how to tune-up a boiler to get the best air-to-fuel ratio. Note that for natural gas, % CO_2 must also be measured, while for fuel oil, smoke spot numbers or visual smoke measurements are used.

The principal method used for improving boiler efficiency involves operating the boiler at the lowest practical excess O_2 level with an adequate margin for variations in fuel properties, ambient conditions and the repeatability and response characteristics of the combustion control system.

These tests should only be conducted with a through understanding of the test objectives and by following a systematic, organized series of tests.

Cautions

Extremely low excess O_2 operation can result in catastrophic results.

Know at all times the impact of the modification on fuel flow, air flow and the control system.

Observe boiler instrumentation, stack and flame conditions while making any changes.

When in doubt, consult the plant engineering personnel or the boiler manufacturer.

Consult the boiler operation and maintenance manual supplied with the unit for details on the combustion control system or for methods of varying burner excess air.

Test Description

The test series begins with baseline tests that document existing "as-found conditions" for several firing rates over the boiler's normal operating range. At each of these firing rates, variations in excess O_2 level from 1 to 2% above the normal operating point to the "minimum O_2" level are noted. Curves of combustibles as a function of excess O_2 level will be constructed similar to those given in Figures

THEORETICAL AIR CURVE

(FUEL OIL)

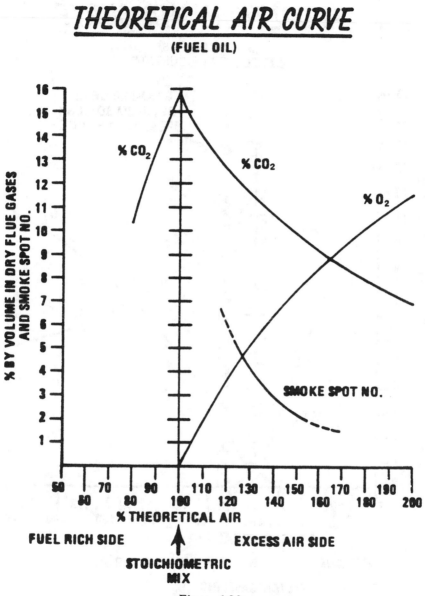

Figure 6-20.

THEORETICAL AIR CURVE

(NATURAL GAS)
COMPLETE COMBUSTION

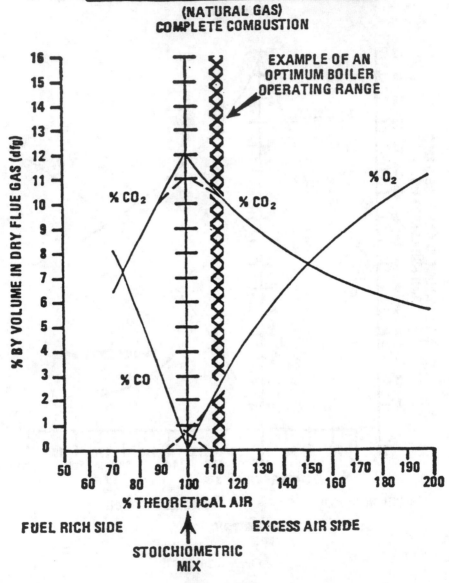

Figure 6-21.

6-22 and 6-23. As illustrated in these figures, high levels of smoke or CO indicating potentially unstable operation can occur with small changes in excess O_2; thus small changes in excess O_2 should be made for conditions near the smoke or CO limit. It is important to note that the boiler may exhibit a gradual smoke or CO behavior at one firing rate and a steep behavior at another. Minimum excess O_2 will be that at which the boiler just starts to smoke, or the CO emissions rise above 400 ppm or the Smoke Spot Number equals the maximum value as given in Table 6-5. Once minimum excess O_2 levels are established, an appropriate O_2 margin or operating cushion ranging from 0.5 to 2.0% O_2 above the minimum point depending on the particular boiler control system and fuels, should be maintained.

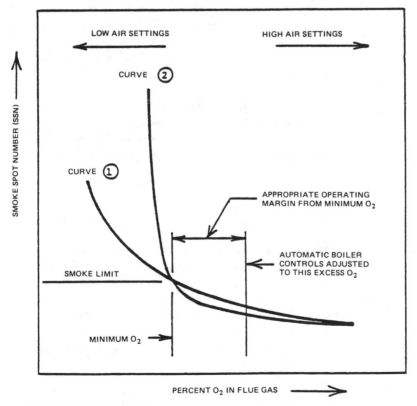

CURVE 1 — GRADUAL SMOKE/O_2 CHARACTERISTIC
CURVE 2 — STEEP SMOKE/O_2 CHARACTERISTIC

Figure 6-22. Typical Smoke-O_2 Characteristic Curves for Coal- or Oil-Fired Industrial Boilers

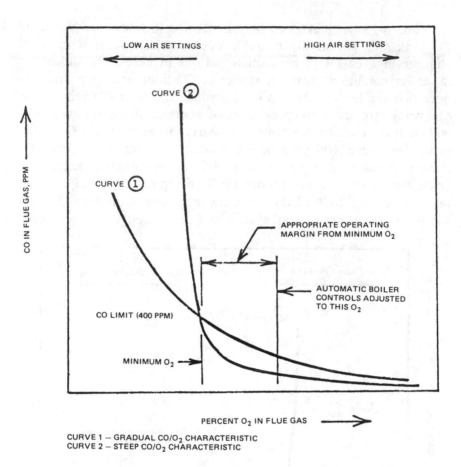

Figure 6-23. Typical CO–O Characteristic Curve for Gas-Fired
Industrial Boilers

Table 6-5. Maximum Desirable Smoke Spot Number

Fuel Grade	Maximum Desirable SSN
No. 2	less than 1
No. 4	2
No. 5 (light and heavy), and low-sulfur resid.	3
No. 6	4

Repeated tests at the same firing condition, approached from both the "high side" and "low side" (i.e., from higher and lower firing rates), can determine whether there is excessive play in the boiler controls.

Record all pertinent data for future comparisons. Readings should be made only after steady boiler conditions are reached and at normal steam operating conditions.

Step-By-Step Boiler Adjustment Procedure
for Low Excess O_2 Operation

1. Bring the boiler to the test firing rate and put the combustion controls on manual.
2. After stabilizing, observe flame conditions and take a complete set of readings.
3. Raise excess O_2 1 to 2%, allowing time to stabilize and take readings.
4. Reduce excess O_2 in small steps while observing stack and flame conditions. Allow the unit to stabilize following each change and record data.
5. Continue to reduce excess air until a minimum excess O_2 condition is reached.
6. Plot data similar to Figures 6-22 and 6-23.
7. Compare the minimum excess O_2 value to the value provided by the boiler manufacturers. High excess O_2 levels should be investigated.
8. Establish the margin in excess O_2 above the minimum and reset the burner controls to maintain this level.
9. Repeat Steps 1 thru 8 for each firing rate to be considered. Some compromise in optimum O_2 settings may be necessary since control adjustments at one firing rate may affect conditions at other firing rates.
10. After these adjustments have been completed, verify the operation of these settings by making rapid load pick-ups and drops. If undesirable conditions are encountered, reset controls.
11. Low fire conditions may require special consideration.
12. Perform tests on any alternate fuel used. Again, some discretion will be required to resolve differences in control settings for optimum conditions on the two fuels.

Evaluation of the New Low O_2 Settings

Extra attention should be given to furnace and flame patterns for the first month or two following implementation of the new operating modes. Thoroughly inspect the boiler during the next shutdown. To assure high boiler efficiency, periodically make performance evaluations and compare with the results obtained during the test program.

Burner Adjustments

Adjustments to burner and fuel systems can also be made in addition to the low excess O_2 test program previously described. The approach in testing these adjustments is a "trial-and-error" procedure with sufficient organization to allow meaningful comparisons with established data.

Items that may result in lower minimum excess O_2 levels include changes in burner register settings, oil gun tip position, oil gun diffuser position, coal spreader position, fuel oil temperature, fuel and atomizing pressure, and coal particle size. Evaluation of each of these items involves the same general procedures, precautions and data evaluation as outlined previously. The effect of these adjustments on minimum O_2 is variable from boiler to boiler and difficult to predict.

Conclusions

1. Combustion modifications are potentially catastrophic unless caution is continuously observed.
2. Test programs must be well thought out and planned in advance and anticipated results formulated and checked with obtained data.
3. Any modifications at low excess O_2 levels must be made slowly in small steps and with continuous evaluation of the flame, furnace and stack gas conditions.
4. All data must be recorded to allow future comparisions to be made.
5. Practical operating excess O_2 levels must be given an adequate excess O_2 margin to allow for load variation and fuel property or ambient air condition changes.
6. New recommended operating conditions should be moni-

tored for a sufficient length of time to gain confidence in their long-term use.

7. Periodic efficiency checks can indicate deviations from optimum performance conditions.

AUXILIARY EQUIPMENT FOR INCREASING EFFICIENCY

The efficiency improvement potential of auxiliary equipment modifications is dependent on the existing boiler conditions.

Stack gas heat recovery equipment (air preheaters and economizers) are generally the most cost-effective auxiliary equipment additions. Addition of turbulators to firetube boilers can improve operating efficiencies and promote balanced gas flows between tube banks. Advanced combustion control systems and burners generally are less beneficial than stack gas heat recovery equipment on industrial sized units. Insulation and sootblowers, judiciously applied, can have beneficial effects on boiler efficiency. Significant energy savings potentials exist in wastewater heat recovery from blowdown water and returned condensate.

Table 6-6 summarizes the options available to improve boiler efficiency.

Air preheaters and economizers are common equipment used. A caution should be noted in that these and other options will reduce stack temperature. In order to avoid corrosion problems, exit temperatures should be as illustrated in Table 6-7. Exit gas temperature is determined by the extent of boiler convective surface or the presence of stack gas heat recovery equipment.

For boilers without heat recovery equipment, the minimum exit gas temperature is fixed by the boiler operating pressure since this determines the steam temperature. Usual design practices result in an outlet gas temperature 150°F above the saturated steam temperature.

It becomes increasingly expensive to approach boiler saturation temperatures by simply adding convective surface area. As operating pressures increase, the stack gas temperature increases to make heat recovery equipment more desirable. Economizers will permit a reduction in exit gas temperatures since the feedwater is at a lower temperature (220°F) than the steam saturation temperature. Stack gas temperatures of 300°F can be achieved with stack gas heat recovery

Table 6-6. Boiler Efficiency Improvement Equipment

DEVICE	PRINCIPLE OF OPERATION	EFFICIENCY IMPROVEMENT POTENTIAL	SPECIAL CONSIDERATIONS
Air Preheaters	Transfer energy from stack gases to incoming combustion air.	2.5 % for each 100 degree F decrease in stack gas temperature.	* Results in improved combustion condition. * Minimum flue gas temperatures limited by corrosion characteristics of the flue gas. * Application limited by space, duct orientation and maximum combustion air temperatures.
Economizers	Transfer energy from stack gases to incoming feedwater.	2.5 % increase for each 100 degree F decrease in stack gas temperature (1 % increase for each 10 degree F increase in feedwater temperature.	* Minimum flue gas temperature limited by corrosion characteristics of the flue gas. * Application limited on low pressure boilers. * Generally preferred over air preheaters for small (50,000 lbs./hr.) units.
Firetube Turbulators	Increases turbulence in the secondary passes of firetube units thereby increasing efficiency	2.5 % increase for each 100 degree F decrease in stack gas temperature.	* Limited to gas and oil fired units. * Properly deployed, they can balance gas flows through the tubes. * Increases pressure drop in the system.
Combustion Control Systems	Regulate the quantity of fuel and air flow	0.25 % increase for each 1 % decrease in excess O_2 depending on the stack gas temperature.	* Vary in complexity from the simplest jackshaft system to cross limited oxygen correction system. * Can operate either pneumatically or electrically. * Retrofit applications must be compatible with existing burner hardware.
Instrumentation	Provide operational data		* Provide records so that efficiency comparisions can be made.
Oil and Gas Burners	Promote flame conditions that result in complete combustion at lower excess air levels.	0.25 % increase for each 1 % decrease in excess O_2 depending on the stack gas temperature.	* Operation with the most elaborate low excess air burners require the use of advanced combustion control systems. * Flame shape and heat release rate must be compatible with furnace characteristics.

			* Flame scanners increase reliability. * Advanced atomizing systems are available. * Close control of oil viscosity improves atomization.
Insulation	Reduce external heat transfer	Dependent on surface temperature	* Mass type insulation has low thermal conductivity and release heat loss by conduction. * Reflective insulation has smooth, metallic surfaces that reduce heat loss by radiation. * Insulation provides several other advantages including structural strength, reduced noise and fire protection.
Sootblowers	Remove boiler tube deposits that retard heat transfer.	Dependent on the gas temperature	* Can use steam or air as the blowing media. * Fixed position systems are used in low temperature regions whereas retractable "losses" are employed in high temperature areas. * The choice of the cleaning media will depend on the characteristics of the deposits.
Blowdown Systems	Transfer energy from expelled blowdown liquids to incoming feedwater.	1 - 3 % dependent on blowdown quantities and operating pressures	* Quantity of expelled blowdown water is dependent on the boiler and makeup water quality. * Continuous blowdown operation not only decreases expelled liquids but also allows the incorporation of heat recovery equipment.
Condensate Return Systems	Reduce hot water requirements by recovering condensate.	12 - 15 %	* Quantity of condensate returned dependent on process and contamination. * Several systems available range from atmospheric (open) to fully pressurized (closed) systems.

Table 6-7. Minimum Exit Gas Temperatures

Oil Fuel ($>$ 2.5% S)	390°F
Oil Fuel ($<$ 1.0% S)	330°F
Bituminous Coal ($>$ 3.5% S)	290°F
Bituminous Coal ($<$ 1.5% S)	230°F
Pulverized Anthracite	220°F
Natural Gas (Sulfur-free)	220°F

equipment. Further reductions are achieved using air preheaters. Present design criteria limit the degree of cooling using stack gas heat recovery equipment to a level which will minimize condensation on heat transfer surfaces. The sulfur content of the fuel has a direct bearing on the minimum stack gas temperature since SO_3 combines with condensed water to form sulfuric acid and since the SO_3 concentration in the flue gas also determines the condensation temperature.

7

Heating, Ventilation, Air Conditioning, and Building System Optimization

This chapter will review the basics of Heating, Ventilation and Air Conditioning (HVAC) and building as related to energy engineering.

DEGREE DAYS

Degree days are the summation of the product of the difference in temperature (ΔT) between the *average outdoor* and hypothetical *average indoor* temperatures (65°F) and the number of days (t) the outdoor temperature is below 65°F. Therefore:

$$DD = \Delta T \times t, \text{ therefore } \Delta T = DD/t \qquad \textit{Formula (7-1)}$$

Degree days divided by the total number of days on which Degree days were accumulated will yield an average ΔT for the season, based on an assumed indoor temperature of 65°F. To find the average outdoor temperature of the season, this figure must be subtracted from 65°F.

Example Problem 7-1

If there are 6750 degree days recorded over a heating season of 270 days, what is the mean outdoor temperature for that season?

Answer

$$\Delta T = DD/t \qquad \Delta T = \frac{6750 DD}{270 \text{ days}} \qquad \Delta T = 25°F$$

The average outdoor temperature can now be found, since
$\Delta T = T$ (avg. indoor) $- T$ (avg. outdoor)
T (avg. outdoor) $= T$ (avg. indoor) $- \Delta T$
T (avg. outdoor) $= 65°F - 25°F$
T (avg. outdoor) $= 40°F$

RESISTANCE (*R*) TO HEAT FLOW AND
CONDUCTANCE (*U*) AND CONDUCTIVITY (*K*)

The rate at which heat flows through a material depends on its characteristics. Some materials transmit heat more readily than others. This characteristic of materials which affects the flow of heat through them can be viewed either as their *resistance* to the flow of heat or as their *conductance* allowing the flow of heat.

For a section of a building, such as a wall, the conductance is expressed as the *U*-value for that wall, that is, the number of Btu's that will pass through a one-square-foot section of a building in one hour with a one-degree temperature difference between the two surfaces.

U = Btu's per square foot per hour per degree Fahrenheit.
or
$$U = Btu/ft^2 \, h°F$$

R-Value = Thermal Resistance = The unit time for a unit area of a particular body or assembly having defined surfaces with a unit average temperature difference established between the two surfaces per unit of Thermal Transmission.

$$\frac{hr \cdot ft^2 \cdot °F}{Btu}$$

$$R = 1/U$$

Where

$$U = \frac{1}{1/h_i + x_1/K_1 + \dfrac{x_2}{K_2} + \ldots + 1/h_o}$$

Formula (7-2)

x   = Respective thickness of different materials
K   = Respective conductivities of layers
h   = Convective heat transfer coefficients

The conductivity of a material as related to conductance and resistance is illustrated by Formula 7-3.

$$U = \frac{K}{d} = \frac{1}{R}$$

Formula (7-3)

where d is the thickness of the material.

In buildings, Table 7-1 may be used to find acceptable conductance values for walls.

Table 7-1. Minimum Conductance Values for Walls

Heating Degree Days	Min. Acceptable U-Value
1000	0.40
2000	0.30
3000	0.30
4000	0.20
5000	0.20
6000	0.15
7000	0.15
8000	0.10
9000	0.10

VOLUME (V) OF AIR

The volume of air within a structure is constant even though the air itself changes—new air enters; old air leaves. Total volume is equal to the volume of space in the conditioned portion of the facility. (Only the volume of conditioned space is considered since air entering and leaving the unconditioned part of the facility does not demand energy to condition it.)

To determine the volume (V) of air, multiply the height (H) of the space times the width (W) of the space times the length (L) of the space.* While this can be done for the facility as a whole, it is more accurate to calculate it for each room and then add these volumes.

AIR CHANGES PER HOUR (AC)

The rate at which the volume of air in a structure changes per hour differs greatly from building to building. The number of air changes per hour (AC/h) has wide variation due to a number of factors such as

---

* This is only appropriate for structures with flat ceilings.

- *The number and size of openings* in the envelope—around doors and windows and in the siding itself
- *The average speed* of the wind blowing against the structure and the protection the structure has from this wind
- *The number and size* of chimneys, vents, and exhaust fans and the frequency of their use
- *The number of times* that doors and windows are opened and
- *How the structure is used.*

Currently, the general standard, or accepted value, for air changes within an average home is one air change (of entire house) per hour, unless for some reason a room contains toxic fumes, etc.

HEATING CAPACITY OF AIR (*HC*)

Air can be heated and cooled. A certain amount of heat is necessary to change the temperature of each cubic foot (ft^3) of air one degree Fahrenheit (F). This amount of heat depends on the density of air which varies with temperature and pressure. This figure will generally be within the range of 0.018-0.022 Btu's/ft^3 °F.

BUILDING DYNAMICS

The building experiences heat gains and heat losses depending on whether the cooling or heating system is present, as illustrated in Figures 7-1 and 7-2. Only when the total season is considered in conjunction with lighting and heating, ventilation and air conditioning (HVAC) can the energy utilization choice be decided. One way of reducing energy consumption of HVAC equipment is to reduce the overall heat gain or heat loss of a building.

CONDUCTION HEAT LOSS

The formula used to determine the amount of heat conducted through the envelope is as follows: Degree days (*DD*) is the product of the difference in temperature (ΔT) and the time (*t*) in days, providing that the days are converted to hours. This is accomplished by multiplying *DD* times 24 hours a day. This will yield the quantity of heat (*Q*) conducted through a particular section of the envelope for the entire heating season.

The formula can be written:

$$Q_{\text{(heating season)}} = U \times DD \times 24 \text{ hrs/day} \qquad Formula\ (7\text{-}4)$$

or

$$Q_{\text{(heating season)}} = \frac{A \times DD \times 24 \text{ hrs/day}}{R} \qquad Formula\ (7\text{-}5)$$

In general, heat flow through a flat surface is defined as

$$Q = U A \Delta T \qquad Formula\ (7\text{-}6)$$

where ΔT is the temperature difference causing the heat flow.

For a composite wall, the heat flow is represented by Figure 7-3. To calculate the overall U of conductance value, the resistance of each material is added in series. This is analogous to an electrical circuit.

$$R = R_1 + R_2 + R_3 + R_4 + R_5 \qquad Formula\ (7\text{-}7)$$

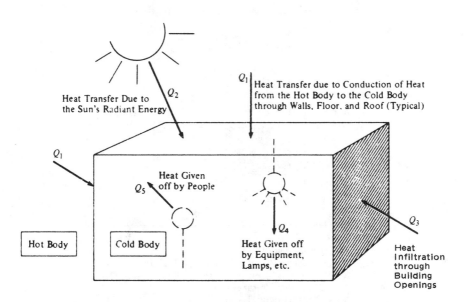

Heat Gain $= Q_1 + Q_2 + Q_3 + Q_4 + Q_5$

Figure 7-1. Heat Gain of a Building

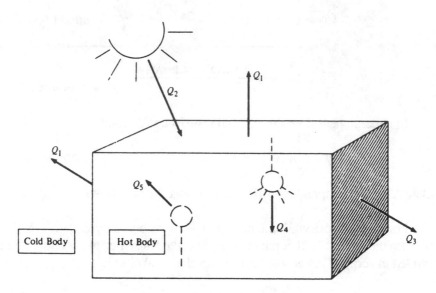

Heat Loss $= Q_1 + Q_3 - Q_2 - Q_4 - Q_5$

Figure 7-2. Heat Loss of a Building

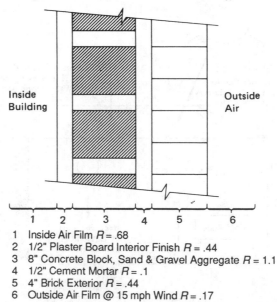

1   Inside Air Film $R = .68$
2   1/2" Plaster Board Interior Finish $R = .44$
3   8" Concrete Block, Sand & Gravel Aggregate $R = 1.1$
4   1/2" Cement Mortar $R = .1$
5   4" Brick Exterior $R = .44$
6   Outside Air Film @ 15 mph Wind $R = .17$

Figure 7-3. Typical Wall Construction

FACTORS IN CONDUCTION

There are four factors which affect the conduction of heat from one area to another. They are

- *The difference in temperature* (ΔT) between the warmer area and the colder area
- *The length of time* (*t*) over which the transfer occurs
- *The area* (*A*) *in common* between the warmer and the colder area
- *The resistance* (*R*) *to heat flow and conduction* (*U*) between the warmer and the colder area.

In a typical house, the major losses due to conduction occur through the walls, the roof, and the floor.

The Wall
— Major heat loss due to conduction, other losses are minor.
— Major concern is about the *U*-values of the walls.
— Minimum acceptable *R*-value for wall varies from 2.5 to 10.0.
— For winter the most important wall to insulate is north.
— For summer the most important wall to insulate is south.

The Roof
—Major heat loss due to conduction, other losses are minor.
— If the funds for insulation are limited, insulate the roof first, then the north wall, then the south wall.
— Use Table 7-2 as a guide for insulating the roof.

Heating Degree Days	Min. Acceptable U - Value
1000	0.30
2000	0.20
3000	0.20
4000	0.15
5000	0.15
6000	0.10
7000	0.10
8000	0.06

Table 7-2. Minimum Conductance Values for Roof

The Floor
— Major heat loss through conduction along the perimeter of the floor.
— Insulating floor should be the last priority compared to the walls and roof.
— Use Table 7-3 as a guide for insulating floors.

Heating Degree Days	Min. Acceptable U - Value
1000	0.40
2000	0.35
3000	0.30
4000	0.22
5000	0.22
6000	0.18
7000	0.18
8000	0.12

Table 7-3. Minimum Conductance Values for Floors

Difference in Temperature

Heat flows (much as water moves downhill) from warm areas to cold ones. The steeper the gradient between its origin and its destination, the faster it will flow. In fact, the rate at which heat is conducted is directly proportionate to the difference in temperature (ΔT) between the warm area and the colder one.

Length of Time

The longer the heat is allowed to flow across the gradient, the more heat will be conducted. The amount of heat (Btu's) is directly proportionate to the time span (t) of the transfer.

Btu/h is the amount of heat transferred in one hour.

The Area (A) in Common

The larger the area common to the warmer and colder surfaces through which the heat flows, the greater is the rate of conducted heat. For the same material, for the same length of time, at the same ΔT, the amount of heat (Btu's) transferred is directly proportionate to the area (A) in common.

Example Problem 7-2

Calculate the heat loss through 20,000 ft^2 of building wall, as indicated by Figure 7-3. Assume a temperature differential of 17°F.

Answer

Description	Resistance
Outside air film at 15 mph	0.17
4" brick	0.44
Mortar	0.10
Block	1.11
Plaster Board	0.44
Inside film	0.68
Total resistance	2.94

$U = 1/R = 0.34$

$Q = U A \Delta T$

$= 0.34 \times 20,000 \times 17 = 115,600 \text{ Btu/h}$

In Figure 7-4 a surface film conductance is introduced.

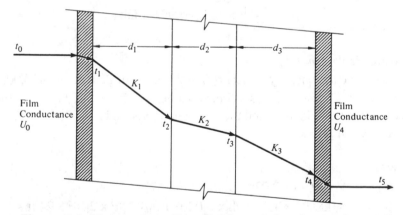

Figure 7-4. Temperature Distribution of the Composite Wall

The surface or film conductance is the amount of heat transferred in Btu per hour from a surface to air or from air to a surface per square foot for one degree difference in temperature. The flow of heat for the composite material can also be specified in terms of the conductivity of the material and the conductance of the air film.

$$Q = \frac{A \ (t_0 - t_5)}{1/U_o + d_1/K_1 + d_2/K_2 + d_3/K_3 + 1/U_4}$$

Formula (7-8)

LATENT HEAT AND SENSIBLE HEAT

The latent heat gain of a space means that moisture is being added to the air in the space. Moisture in the air is really in the form of superheated steam. Removing sensible heat from a space through air-conditioning equipment lowers the dry-bulb temperature of the air. On the other hand, removing latent heat from a space changes the substance state from a vapor to a liquid. The latent heat gain of a space is expressed in terms of moisture, heat units (Btu) or grains of moisture per hour (7,000 gains equals one pound). The average value for the latent heat or vaporization for superheated steam in air is 1050.

Example Problem 7-3

2000 grains of moisture are released in a conditioned room each hour. Calculate the heat that must be removed in order to condense this moisture at the cooling coils.

Answer $\dfrac{2000}{7000}$ x 1050 = 299.9 Btuh

Example Problem 7-4

Calculate the quantity of heat (Q) required by infiltration in an 8,000-cubic foot (ft^3) home that has 1.7 air changes per hour (AC/h) when the outside temperature is 48°F and the inside temperature is 68°F ($\Delta T = 20°F$) for one day (24 hours).

Answer

$$Q = V \text{ x } AC/h \text{ x } 0.020^* \text{ Btu/ft}^3 \text{ °F x } \Delta T \text{ x } t$$
$$Q = 8,000 \text{ ft}^3 \text{ x } 1.7 \text{ } AC/h \text{ x } 0.020^* \text{ Btu/ft}^3 \text{ °F x } 20°F \text{ x } 24 \text{ hrs}$$
$$Q = 8,000 \text{ x } 1.7 \text{ x } 0.020° \text{ x } 20 \text{ x } 24$$
$$Q = 130,560 \text{ Btu's}^{**}$$

INFILTRATION

Leakage or infiltration of air into a building is similar to the effect of additional ventilation. Unlike ventilation, it cannot be controlled or turned

---

* This is a regional variable.
** Once again, all of the units in the formula cancel except Btu's, leaving the units for Q as Btu's.

off at night. It is the result of cracks, openings around windows and doors, and access openings. Infiltration is also induced into the building to replace exhaust air unless the HVAC balances the exhaust. Wind velocity increases infiltration, and stack effects are potential problems. Air that is pushed out the window and door cracks is referred to as exfiltration.

To estimate infiltration, the Air Change Method or Crack Method is used.

Air Change Method

The five factors which determine the amount of energy lost through infiltration can be assembled in a formula that states:

The quantity of heat (Q) equals the volume of air (V) times the number of air changes per hour (AC/h) times the amount of heat required to raise the temperature of air one degree Fahrenheit (0.018-0.022 Btu's) times the temperature difference (ΔT) times the length of time (t). This is expressed as follows:

$$Q = V \times AC/h \times 0.020 \text{ Btu's} \cdot /\text{ft}^3 \, °F \times \Delta T \times t \qquad Formula\ (7\text{-}9)$$

The Air Change Method is considered to be a quick estimation method and is not usually accurate enough for air-conditioning design. A second method used to determine infiltration is the Crack Method.

Crack Method

When infiltration enters a space, it adds sensible and latent heat to the room load. To calculate this gain, the following equations (7-10 and 7-11) are used.

Sensible Heat Gain $\qquad Q_S = 1.08 \text{ CFM } \Delta T \qquad Formula\ (7\text{-}10\_$

Latent Heat Gain $\qquad Q_L = .7 \text{ CFM } (HR_O - HR_i) \qquad Formula\ (7\text{-}11)$
Where
$\quad Q_S \quad = \quad$ Sensible heat gain Btuh
$\quad Q_L \quad = \quad$ Latent heat gain Btuh
$\quad$ CFM $\ = \quad$ Air Flow Rate
$\quad \Delta T \quad = \quad$ Temperature differential between outside and inside air, F
$\quad HR_O \quad = \quad$ Humidity ratio of outside air, grains per lb
$\quad HR_i \quad = \quad$ Humidity ratio of room air, grains per lb

BODY HEAT

The human body releases sensible and latent heat depending on the degree of activity. Heat gains for typical applications are summarized in Table 7-4.

Table 7-4. Heat Gain from Occupants

Activity	Sensible Heat Btuh	Latent Heat Btuh	Total Heat Gain Btuh
Very Light Work – Seated			
(Offices, Hotels, Apartments)	215	185	400
Moderately Active Work			
(Offices, Hotels, Apartments)	220	230	450
Moderately Heavy Work			
(Manufacturing)	330	670	1000
Heavy Work (Manufacturing)	510	940	1450

Source: ASHRAE—Guide & Data Book

EQUIPMENT, LIGHTING AND MOTOR HEAT GAINS

It is important to include heat gains from equipment, lighting systems and motor heat gain in the overall calculations.

For a manufacturing facility, the major source of heat gains will be from the process equipment. Consideration must be given to all equipment including motors driving supply and exhaust fans.

To convert motor horsepower to heat gain in Btuh, Formula 7-12 is used.

$$Q = \frac{hp \times .746}{\eta} \times 3412$$

Formula (7-12)

Where

hp   is the running motor horsepower

η is the efficiency of the motor

Q is the heat gain from the motor Btuh

Similarly, the kilowatts of the lighting system can be converted to heat gain.

$$Q = (KW_F + KW_B) \times 3412$$

Formula (7-13)

Where

KW_F is the kilowatts of the lighting fixtures

KW_B if the kilowatts of the ballast

Q is the heat gain from the lighting system Btuh

RADIANT HEAT GAIN

Heat from the sun's rays greatly increases heat gain of a building. If the building energy requirements were mainly due to cooling, then this gain should be minimized. Solar energy affects a building in the following ways:

1. *Raises the surface temperature*; thus a greater temperature differential will exist at roofs than at walls.
2. A large percentage of direct solar radiation and diffuse sky radiation *passes through* transparent materials, such as glass.
3. Envelopes can be designed to have individual control loops to stop or enhance the solar radiation into the building.

SURFACE TEMPERATURES

The temperature of a wall of roof depends upon

(a) the angle of the sun's rays
(b) the color and roughness of the surface
(c) the reflectivity of the surface
(d) the type of construction

When an engineer is specifying building materials, he should consider the above factors. A simple example is color. The darker the surface, the more solar radiation will be absorbed. Obviously, white surfaces have a lower temperature than black surfaces after the same period of solar heating. Another factor is that smooth surfaces reflect more radiant heat than do rough ones.

In order to properly take solar energy into account, the angle of the sun's rays must be known. If the latitude of the facility is known, the angle can be determined.

SUNLIGHT AND GLASS CONSIDERATIONS

A danger in the energy conservation movement is to take steps backward. A simple example would be to exclude glass from building designs because of the poor conductance and solar heat gain factors of clear glass. The engineer needs to evaluate various alternate glass constructions and coating in order to maintain and improve the aesthetic qualities of good design while minimizing energy inefficiencies. It should be noted that the method to reduce heat gain of glass due to conductance is to provide an insulating air space.

To reduce the solar radiation that passes through glass, several techniques are available. Heat absorbing glass (tinted glass) is very popular. Reflective glass is gaining popularity, as it greatly reduces solar heat gains.

To calculate the relative heat gain through glass, a simple method is illustrated below.

$$Q = \pm\, U A\, (t_0 - t_1) + A \times S_1 \times S_2 \qquad\qquad Formula\ (7\text{-}14)$$

Where

Q	is the total heat gain for each glass orientation (Btuh)
U	is the conductance of the glass (Btu/h•ft^2•°F)
A	is the area of glass; the area used should include framing, since it will generally have a poor conductance compared with the surrounding material. (ft^2)
$t_0 - t_1$	is the temperature difference between the inside temperature and outside ambient. (°F)
S_1	is the shading coefficient; S_1 takes into account external shades, such as venetian blinds and draperies, and the qualities of the glass, such as tinting and reflective coatings.
S_2	is the solar heat gain factor. This factor takes into account direct and diffused radiation from the sun. Diffused radiation is basically caused by reflections from dust particles and moisture in the air.

Optimizing Windows
— Use storm windows in winter.
— Use a clear plastic cover to cover windows in the winter so as to allow radiation but stop infiltration.
— Use double or triple glass windows.
— Promote thermal barriers.
— Use solar control devices.

THE PSYCHROMETRIC CHART
Just as the steam table and the Mollier Diagram are used to relate the properties of steam, the psychrometric chart is used to illustrate the properties of moist air. The psychrometric chart is a very important tool in the design of air-conditioning systems.

PROPERTIES OF MOIST AIR

Air expands and contracts with temperature. If pressure is held constant, then air expands or contracts at a specified rate with change in temperature as defined by Formula 7-15.

$$V_2 = V_1 \ \frac{T_2}{T_1}$$

Formula (7-15)

Where
V_1 = initial volume of air
V_2 = final volume of air
T_1 = absolute temperature
T_2 = final absolute temperature

$$V_2 = V_1 \ \frac{P_2}{P_1}$$

At constant temperature,

Formula (7-15a)

Where
P_2 = initial pressure, psia
P_1 = final pressure, psia

The change in the volume occupied by air at any temperature can be found by first using Formula 7-15a to calculate the change in volume with pressure and then using Formula 7-15 to calculate the change in volume with temperature.

The temperature at which the water vapor in the atmosphere begins to condense is the *Dew Point Temperature*. It should be noted that the mass of moisture per pound of dry air in a mixture of air and water vapor depends on the dew point temperature alone. If there is no condensation of moisture, the dew point temperature remains constant.

Humidity Ratio is defined as the mass of water vapor mixed with one pound of dry air.

Degree of Saturation is defined as the actual humidity ratio divided by the humidity ratio at saturation.

Relative Humidity is defined as the vapor pressure of air divided by the saturation pressure of pure water at the same temperature.

Sensible Heat of an Air-Vapor Mixture is defined as the heat which affects the dry-bulb temperature of the mixture only.

Wet-Bulb Temperature can be determined by covering the bulb of a thermometer with a wet wick and holding it in a stream of swiftly moving air. At first the temperature will drop quickly and then reach a stationary point referred to as the wet-bulb temperature. The wet-bulb temperature is lower than the dry-bulb temperature. The amount of water which evaporates from the wet wick into the air depends on the amount of water vapor initially in the air flowing past the wet bulb. A sling psychrometer is a convenient instrument used to measure wet-bulb temperatures.

Total Heat is defined as the sum of the sensible and latent heat. Sensible heat depends only upon the dry-bulb temperature, while the latent heat content depends only upon its dew point.

The *Enthalpy of an Air-Water Vapor Mixture* can be calculated by Formula 7-16.

$$h_{(mix)} = h_{(dry\ air)} + h_{(water\ vapor)}$$

or

$$h_{(mix)} = Cp \times T_{DB} + HR\ hg \qquad Formula\ (7\text{-}16)$$

Where

$h_{(mix)}$	= enthalpy of the mixture of dry air and water vapor, Btu per lb
Cp	= specified heat
T_{DB}	= dry bulb temperature
HR	= humidity ratio of the mixture
hg	= enthalpy of saturated vapor (steam) at the dew point temperature

To determine the properties of air such as the humidity ratio, relative humidity and enthalpy, the psychrometric chart is frequently used. Figure 7-5 illustrates the psychrometric chart.

Example Problem 7-5

Given air at 70°F DB and 50% relative humidity, for the air vapor mixture find

- Wet-Bulb Temperature
- Specific Volume
- Enthalpy
- Vapor Pressure
- Humidity Ratio
- Percentage Humidity
- Dew Point Temperature

Answer

From Figure 7-5 find the intersection of 70°F DB and 50% *HR*, point "A."

The WB temperature is found as 58.6 WB, point "B."

The Enthalpy is found as 25.5 Btu/lb, point "C."

The Humidity Ratio is found to be 56 grains of moisture per pound of dry air, point "D," and the Dew Point temperature is 53°F, point "E."

The Specific Volume is found to be 13.5 cubic feet per pound of air, point "F," with a Vapor Pressure of .38 inches of mercury, point "G."

The percentage humidity equals the actual humidity 56, point "D," divided by the humidity ratio at saturation (100% RH) which is found to be 110, point "H." Thus % humidity = 56/110 = .50.

Example Problem 7-6

Given 8000 CFM of chilled air at 55°F DB and 50°F WB mixed with 3000 CFM of outside air at 90°F DB and 80°F WB, compute the properties of the mixture.

Answer

From Figure 7-6, the intersection of 55°F DB and 50°F WB is point "A." The specific volume is then 13.1 cubic feel/lb, point "B."

Similarly, for the outside air, the specific volume is 14.3 cubic feet/lb, point "D."

The total weight and dry-bulb temperature of the mixture can be found by the following ratios:

$$\frac{8,000}{13.1} = 610.6 \text{ lb/min.}$$

$$\frac{3,000}{14.3} = \underline{209.7 \text{ lb/min.}}$$
$$\text{Total} \quad 820.3 \text{ lb/min. for total weight}$$

The dry-bulb temperature is

$$\frac{610.6}{820.3} \times 55 = 40.93°F$$

$$\frac{209.7}{820.3} \times 90 = \underline{23.0°F}$$
$$63.9°F \text{ DB}$$

The properties of the mixture, point "E," can now be determined from the chart.

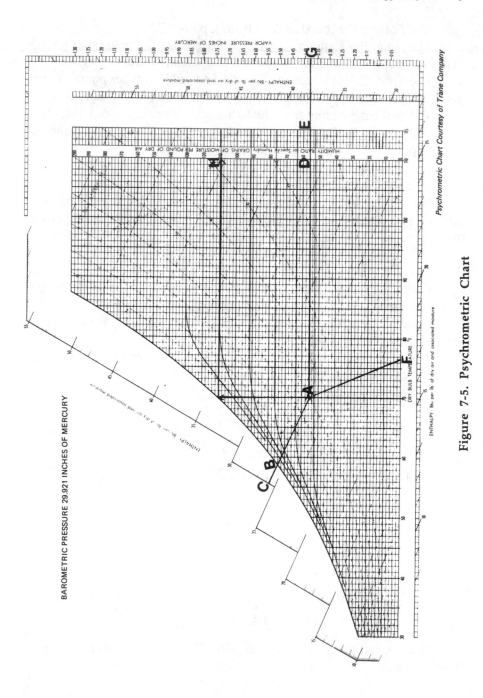

Figure 7-5. Psychrometric Chart

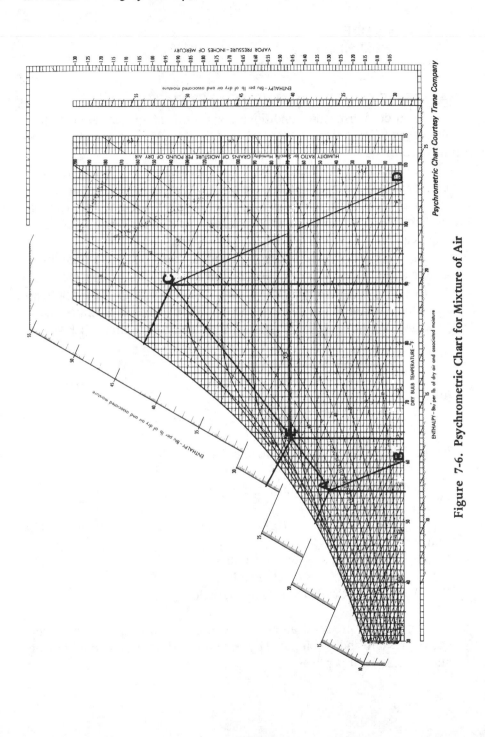

Figure 7-6. Psychrometric Chart for Mixture of Air

WB   = 59.8°F
  h = 26.6 Btu/lb
Humidity Ratio = 70 gr of moisture/lb of dry air.

BASICS OF FAN DISTRIBUTION SYSTEMS

In order to distribute conditioned or ventilated air, fans are the chief vehicle used. Several basic types of fans commonly used in industry are illustrated in Figure 7-7 and are listed below.

Centrifugal, airfoil blade—used on large heating, ventilating, and air-conditioning systems. Airfoil fans are used where clean air is handled.

Centrifugal, backward curved blade—used for general heating, ventilating, and air-conditioning systems. Air handling need not be as clean as above.

Centrifugal, radial blade—a rugged, heavy duty fan for high pressure applications. It is designed to handle sand, wood chips, etc.

Centrifugal, forward curved blade—ideal for low pressure applications, such as domestic furnaces or room and packaged air conditioners.

The brake horsepower of a fan is illustrated by Formula 7-17.

$$\text{Brake Horsepower} = \frac{\text{CFM x Fan PS}}{6356 \times \eta_F}$$ *Formula (7-17)*

Where
  CFM   is the quantity of air in CFM
  η_F is the fan static efficiency
  Fan PS   is the fan static pressure in inches

To compute the fan static pressure

$$\text{Fan PS} = P_T(O) - P_T(i) - P_v(O)$$ *Formula (7-18)*

Where
  $P_T(O)$ is the total pressure at fan outlet
  $P_T(i)$ is the total pressure at fan inlet
  $P_v(O)$ is the velocity pressures at fan outlet

The excess pressure above the static pressure is known as the velocity pressure and is computed by Formula 7-19 for standard air having value of 13.33 cu ft per lb as

Figure 7-7. Fan Types: (A) Vaneaxial; (B) Backward Curved Blade; (C) Tubeaxial; (D) Radial; (E) Radial Tip Blade; (F) Airfoil Blade. *(Courtesy of Buffalo Forge Company.)*

$$P_v = \left(\frac{V}{4005}\right)^2$$

Formula (7-19)

where V is the velocity of air in FPM.

Note that the pressure of air in sheet metal ducts is so low that ordinary pressure gage (Bourdon type) cannot be used; thus a V-tube or manometer is used, which measures pressure in inches of water. A pressure of 1 psi will support a column of water 2.31 feet high or 27.7 inches.

Fan Laws

The performance of a fan at varying speeds and air densities may be predicted by certain basic fan laws as illustrated in Table 7-5.

Table 7-5. Fan Laws

1.   Fan Law for variation in fan speed at constant air density with a constant system
1.1   Air volume, CFM varies as fan speed
1.2   Static velocity or total pressure varies as the square of fan speed
1.3   Power varies as cube of fan speed
2.   Fan Law variation in air density at constant fan speed with a constant system
2.1   Air volume is constant
2.2   Static velocity or pressure varies as density
2.3   Power varies as density

Example Problem 7-7

An energy audit indicates that the ventilation requirements of a space can be reduced from 15,000 CFM to 12,000 CFM. Comment on the savings in brake horsepower if the fan pulley is changed to reduce the fan speed accordingly.

Answer

From the Fan Laws

$$hp_1 = hp_2 \times \left(\frac{CFM\ new}{CFM\ old}\right)^3$$

$$= hp_2 \times \left(\frac{12,000^3}{15,000}\right) = hp_2\ (.8)^3 = .512\ hp_2$$

or a 48.8% savings.

The fan performance is affected by the density of the air that the fan is handling. All fans are rated at standard air with a density of .075 lb per cu ft and a specific volume of 13.33 cu ft per lb. When a fan is tested in a laboratory at different than standard air, the brake horsepower is corrected by using the Fan Laws.

Fan Performance Curves

Fan performance curves are used to determine the relationship between the quantity of air that a fan will deliver and the pressure it can discharge at various air quantities. For each fan type, the manufacturer can supply fan performance curves which can be used in design and as a tool of determining the fan efficiency.

As illustrated by Ex. Prob. 7-6, one energy engineering technique to reduce fan horsepower is to reduce fan speed. An alternate way is to throttle the air flow by a damper. The fan performance curves can be used to illustrate the best choice of these options. The system characteristics can be plotted on the fan curves to show the static pressure required to overcome the friction loss in the duct system. From the Fan Laws the system friction loss varies with the square of fan speed; thus, as the air quantity increases, the friction loss will vary as illustrated by Figure 7-8.

For a detailed analysis, the fan performance curve should be used to predict how a specific fan will perform in a desired application.

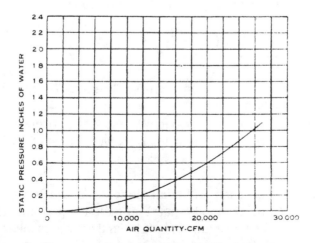

Figure 7-8. System Characteristic Curve

Example Problem 7-8

Given Fan Performance Curve Figure 7-9. The fan delivers 21,500 CFM at 600 rpm at a brake horsepower of 12.3. Comment on the savings in brake horsepower by reducing air flow to 14,400 CFM by each of the following methods: (a) reducing fan speed to 400 rpm, (b) throttling the air flow by a damper.

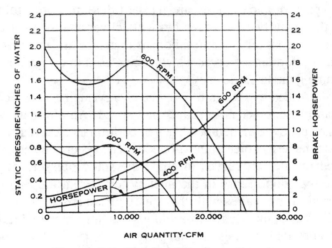

Figure 7-9. Fan Performance Curves

(a) From the Fan Laws the brake horsepower is reduced as follows:

$$hp = (12.3) \left(\frac{400}{600}\right)^3 = 3.64$$

Using the fan performance curve, Figure 7-10, the system characteristic curve "A" is plotted.

By reducing the rpm from 600 to 400, the system operates at point 1 and then moves to point 2. The brake horsepower is found from Figure 7-9 to be 3.7.

(b) By closing the air damper, the air flow is reduced to 14,400 CFM while still running the fan at 600 rpm. Using the fan performance curve, Figure 7-10, the system operates at point 1 and then moves to point 3. The power to operate the fan at point 3 is 7.2 hp from Figure 7-9. Thus, if the fan speed can be reduced, it is more efficient than throttling the air flow damper.

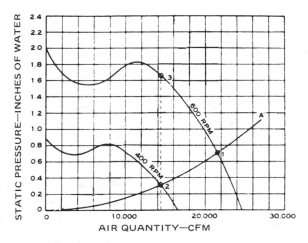

Figure 7-10. Fan Performance and System Characteristic Curves

FLUID FLOW

Pump and piping considerations are extremely important due to the fact that energy transport losses are a part of any distribution system. Losses occur due to friction, and that lost energy must be supplied by pump horsepower.

Centrifugal pumps are commonly used in heating, ventilating and air-conditioning applications as well as utility systems. The output torque for the pump is supplied by a driver such as a motor. Liquid enters the eye of the impeller which rotates. Pressure energy builds up by the action of centrifugal force, which is a function of the impeller vane peripheral velocity.

As with fan systems, Pump Laws and curves can be used to predict system responsiveness. The affinity laws of a pump are illustrated in Table 7-6.

The horsepower required to operate a pump is illustrated by Formula 7-20.

$$hp = \frac{\Delta P \ GPM}{1715 \ \eta}$$

Formula (7-20)

Table 7-6. Affinity Laws for Pumps

Impeller Diameter	Speed	Specific Gravity (SG)	To Correct for	Multiply by
Constant	Variable	Constant	Flow	$\left(\dfrac{\text{New Speed}}{\text{Old Speed}}\right)$
			Head	$\left(\dfrac{\text{New Speed}}{\text{Old Speed}}\right)^2$
			BHP (or kW)	$\left(\dfrac{\text{New Speed}}{\text{Old Speed}}\right)^3$
Variable	Constant		Flow	$\left(\dfrac{\text{New Diameter}}{\text{Old Diameter}}\right)$
			Head	$\left(\dfrac{\text{New Diameter}}{\text{Old Diameter}}\right)^2$
			BHP (or kW)	$\left(\dfrac{\text{New Diameter}}{\text{Old Diameter}}\right)^3$
	Constant	Variable	BHP (or kW)	$\left(\dfrac{\text{New SG}}{\text{Old SG}}\right)$

Where

ΔP is the differential pressure across a pump in psi

GPM   is the required flow rate in gallons per minute

η is the pump efficiency

To convert psi to read in feet use Formula 7-21.

$$\text{Head in Feet} = \frac{\text{psi} \times 2.31}{\text{Specific Gravity of Fluid}}$$

Formula (7-21)

Basically, the size of discharge line piping from the pump determines the friction loss through the pipe that the pump must overcome. The greater the line loss, the more pump horsepower required. If the line is short or has a small flow, this loss may not be significant in terms of the total system head requirements. On the other hand, if the line is long and has a large flow rate, the line loss will be significant.

To calculate the pressure loss for water system piping and the corresponding velocity, Formulas 7-22 and 7-23 are used.

$$\Delta P = \frac{0.55\, CF^{1.85}}{d^{4.87}}$$

Formula (7-22)

$$V = \frac{.41\,F}{d^2}$$

Formula (7-23)

Where

ΔP = pressure loss per 100 feet of pipe, psi
V = velocity of fluid, ft/sec.
C = roughness factor
    1 for copper tubing
    1.62 for steel pipe
    .77 for plastic pipe
F = flow rate in gallons per minute
d = inside diameter of pipe, inches

The pressure loss due to fittings is determined by Formula 7-24.

$$\Delta P = .0067\,KV^2$$

Formula (7-24)

where K is the loss coefficient.

Options to Reduce Pump Horsepower

There are several alternatives that will significantly reduce pump horsepower. Summarized below are some of the options available.

1. Many pumps are oversized to very conservative design practices. If the pump is oversized, install a smaller impeller to match the load.
2. In some instances heating or cooling supply flow rates can be reduced. To save on pump horsepower, either reduce motor speed or change the size of the motor sheave.
3. Check economics of replacing corroded pipe with a large pipe diameter to reduce friction losses.
4. Consider using variable speed pumps to better match load conditions. Motor drive speed can be varied to match pump flow rate or head requirements.
5. Consider adding a smaller auxiliary pump. During part load situations a larger pump can be shut down and a smaller auxiliary pump used.

The distribution and overall load that is placed on HVAC transportation devices can additionally be optimized by using the following guidelines.
1. Reduce the resistance to flow in ducts and piping.

2. Eliminate leaks in ducts or piping.
3. Improve the efficiency of terminal devices. (Heat exchangers, diffusers, etc.)
4. Reduce or eliminate the opportunity for heat transfer during distribution.
5. Reduce the transport rate. (Flow velocity)

Remember that improvements in the envelope efficiency will automatically reduce the distribution load. This effect should always be considered and optimized by sizing distribution components accordingly.

Common Sense Methods

Do not underestimate the effects of energy saving techniques which require only a change of habit or life style.

Some of the following guidelines can lead to significant savings with little or no initial cost.

1. Set back the thermostat. Some suggested guidelines are 68°F during the heating season and 78°F during the cooling season. When possible, reduce thermostat in unoccupied areas. This is easily done with baseboard electric heat.

2. Use or adjust thermostats to allow for a temperature deadband or range in which no action is taken by HVAC systems.

3. Regulate thermostat settings. (Hide or lock where necessary.)

4. Turn off lights in unoccupied areas and reduce overall lighting levels by illuminating the required area (task area). A good example of the potential for energy savings is illustrated on pages 310-312 of this text.

5. Reduce levels of relative humidity during the heating season. The energy required to vaporize water for humidification purposes can be significant.

6. Shut down ventilation systems during unoccupied hours and reduce levels to acceptable limit during occupied hours.

7. Check and ensure that ventilation dampers open and close as desired by HVAC system.

8. Filter indoor air to reduce outdoor air ventilation requirements. Filtering systems are available to remove cigarette smoke, cooking smoke and odors, and other common air contaminants.

9. Use storm windows or plastic to reduce infiltration as well as conduction and convection heating/cooling losses of windows.

10. Correct duct inadequacies which cause leakage, obstruction, or allow unnecessary heat transfer.

HVAC SYSTEMS

The summary below illustrates the types of systems frequently encountered in heating and air-conditioning systems.

Single Zone System

Single zone systems consist of a mixing, conditioning and fan section. The conditioning section may have heating, cooling, humidifying or a combination of capabilities. Single zone systems can be factory assembled roof-top units or built up from individual components and may or may not have distributing duct work.

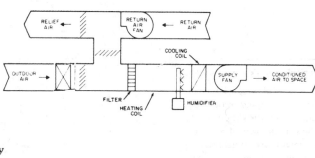

Terminal Reheat System

Reheat systems are modifications of single zone systems. Fixed cold temperature air is supplied by the central conditioning system and reheated in the terminal units when the space cooling load is less than maximum. The reheat is controlled by thermostats located in each conditioned space.

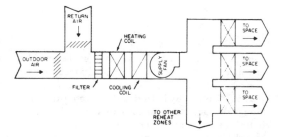

Multizone Systems

Multizone systems condition all air at the central system and mix heated and cooled air at the unit to satisfy various zone loads as sensed by zone thermostats. These systems may be packaged roof-top units or field-fabricated systems.

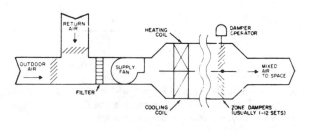

Dual Duct Systems

Dual duct systems are similar
to multizone systems
except heated and cooled
air is ducted to the conditioned spaces and mixed as
required in terminal
mixing boxes.

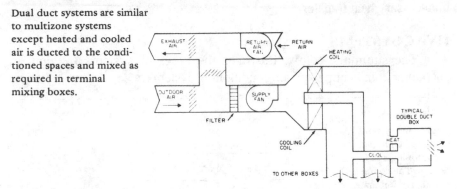

Variable Air Volume Systems

A variable air volume system
delivers a varying amount of air as
required by the conditioned spaces.
The volume control may be by
fan inlet (vortex) damper,
discharge damper or fan speed
control. Terminal sections may be
single duct variable volume units
with or without reheat, controlled
by space thermostats.

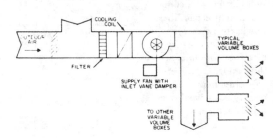

Induction Systems

Induction systems generally have units at the outside
perimeter of conditioned spaces. Conditioned primary
air is supplied to the units where it passes through
nozzles or jets and by induction draws room air
through the induction unit coil. Room temperature
control is accomplished by modulating water flow
through the unit coil.

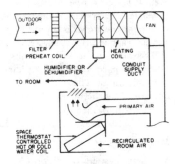

Fan Coil Units

A fan coil unit consists of a cabinet with heating and/or cooling coil,
motor and fan and a filter. The unit may be floor or ceiling mounted
and uses 100% return air to condition a space.

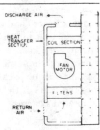

Unit Ventilator

A unit ventilator consists of a cabinet with heating and/or cooling coil, motor and fan, a filter and a return air—outside air mixing section. The unit may be floor or ceiling mounted and uses return and outside air as required by the space.

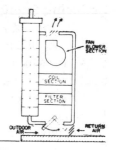

Unit Heater

Unit heaters have a fan and heating coil which may be electric, hot water or steam. They do not have distribution duct work but generally use adjustable air distribution vanes. Unit heaters may be mounted overhead for heating open areas or enclosed in cabinets for heating corridors and vestibules.

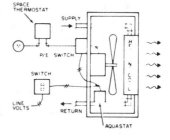

Perimeter Radiation

Perimeter radiation consists of electric resistance heaters or hot water radiators usually within an enclosure but without a fan. They are generally used around the conditioned perimeter of a building in conjunction with other interior systems to overcome heat losses through walls and windows.

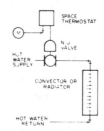

Hot Water Converters

A hot water converter is a heat exchanger that uses steam or hot water to raise the temperature of heating system water. Converters consist of a shell and tubes with the water to be heated circulated through the tubes and the heating steam or hot water circulated in the shell around the tubes.

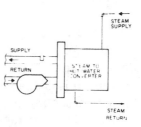

Source: "Energy Conservation with Comfort," Honeywell

Reducing Energy Consumption in HVAC Systems

Variable Air Volume (VAV) System—A variable air volume system provides heated or cooled air at a constant temperature to all zones served. VAV boxes located in each zone or in each space adjust the quantity of air reaching each zone or space depending on its load requirements. Methods for conserving energy consumed by this system include:

1. Reduce the volume of air handled by the system to that point which is minimally satisfactory.
2. Lower hot water temperature and raise chilled water temperature in accordance with space requirements.
3. Lower air supply temperature to that point which will result in the VAV box serving the space with the most extreme load being fully open.
4. Consider installing static pressure controls for more effective regulation of pressure bypass (inlet) dampers.
5. Consider installing fan inlet damper control systems if none now exists.

Constant Volume System—Most constant volume systems either are part of another system—typically dual duct systems—or serve to provide precise air supply at a constant volume. Opportunities for conserving energy consumed by such systems include:

1. Determine the minimum amount of airflow which is satisfactory and reset the constant volume device accordingly.
2. Investigate the possibility of converting the system to variable (step controlled) constant volume operation by adding the necessary controls.

Induction System—Induction systems comprise an air handling unit which supplies heated or cooled primary air at high pressure to induction units located on the outside walls of each space served. The high pressure primary air is discharged within the unit through nozzles inducing room air through a cooling or heating coil in the unit. The resultant mixture of primary air and induced air is discharged to the room at a temperature dependent upon the cooling and heating load of the space. Methods for conserving energy consumed by this system include:

1. Set primary air volume to original design values when adjusting and balancing work is performed.

2. Inspect nozzles. If metal nozzles, common on most older models, are installed, determine if the orifices have become enlarged from years of cleaning. If so, chances are that the volume/pressure relationship of the system has been altered. As a result, the present volume of primary air and the appropriate nozzle pressure required must be determined. Once done, rebalance the primary air system to the new nozzle pressures and adjust individual induction units to maintain airflow temperature. Also, inspect nozzles for cleanliness. Clogged nozzles provide higher resistance to air flow, thus wasting energy.

3. Set induction heating and cooling schedules to minimally acceptable levels.

4. Reduce secondary water temperatures during the heating season.

5. Reduce secondary water flow during maximum heating and cooling periods by pump throttling or, for dual-pump systems, by operating one pump only.

6. Consider manual setting of primary air temperature for heating, instead of automatic reset by outdoor or solar controllers.

Dual-Duct System—The central unit of a dual-duct system provides both heated and cooled air, each at a constant temperature. Each space is served by two ducts, one carrying hot air, the other carrying cold air. The ducts feed into a mixing box in each space which, by means of dampers, mixes the hot and cold air to achieve that air temperature required to meet load conditions in the space or zone involved. Methods for improving the energy consumption characteristics of this system include:

1. Lower hot deck temperature and raise cold deck temperature.

2. Reduce air flow to all boxes to minimally acceptable level.

3. When no cooling loads are present, close off cold ducts and shut down the cooling system. Reset hot deck according to heating loads and operate as a single duct system. When no heating loads are present, follow the same procedure for heating ducts and hot deck. It should be noted that operating a dual-duct system as a single-duct system reduces air flow, resulting in increased energy savings through lowered fan speed requirements.

Single Zone System—A zone is an area or group of areas in a building which experiences similar amounts of heat gain and heat loss. A single zone system is one which provides heating and cooling to one zone controlled by the zone thermostat. The unit may be installed within or remote from the space it serves, either with or without air distribution ductwork.

1. In some systems air volume may be reduced to the minimum required, therefore reducing fan power input requirements. Fan brake horsepower varies directly with the cube of air volume. Thus, for example, a 10% reduction in air volume will permit a reduction in fan power input by about 27% of original. This modification will limit the degree to which the zone serviced can be heated or cooled as compared to current capabilities.
2. Raise supply air temperatures during the cooling season and reduce them during the heating season. This procedure reduces the amount of heating and cooling which a system must provide, but, as with air volume reduction, limits heating and cooling capabilities.
3. Use the cooling coil for both heating and cooling by modifying the piping. This will enable removal of the heating coil, which provides energy savings in two ways. First, air flow resistance of the entire system is reduced so that air volume requirements can be met by lowered fan speeds. Second, system heat losses are reduced because the surface area of cooling coils is much larger than that of heating coils, thus enabling lower water temperature requirements. Heating coil removal is not recommended if humidity control is critical in the zone serviced and alternative humidity control measures will not suffice.

Multizone System—A multizone system heats and cools several zones—each with different load requirements—from a single, central unit. A thermostat in each zone controls dampers at the unit which mix the hot and cold air to meet the varying load requirements of the zone involved. Steps which can be taken to improve energy efficiency of multizone systems include:

1. Reduce hot deck temperatures and increase cold deck temperatures. While this will lower energy consumption, it also will reduce the system's heating and cooling capabilities as compared to current capabilities.
2. Consider installing demand reset controls which will regulate hot and cold deck temperatures according to demand. When properly installed, and with all hot deck or cold deck dampers partially closed, the control

will reduce hot and raise cold deck temperature progressively until one or more zone dampers is fully open.
3. Consider converting systems serving interior zones to variable volume. Conversion is performed by blocking ·off the hot deck, removing or disconnecting mixing dampers, and adding low pressure variable volume terminals and pressure bypass.

Terminal Reheat System—The terminal reheat system essentially is a modification of a single zone system which provides a high degree of temperature and humidity control. The central heating/cooling unit provides air at a given temperature to all zones served by the system. Secondary terminal heaters then reheat air to a temperature compatible with the load requirements of the specific space involved. Obviously, the high degree of control provided by this system requires an excessive amount of energy. Several methods for making the system more efficient include:
1. Reduce air volume of single zone units.
2. If close temperature and humidity control must be maintained for equipment purposes, lower water temperature and reduce flow to reheat coils. This still will permit control but will limit the system's heating capabilities somewhat.
3. If close temperature and humidity control is not required, convert the system to variable volume by adding variable volume valves and eliminating terminal heaters.

THE ECONOMIZER CYCLE
The basic concept of the economizer cycle is to use outside air as the cooling source when it is cold enough. There are several parameters which should be evaluated in order to determine if an economizer cycle is justified. These include:
- Weather
- Building occupancy
- The zoning of the building
- The compatibility of the economizer with other systems
- The cost of the economizer

What Are the Costs of Using
The Economizer Cycle
Outside air cooling is accomplished usually at the expense of an additional return air fan, economizer control equipment, and an additional

burden on the humidification equipment. Therefore, economizer cycles must be carefully evaluated based on the specific details of the application.

Using outside air to cool a building can result in lower mechanical refrigeration cost whenever outdoor air has a lower total heat content (enthalpy) than the return air. This can be accomplished by an "integrated economizer" or enthalpy control. See Figure 7-11 for a comparison of controls.

Dry-Bulb Economizer

Operation of the "integrated economizer" can be made automatic by providing (1) dampers capable of providing 100% outdoor air, and (2) local controls that sequence the chilled water or DX (direct expansion) coil and dampers so that during economizer operation, on a rise in discharge (or space) temperature, the outdoor damper opens first; then on a further rise, the cooling coil is turned "on."

Economizer operation is activated by outside air temperature, for example, 72°F DB*. If outside air is below 72°F, the above described economizer sequence occurs. Above 72°F, outside air cooling is not economical, and the outdoor air damper closes to its minimum position to satisfy ventilation requirements only.

Enthalpy Control

If an economizer system is equipped with enthalpy control, savings will accrue due to a more accurate changeover point. The load on a cooling coil for an air handling system is a function of the *total heat* of air entering the coil. Total heat is a function of *two* measurements, dry-bulb (DB) and relative humidity (RD) or dew point (DP). The enthalpy control measures both conditions (DB and RH) in the return air duct and outdoors. It then computes which air source would impose the lowest load on the cooling system. If outside air is the smallest load, the controller enables the economizer cycle. (Dry-bulb economizer control savings will be less than those shown for enthalpy control, except in dry climates.)

Enthalpy Control Savings Calculations

Savings are based on the assumption that the system previously had either (a) no 100% O.A. damper or (b) a fixed minimum outdoor air

----

* This varies according to location.

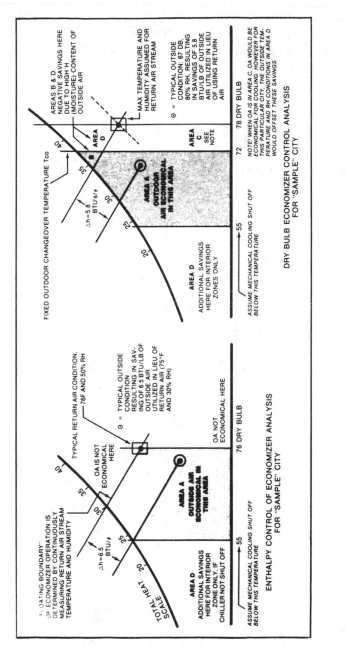

Figure 7-11. Comparative Economizer Controls

setting (whenever the fan operated) sufficient for ventilation purposes; it is also assumed that (c) minimum outdoor air has already been reduced, and the minimum damper opening will be at the *new value*.

   Step 1. Determine minimum CFM of outdoor air to be used during occupied hours.

   Step 2. Calculate annual savings.

$$A \frac{ft^3}{min.} \left(1 - \frac{B\%}{100}\right) \times K \frac{10^6 \, Btu}{yr \, 1000 \, CFM} \times \frac{operating \, hrs/wk}{50} \times J \frac{\$}{10^6 \, Btu}$$

$$= \$ \, SAVED \, PER \, YEAR \qquad\qquad Formula \, (7\text{-}25)$$

Where

A = air handling capacity $\left(\frac{ft^3}{min.}\right)$

B = present ventilation air (%)

J = cost of cooling $\left(\frac{\$}{10^6 \, Btu}\right)$

K = seasonal cooling savings $\left(\frac{10^6 \, Btu}{yr \, 1000 \, CFM}\right)$

   Formula 7-25 is used to calculate the savings resulting from enthalpy control of outdoor air. The calculated savings generally will be greater than the savings resulting from a dry-bulb economizer. To estimate dry-bulb economizer savings, multiply the enthalpy savings by .93.

8

HVAC Equipment

PERFORMANCE RATIOS

Measuring Efficiency by Using the Coefficient of Performance

The coefficient of performance *(COP)* is the basic parameter used to compare the performance of refrigeration and heating systems. *COP* for cooling and heating applications is defined as follows:

$$COP \text{ (Cooling)} = \frac{\text{Rate of Net Heat Removal}}{\text{Total Energy Input}} \qquad \textit{Formula (8-1)}$$

$$COP \text{ (Heating, Heat Pump*)} = \frac{\text{Rate of Useful Heat Delivered*}}{\text{Total Energy Input}}$$
$$\textit{Formula (8-2)}$$

Measuring System Efficiency Using the Energy Efficiency Ratio

The energy efficiency ratio *(EER)* is used primarily for air-conditioning systems and is defined by Formula 8-3.

$$EER = \frac{COP\,(3412)}{1000} \text{ Btu/watt-hr} \qquad \textit{Formula (8-3)}$$

---

*For Heat Pump Applications, exclude supplemental heating.

THE HEAT PUMP

The heat pump has gained wide attention due to its high potential *COP*. The heat pump in its simplest form can be thought of as a window air conditioner. During the summer, the air on the room side is cooled while air is heated on the outside-air side. If the window air conditioner is turned around in the winter, some heat will be pumped into the room. Instead of switching the air conditioner around, a cycle reversing valve is used to switch functions. This valve switches the function of the evaporator and condenser, and refrigeration flow is reversed through the device. Thus, *the heat pump is heat recovery through a refrigeration cycle.* Heat is removed from one space and placed in another. In Chapter 7, it was seen that the direction of heat flow is from hot to cold. Basically, energy or pumping power is needed to make heat flow "up hill." The mechanical refrigeration compressor "pumps" absorbed heat to a higher level for heat rejection. The refrigerant gas is compressed to a higher temperature level so that the heat absorbed by it, during the evaporation or cooling process, is rejected in the condensing or heating process. Thus, the heat pump provides cooling in the summer and heating in the winter. The source of heat for the heat pump can be from one of three elements: air, water or the ground.

Air to Air Heat Pumps

Heat exists in air down to 460°F below zero. Using outside air as a heat source has its limitations, since the efficiency of a heat pump drops off as the outside air level drops below 55°F. This is because the heat is more dispersed at lower temperatures, or more difficult to capture. Thus, heat pumps are generally sized on cooling load capacities. Supplemental heat is added to compensate for declining capacity of the heat pump. This approach allows for a realistic first cost and an economical operating cost.

An average of two to three times as much heat can be moved for each kW input compared to that produced by use of straight resistance heating. Heat pumps can have a COP of much more than 2 or 3 in industrial processes, depending on the temperature involved. Commercially available heat pumps range in size from 2 to 3 tons for residences up to 40 tons for commercial and industrial users. Figure 8-1 illustrates a simple scheme for determining the supplemental heat required when using an air-to-air heat pump.

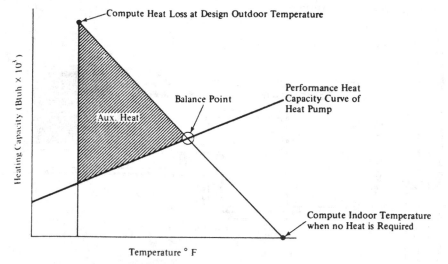

Figure 8-1. Determining Balance Point of Air to Air Heat Pump

Hydronic Heat Pump

The hydronic heat pump is similar to the air to air unit, except the heat exchange is between water and refrigerant instead of air to refrigerant, as illustrated in Figure 8-2. Depending on the position of the reversing valve, the air heat exchanger either cools or heats room air. In the case of cooling, heat is rejected through the water-cooled condenser to the building water. In the case of heating, the reversing valve causes the water to refrigerant heat exchanger to become an evaporator. Heat is then absorbed from the water and discharged to the room air.

Imagine several hydronic heat pumps connected to the same building water supply. In this arrangement, it is conceivable that while one unit is providing cool air to one zone, another is providing hot air to another zone; the first heat pump is providing the heat source for the second unit, which is heating the room. This illustrates the principle of energy conservation. In practice, the heat rejected by the cooling units does not equal the heat absorbed. An additional evaporative cooler is added to the system to help balance the loads. A better heat source would be the water from wells, lakes or rivers which is thought of as a constant heat source. Care should be taken to insure that a heat pump connected to such a heat source does not violate ecological interests.

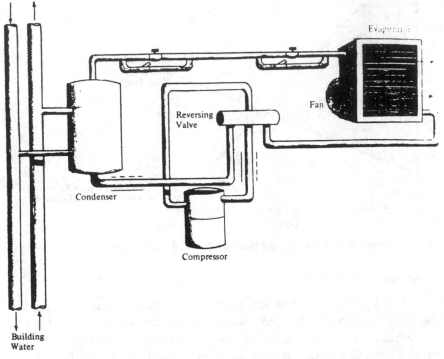

Figure 8-2. Hydronic Heat Pump

Liquid Chiller

A liquid chilling unit (mechanical refrigeration compressor) cools water, brine, or any other refrigeration liquid, for air-conditioning or refrigeration purposes. The basic components include a compressor, liquid cooler, condenser, the compressor drive, and auxiliary components. A simple liquid chiller is illustrated in Figure 8-3. The type of chiller usually depends on the capacity required. For example, small units below 80 tons are usually reciprocating, while units above 350 tons are usually centrifugal.

Factors which affect the power usage of liquid chillers are the percent load and the temperature of the condensing water. *A reduced condenser water temperature saves energy.* In Figure 8-4, it can be

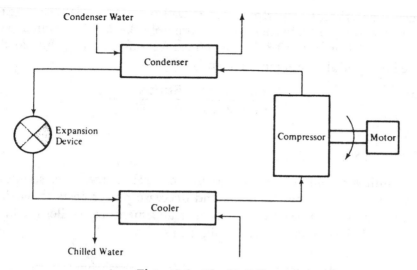

Figure 8-3. Liquid Chiller

seen that by reducing the original condenser water temperature by 10 degrees, the power consumption of the chiller is reduced. Likewise, a chiller operating under part load consumes less power. The

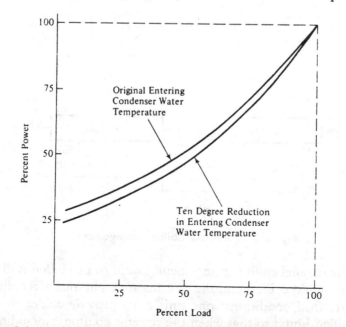

Figure 8-4. Typical Power Consumption Curve for Centrifugal Liquid Chiller

"ideal" coefficient of performance *(COP)* is used to relate the measure of cooling effectiveness. Approximately 0.8 kW is required per ton of refrigeration (0.8 kW is power consumption at full load, based on typical manufacturer's data).

$$COP = \frac{1 \text{ ton} \times 12{,}000 \text{ Btu/ton}}{0.8 \text{ kW} \times 3412 \text{ Btu/kW}} = 4.4$$

Chillers in Series and in Parallel

Multiple chillers are used to improve reliability, offer standby capacity, reduce inrush currents and decrease power costs at partial loads. Figure 8-5 shows two common arrangements for chiller staging namely, chillers in parallel and chillers in series.

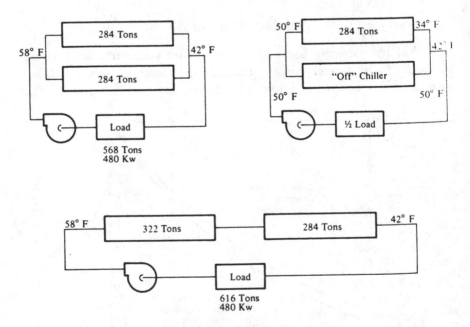

Figure 8-5. Multiple Chiller Arrangements

In the parallel chiller arrangement, liquid to be chilled is divided among the liquid chillers and the streams are combined after chilling. Under part load conditions, one unit must provide colder than designed chilled liquid so that when the streams combine, including the one from the off chiller, the supply temperature is provided. The parallel chillers have a lower first cost than the series chillers counterparts but usually consume more power.

In the series arrangement, a constant volume of flow of chilled water passes through the machines, producing better temperature control and better efficiency under part load operation; thus, the upstream chiller requires less kW input per ton output. The waste of energy during the mixing aspect of the parallel chiller operation is avoided. The series chillers, in general, require higher pumping costs. The energy conservation engineer should evaluate the best arrangement, based on load required and the partial loading conditions.

The Absorption Refrigeration Unit

Any refrigeration system uses external energy to "pump" heat from a low temperature level to a high temperature. Mechanical refrigeration compressors pump absorbed heat to a higher temperature level for heat rejection. Similarly, absorption refrigeration changes the energy level of the refrigerant (water) by using lithium bromide to alternately absorb it at a low temperature level and reject it at a high level by means of a concentration-dilution cycle.

The single-stage absorption refrigeration unit uses 10 to 12 psig steam as the driving force. Whenever users can be found for low pressure steam, energy savings will be realized. A second aspect for using absorption chillers is that they are compatible for use with solar collector systems. Several manufacturers offer absorption refrigeration equipment which uses high temperature water (160°–200°F) as the driving force.

A typical schematic for a single-stage absorption unit is illustrated in Figure 8-6. The basic components of the system are the evaporator, absorber, concentrator and condenser. These components can be grouped in a single or double shell. Figure 8-6 represents a single-stage arrangement.

Evaporator—Refrigerant is sprayed over the top of the tube bundle to provide for a high rate of transfer between water in the tubes and the refrigerant on the outside of the tubes.

Absorber—The refrigerant vapor produced in the evaporator migrates to the bottom half of the shell where it is absorbed by a

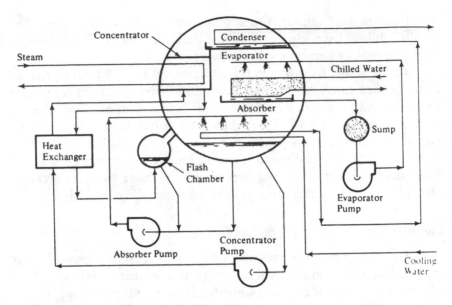

Figure 8-6. One-Shell Lithium Bromide Cycle Water Chiller
(Source: Trane Air Conditioning Manual)

lithium bromide solution. Lithium bromide is basically a salt solution which exerts a strong attractive force on the molecules of refrigerant (water) vapor. The lithium bromide is sprayed into the absorber to speed up the condensing process. The mixture of lithium bromide and the refrigerant vapor collects in the bottom of the shell; this mixture is referred to as the dilute solution.

Concentrator—The dilute solution is then pumped through a heat exchanger where it is preheated by a hot solution leaving the concentrator. The heat exchanger improves the efficiency of the cycle by reducing the amount of steam or hot water required to heat the dilute solution in the concentrator. The dilute solution enters the upper shell containing the concentrator. Steam coils supply heat to boil away the refrigerant from the solution. The absorbent left in the bottom of the concentrator has a higher percentage of absorbent than it does refrigerant; thus it is referred to as concentrated.

Condenser—The refrigerant vapor boiling from the solution in the concentrator flows upward to the condenser and is condensed. The condensed refrigerant vapor drops to the bottom of the condenser and from there flows to the evaporator through a regulating orifice. This completes the refrigerant cycle.

The single-stage absorption unit consumes approximately 18.7 pounds of steam per ton of capacity (steam consumption at full load based on typical manufacturer's data). For a single-state absorption unit

$$COP = \frac{1 \text{ ton} \times 12{,}000 \text{ Btu/ton}}{18.7 \text{ lb} \times 955 \text{ Btu/lb}} = 0.67$$

The single-stage absorption unit is not as efficient as the mechanical chiller. It is usually justified based on availability of low pressure steam, equipment considerations or use with solar collector systems.

Example Problem 8-1

Compute the energy wasted when 15 psig steam is condensed prior to its return to the power plant. Comment on using the 15 psig steam directly for refrigeration.

Answer

From Steam Table 15-14 for 30 psia steam, hfg is 945 Btu per pound of steam; thus, 945 Btu per pound of steam is wasted. In this case where *excess low pressure* steam cannot be used, absorption units should be considered in place of their electrical-mechanical refrigeration counterparts.

Example Problem 8-2

2000 lb/hr of 15 psig steam is being wasted. Calculate the yearly (8000 hr/yr) energy savings if a portion of the centrifugal refrigeration system is replaced with single-stage absorption. Assume 20 kW additional energy is required for the pumping and cooling tower cost associated with the single-stage absorption unit. Energy rate is $.09 kWh and the absorption unit consumes 18.7 lb of steam per ton of capacity.

The centrifugal chiller system consumes 0.8 kWh per ton of refrigeration.

Answer

Tons of mechanical chiller capacity replaced = 2000/18.7 = 106.95 tons. Yearly energy savings = 2000/18.7 × 8000 × 0.8 × $.09 = $61,602.

Two-Stage Absorption Unit

The two-stage absorption refrigeration unit (Figure 8-7) uses steam at 125 to 150 psig as the driving force. In situations where excess medium pressure steam exists, this unit is extremely desirable. The unit is similar to the single-stage absorption unit. The two-stage absorption unit operates as follows:

Medium pressure steam is introduced into the first-stage concentrator. This provides the heat required to boil out refrigerant from the dilute solution of water and lithium bromide salt. The liberated refrigerant vapor passes into the tubes of the second-stage concentrator, where its temperature is utilized to again boil a lithium bromide solution, which in turn further concentrates the solution and liberates additional refrigerant. In effect, the concentrator frees an increased amount of refrigerant from solution with each unit of input energy.

The condensing refrigerant in the second-stage concentrator is piped directly into the condenser section. The effect of this is to reduce the cooling water load. A reduced cooling water load decreases the size of the cooling tower which is used to cool the water. The remaining portions of the system are basically the same as the single-stage unit.

The two-stage absorption unit consumes approximately 12.2 pounds of steam per ton of capacity; thus, it is more efficient than its single-stage counterpart. The associated COP is

$$COP = \frac{1 \text{ ton} \times 12,000 \text{ Btu/ton}}{12.2 \text{ lb} \times 860 \text{ Btu/lb}} = 1.14$$

Either type of absorption unit can be used in conjunction with centrifugal chillers when it is desirable to reduce the peak electrical demand of the plant or to provide for a solar collector addition at a later date.

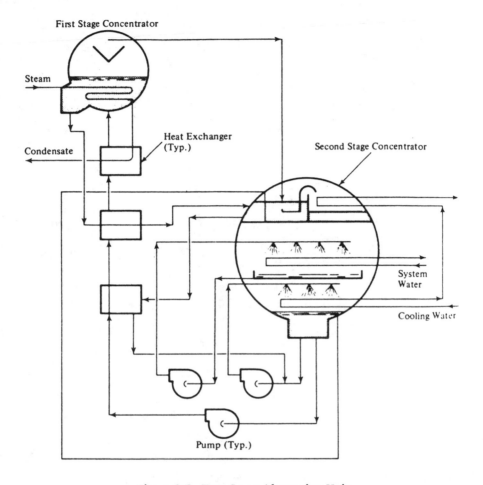

Figure 8-7. Two-Stage Absorption Unit

9

Cogeneration:
Theory and Practice

Because of its enormous potential, it is important to understand and apply cogeneration theory. In the overall context of Energy Management Theory, cogeneration is just another form of the conservation process. However, because of its potential for practical application to new or existing systems, it has carved a niche that may be second to no other conservation technology.

This chapter is dedicated to development of a sound basis of current theory and practice of cogeneration technology. It is the blend of theory and practice, or praxis of cogeneration, that will form the basis of the most workable conservation technology during the remainder of the 20th century.

DEFINITION OF "COGENERATION"

Cogeneration is the sequential production of thermal and electric energy from a single fuel source. In the cogeneration process, heat is recovered that would normally be lost in the production of one form of energy. That heat is then used to generate the second form of energy. For example, take a situation in which an engine drives a generator that produces electricity: With cogeneration, heat would be recovered from the engine exhaust and/or coolant, and that heat would be used to produce, say, hot water.

Making use of waste heat is what differentiates cogeneration facilities from central station electric power generation. The overall fuel utilization efficiency of cogeneration plants is typically 70-80% versus 35-40% for utility power plants.

This means that in cogeneration systems, rather than using energy in the fuel for a single function, as typically occurs, the available energy is cascaded through at least two *useful* cycles.

To put it in simpler terms: Cogeneration is a very efficient method of making use of *all* the available energy expended during any process generating electricity (or shaft horsepower) and then utilizing the waste heat.

A more subjective definition of cogeneration calls upon current practical applications of power generation and process needs. Nowhere more than in the United States is an overall system efficiency of only 30% tolerated as "standard design." In the name of limited *initial* capital expenditure, all of the waste heat from most processes is rejected to the atmosphere.

In short, present design practices dictate that of the useful energy in one gallon of fuel, only 30% of that fuel is put to useful work. The remaining 70% is rejected randomly.

If one gallon of fuel goes into a process, the designer may ask, "How much of that raw energy can I make use of within the constraints of the overall process?"

In this way, cogeneration may be taken as a way to use a maximum amount of available energy from any raw fuel process. Thus, cogeneration may be thought of as *just good design.*

COMPONENTS OF A
COGENERATION SYSTEM

The basic components of any cogeneration plant are

- A prime mover
- A generator
- A waste heat recovery system
- Operating control systems

The prime mover is an engine or turbine which, through combustion, produces mechanical energy. The generator converts the mechanical energy to electrical energy. The waste heat recovery system is one or more heat exchangers that capture exhaust heat or engine coolant heat and convert that heat to a useful form. The operating control systems insure that the individual system components function together.

The prime mover is the heart of the cogeneration system. The three basic types are steam turbines, combustion gas turbines and internal combustion engines. Each has advantages and disadvantages, as explained below.

Steam Turbines

Steam turbine systems consist of a boiler and turbine. The boiler can be fired by a variety of fuels such as oil, natural gas, coal and wood. In many installations, industrial by-products or municipal wastes are used as fuel. Steam turbine cogeneration plants produce high-pressure steam that is expanded through a turbine to produce mechanical energy which, in turn, drives a device such as an electric generator. Thermal energy is recovered in several different ways, which are discussed in Chapter 5. Steam turbine systems generally have a high fuel utilization efficiency.

Combustion Gas Turbines

Combustion gas turbine systems are made up of one or more gas turbines and a waste heat recovery unit. These systems are fueled by natural gas or light petroleum products. The products of combustion drive a turbine which generates mechanical energy. The mechanical energy can be used directly or converted to electricity with a generator. The hot exhaust gases of the gas turbine can be used directly in process heating applications, or they can be used indirectly, with a heat exchanger, to produce process steam or hot water.

A variation on the combustion gas turbine system is one that uses high-pressure steam to drive a steam turbine in conjunction with the cogeneration process. This is referred to as a combined cycle.

Internal Combustion Engines

Internal combustion engine systems utilize one or more reciprocating engines together with a waste heat recovery device. These are fueled by natural gas or distillate oils. Electric power is produced by a generator which is driven by the engine shaft. Thermal energy can be recovered from either exhaust gases or engine coolant. The engine exhaust gases can be used for process heating or to generate low-pressure steam. Waste heat is recovered from the engine cooling jacket in the form of hot water.

Cogeneration plants that use the internal combustion engine generate the greatest amounts of electricity for the amount of heat produced. Of the three types of prime movers, however, the fuel utilization efficiency is the lowest, and the maximum steam pressure that can be produced is limited.

AN OVERVIEW OF
COGENERATION THEORY

As discussed in the introduction, and as may be seen from Figure 9-1, standard design practices make use of, at best, 30% of available energy from the raw fuel source (gas, oil, coal).

Of the remaining 70% of the available energy, approximately 30% of the heat is rejected to the atmosphere through a condenser (or similar) process. An additional 30% of the energy is lost directly to the atmosphere through the stack, and finally, approximately 7% of the available energy is radiated to the atmosphere because of the high relative temperature of the process system.

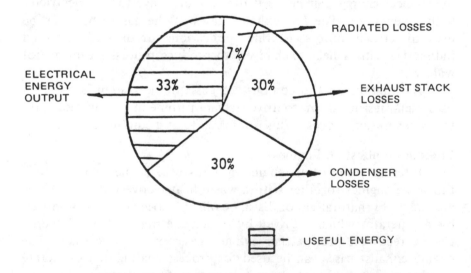

Figure 9-1. Energy Balance Without Heat Recovery

With heat recovery, however, potential useful application of available energy more than doubles. Although in a "low quality" form, *all* of the condenser-related heat may be used, and 40% of the stack heat may be recovered. This optimized process is depicted (in theory) as Figure 9-2.

Thus, it may be seen that effective use of all available energy may more than double the "worth" of the raw fuel. System efficiency is increased from 30% to 75%.

This higher efficiency allows the designer to use low grade energy for various cogeneration cycles.

Example Problem 9-1

A cogeneration system vendor recommends the installation of a 20-megawatt cogeneration system for a college campus. Determine the approximate *range* of useful *thermal* energy. Use the energy balance of Figure 9-2.

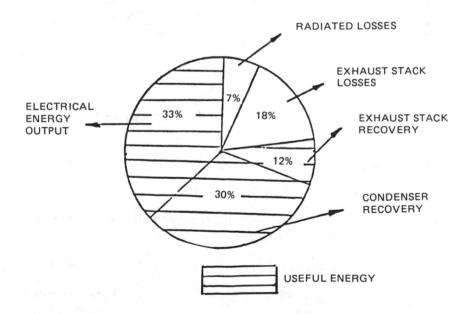

Figure 9-2. Energy Balance With Heat Recovery

Analysis

Step 1:   "Range" is defined by the best and worst operating *times* of the installed system:

At best, system will operate 365 days per year, 24 hours per day.

At worst, system will operate 5 days per week, 10 hours per day.

Step 2:   Perform heat balance. See diagram.

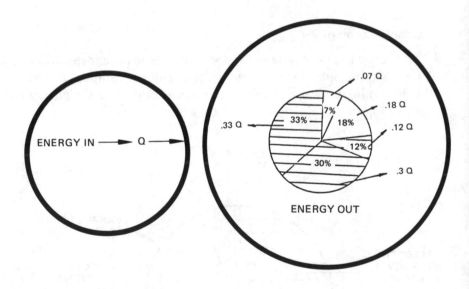

$$Q = E_{out} = E_{electricity} + E_{condenser} + E_{stack} + E_{radiated}$$ *Formula (9-1)*

Step 3:   Calculate *available thermal* energy.

From Figure 9-2, available energy equals condenser energy and 40% stack energy.

$$E_{available} = E_{condenser} + E_{stack}$$ *Formula (9-2)*

$$E_{condenser} = .3Q$$ *Formula (9-3)*

$$E_{stack} = 0.12Q$$

$$E_{available} = 0.30Q + 0.12Q \qquad \text{Formula (9-4)}$$

$$E_{electrical} = .33Q \qquad \text{Formula (9-5)}$$

$$Q = \frac{E_{electrical}}{.33} \qquad \text{Formula (9-6)}$$

By substitution $E_{available} = 0.3\dfrac{E_{electrical}}{0.33} + 0.12\dfrac{E_{electrical}}{0.33}$

$$E_{available} = 1.272 \text{ Electrical} \qquad \text{Formula (9-7)}$$

Calculate available energy

$$Q = Q_{1 \times t} = K_1 \, E_{available} \times K_2 \times t$$

K_1 = 3413 Btu/kWh
K_2 = 1000 Kilowatts per megawatt
t = equipment hours of operation

$$Q_1 = 3413 \text{ Btu/kWh} \times (1.272 \times 20 \text{ megawatt}) \times 1000 \frac{kW}{MWatt}$$

$$Q_1 = 86.82 \times 10^6 \text{ Btu/hr}$$

Worst Case: System operates
5 days/wk × 52 wks/yr × 10 hrs/day = 2600 hrs/yr

Best Case: System operates
365 days/yr × 24 hrs/day = 8670 hrs/yr

Thus range is

$$Q = 86.82 \times 10^6 \times 2600 = 225.7 \times 10^9 \text{ Btu/yr}$$
$$Q = 86.82 \times 10^6 \times 8760 = 760.5 \times 10^9 \text{ Btu/yr}$$

APPLICATION OF THE COGENERATION CONSTANT

The *cogeneration constant* may be used as a fast check on any proposed cogeneration installation. Notice from the sample problem which follows, the ease with which a thermal vs. electrical comparison of end needs may be made.

Example Problem 9-2

A cogeneration system vendor recommends a 20 megawatt installation. Determine the approximate rate of *useful* thermal energy.

Analysis

$$E_{available} = E_{electrical} \times K_c \qquad \qquad Formula\ (9\text{-}8)$$

E is the cogeneration system electrical rated capacity
K_c is the cogeneration constant

$E_{available}$ = 20 MW $\times$ 1.272
        = 25.4 MW of *useful* heat
or

$$25.4\ MW \times 1000\ \frac{kW}{MW} \times \frac{3413\ Btu}{kWh} \times \frac{Therm}{100{,}000\ Btu}$$

$E_{available}$ = 866.9 therms/hour

APPLICABLE SYSTEMS

To ease the complication of matching power generation to load, and because of newly established laws, it is most advantageous that the generator operate in parallel with the utility grid which thereby "absorbs" all generated electricity. The requirement for "qualify facility (QF) status," and the consequent utility rate advantages which are available when "paralleling the grid," is that a significant portion of the thermal energy produced in the generation process must be recovered. Specifically, Formula 9-10 must be satisfied:

$$\frac{Power\ Output + \frac{1}{2}\ Useful\ Thermal\ Output}{Energy\ Input} \geq 42.5\%\ \text{(for any calendar year)}$$

$$Formula\ (9\text{-}9)$$

Note that careful application of the *cogeneration constant* will generally assure that the qualifying facility status is met.

BASIC THERMODYNAMIC CYCLES

Bottoming and Topping Cycles

Cogeneration systems can be divided into "bottoming cycles" and "topping cycles."

Bottoming Cycles

In a bottoming cycle system, thermal energy is produced directly from the combustion of fuel. This energy usually takes the form of steam that supplies process heating loads. Waste heat from the process is recovered and used as an energy source to produce electric or mechanical power.

Bottoming cycle cogeneration systems are most commonly found in industrial plants that have equipment with high-temperature heat requirements such as steel reheat furnaces, clay and glass kilns and aluminum remelt furnaces. Some bottoming cycle plants operating in Georgia are Georgia Kraft Company, Brunswick Pulp and Paper Company and Burlington Industries.

Topping Cycles

Topping cycle cogeneration systems reverse the order of bottoming cycle systems: Electricity or mechanical power is produced first; then heat is recovered to meet the thermal loads of the facility. Topping cycle systems are generally found in facilities which do not have extremely high process temperature requirements.

Figure 9-3 on the following page shows schematic examples of these two cogeneration operating cycles.

A sound understanding of basic cogeneration principles dictates that the energy manager should be familiar with two standard thermodynamic cycles. These cycles are

1. Brayton Cycle
2. Rankine Cycle

The Brayton Cycle is the basic thermodynamic cycle for the simple gas turbine power plant. The Rankine Cycle is the basic cycle for a vapor-liquid system typical of steam power plants. An excellent theoretical discussion of these two cycles appears in Reference 10.

Figure 9-3. Cogeneration Operating Cycles

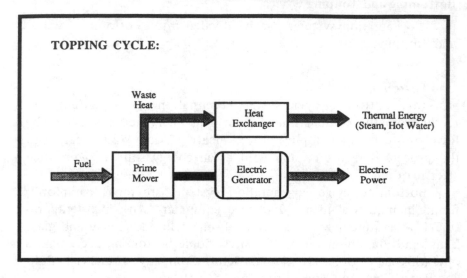

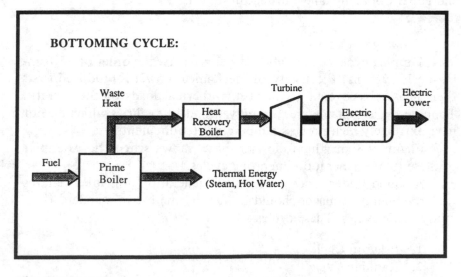

The Brayton Cycle

In the open Brayton Cycle plant, energy input comes from the fuel that is injected into the combustion chamber. The gas/air mixture drives the turbine with high-temperature waste gases exiting to the atmosphere. (See Figure 9-4.)

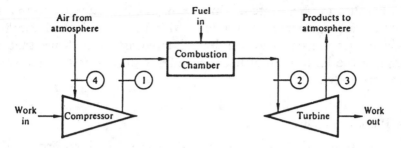

Figure 9-4. Open Gas-Turbine Power Plant (Brayton Cycle)

The basic Brayton Cycle, as applied to cogeneration, consists of a gas turbine, waste heat boiler and a process or "district" heating load. This cycle is a *full-load* cycle. At part loads, the efficiency of the gas turbine goes down dramatically. A simple process diagram is illustrated in Figure 9-5.

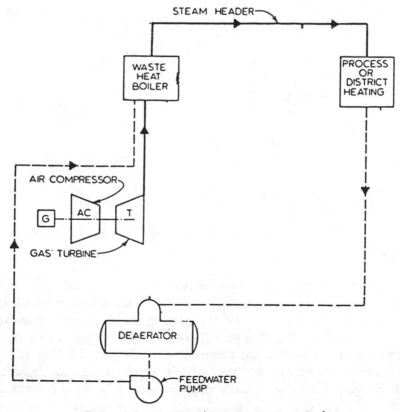

Figure 9-5. Process Diagram — Brayton Cycle

Because the heat rate of this arrangement is superior to that of all other arrangements at full load, this simple, standard Brayton Cycle merits consideration under all circumstances. Note that to complete the loop in an efficient manner, a deaerator and feedwater pump are added.

The Rankine Cycle

The Rankine Cycle is illustrated in the simplified process diagram, Figure 9-6. Note that this is the standard boiler/steam turbine arrangement found in many power plants and central facility plants throughout the world.

| The Rankine Cycle, or steam turbine, provides a real-world outlet for waste heat recovered from any process or generation situation. Hence, it is the steam turbine which is generally referred to as the topping cycle.

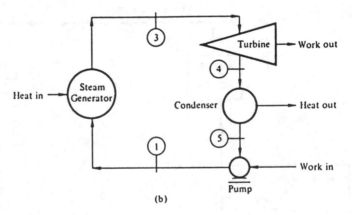

Figure 9-6. Simple Rankine Cycle

Combined Cycles

Of major interest and importance for the serious central plant designer is the Combined Cycle. This cycle forms a hybrid which includes the Brayton Cycle on the "bottoming" portion and a standard Rankine Cycle on the "topping" portion of the combination. A process diagram with standard components is illustrated in Figure 9-7.

The Combined Cycle, then, greatly approximates the cogeneration Brayton Cycle but makes use of a knowledge of the plant requirements and an understanding of Rankine Cycle theory. Note also that the ideal mix of power delivered from the Brayton and Rankine portions of the Combined Cycle is 70% and 30%, respectively.

Even within the seemingly limited set of situations defined as "Combined Cycle," many variations and options become available. These options are as much dependent on any local plant requirements and conditions as they are on available equipment.

Some examples of Combined Cycle variations are:

1. Gas turbine exhaust used to produce 15 psi steam for Rankine Cycle turbine with no additional fuel burned. This situation is shown in Figure 9-7.

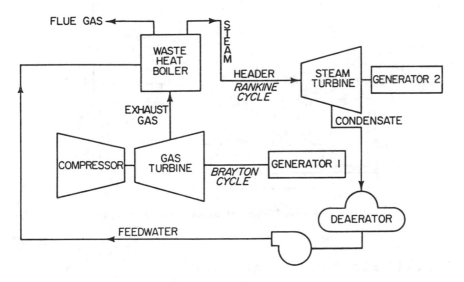

Figure 9-7. Combined Cycle Operation

2. Gas turbine exhaust fired in the duct with additional fuel. This provides a much greater amount of produced power with a correspondingly greater amount of fuel consumption. This situation generally occurs with a steam turbine pressure range of 900–1259 psig.

3. Gas turbine exhaust fired directly and used directly as combustion air for a conventional power boiler. Note here that

the boiler pressure range may vary between 200 and 2600 psig. (See Figure 9-8.)

One other note to keep in mind is that in any Combined Cycle case, the primary or secondary turbine may supply direct mechanical energy to a refrigerant compressor. As discussed, the variations are endless. However, a thorough understanding of the end process generally will result in a final, and best, cogeneration system selection.

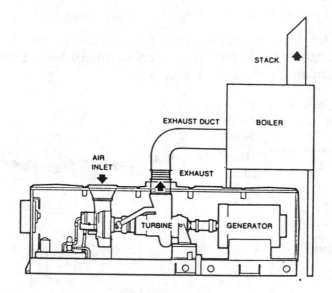

Figure 9-8. Gas Turbine Used As Combustion Air

DETAILED FEASIBILITY EVALUATION[1]

This section will introduce the parameters affecting the evaluation, selection, sizing and operation of a cogeneration plant and shows a means of evaluating those parameters where it counts—on the bottom line.

This section will enable you to answer two basic questions:

1. Is cogeneration technically feasible for us, given our situation?

2. If it is technically feasible, is it also economically feasible,

---
[1]*Georgia Cogeneration Handbook*, Governor's Office of Energy Resources, August 1988.

considering estimated costs, energy savings, current utility rates and regulatory conditions in Georgia?

Keep in mind that this chapter is not designed to take the place of a full-scale feasibility study. It is designed to help you and your engineering staff decide—in-house and at minimal expense—whether such a study is warranted. You will reach one of the following conclusions:

- No, cogeneration is not feasibile in our case.

- Yes, cogeneration looks promising and a feasibility study is warranted.

- Maybe, but I have some questions.

If your conclusion is "no," you will have saved yourself the expense of a feasibility study. If your conclusion is "yes," you will be assured that the expense of employing outside consultants to perform the study is justified by the potential benefits. If your conclusion is "maybe," you may wish to discuss your situation with qualified engineers before deciding whether to proceed.

Utility Data Analysis

The first step in the feasibility evaluation is the gathering and analysis of utility data. These data are necessary to determine the limits of the thermal and electric energy consumed at your facility. They are also necessary for determining the timing of maximum use of each of these energy sources. From this information, you will also determine the thermal and electric load factors for your facility.

Load Factor Defined

A load factor is defined as the average energy consumption rate for a facility divided by the peak energy consumption rate over a given period of time, and it is an important simplification of energy use data. These values are usually expressed as either a decimal number or a percent. To calculate a load factor you need two pieces of data: the total energy consumption for a given time period and the maximum energy demand observed during that time period.

In the electric power industry, load factors serve as a measure of the utilization of generation equipment. A utility load factor of 0.90, or 90%, indicates very good utilization of power generation facilities, while a load factor of 0.30 or 30% indicates poor utilization. The lower the load factor, the larger and more expensive your generating equipment must be merely to meet peak power demands for short time periods.

The thermal and electric load factors of your facility are of great importance in sizing a cogeneration plant. In fact, they are even more important to a cogeneration plant than to an electric utility because the cogeneration plant does not have the advantage of an electric power grid to diversify the variations in energy use.

A high thermal or electrical load factor generally indicates that a cogeneration plant would be utilized a major portion of the time and therefore would provide a favorable return on your investment. An ideal situation exists when both load factors are high, indicating that a properly sized cogeneration plant would efficiently utilize most of its output. If both the thermal and electrical load factors are small, you may not be a practical candidate for cogeneration.

In the analysis of your electric and thermal energy consumption data, you will calculate the load factors for your facility. In this evaluation, we will concentrate on annual load factors. It is important to note, however, that monthly and daily load factors may be important in your final analysis.

Electric Energy Consumption Analysis

In order to obtain the necessary energy-use information, you must refer to your monthly electric utility bills for one complete year. From those bills, calculate the following information:

- Determine your Annual kWh consumption by tabulating and adding the monthly kWh consumption.

- Tabulate the actual kW demand metered for each month. Be sure to use the actual metered demand and not the billed demand, as the billed demand may be based on time of year and a percent ratchet of the previous year's peak demand.

- Identify the maximum monthly kW demand value as the Annual Peak kW.

- Determine the Annual Average kW Demand by dividing the Annual kWh consumption by 8,760 hours per year.

$$\text{Annual Average kW} = \frac{\text{Annual kWh}}{8,760 \text{ hours/year}}$$

- Determine the Annual Electric Load Factor by dividing the Annual Average kW demand by the Annual Peak kW demand.

$$\text{Annual Electric Load Factor} = \frac{\text{Annual Average kW}}{\text{Annual Peak kW}}$$

Thermal Energy Consumption Analysis

If your boiler plant has an accurate steam or Btu metering system, the following data can be gathered directly from your metered output data. If you do not have this degree of metering, your boiler plant fuel bills for one complete year will be required to obtain the necessary data. Monthly natural gas bills can be analyzed with the same method used for analyzing electrical consumption. If you use fuel oil, propane or coal, make sure that accurate measurements of the reserve supply were taken at the beginning and end of the year.

- Determine your annual fuel consumption. From Figure 9-9, select the Btu value per unit of fuel for the type of fuel you use. To obtain the Annual Fuel Btu Input, multiply the annual fuel consumption by this value.

- Determine your boiler's Fuel-to-Steam Efficiency. An estimated fuel-to-steam efficiency of 78% may be used for a well-maintained boiler plant that operates only enough boilers to keep them well loaded. A poorly maintained boiler plant with some leaks, missing insulation, and oversized boilers that cycle frequently may have a Fuel-to-Steam Efficiency of 60% or lower.

- Determine your Annual Btu Output by multiplying your Annual Fuel Btu Input by your Fuel-to-Steam Efficiency.

$$\text{Annual Btu Output} = \frac{\text{Annual Fuel}}{\text{Btu Input}} \times \frac{\text{Fuel-to-Steam}}{\text{Efficiency}}$$

- Determine the Annual Average Btu/Hour Demand by dividing the Annual Btu Output by 8,760 hours per year.

$$\text{Annual Average Btu/Hour Demand} = \frac{\text{Annual Btu Output}}{8,760 \text{ hours/year}}$$

Figure 9-9. Typical Fuel Caloric Values (Btu/CF)

Fuel Type	HHV Higher Heating Value (Approximate)	LHV/HHV Lower/Higher Heating Value (Approximate)
Natural Gas (Dry)	1,000	0.90
Butane	3,200	0.92
Propane	2,500	0.92
Sewage Gas	300-600	0.90
Landfill Gas	300-600	0.90
No. 2 Oil	139,000 Btu/Gal	0.93
No. 6 Oil	154,000 Btu/Gal	0.96
Coal, Bituminous	14,100 Btu/lb	

If you have metered steam production data, select the maximum value for Btu/hour output that occurred last year as the Annual Peak Btu/hour Demand. If these data are not available, estimate your Annual Peak Btu/hour Demand as a percentage of the maximum possible output of your boiler plant.

NOTE: Boilers that operate at approximately 150 psig and below use the terms "lbs per hour" and "mbh output" interchangeably to represent 1,000 Btu/hour. We will use the unit "mbh" to simplify notation when referring to 1,000 Btu/hour units.

- Determine the Annual Thermal Load Factor by dividing the Annual Average Btu/hour Demand by the Annual Peak Btu/hour Demand.

$$\frac{\text{Annual Thermal}}{\text{Load Factor}} = \frac{\text{Annual Average Btu/hour Demand}}{\text{Annual Peak Btu/hour Demand}}$$

If your Annual Minimum Btu/Hour Demand is extremely low or your boiler plant actually shuts down for several months during the summer, your Annual Thermal Load Factor is not an accurate indicator of your cogeneration potential. It is probably too large, and a closer examination of the number of hours at the Annual Minimum Btu/hour Demand may be required to assess feasibility. This can be done using monthly and daily load profiles.

If your Annual Minimum Btu/hour Demand is zero for a large number of hours per year (for example, if your boiler plant shuts down during the summer months), you may not be a good candidate for cogeneration.

Thermal/Electric Load Ratio

For cogeneration to be feasible, the demands for thermal and electric energy must overlap much of the time. Therefore, once the thermal and electric demands of your facility are known, you must determine the ratio of heat demand to electric demand that may be expected to occur together. This is done by using the Thermal/Electric (T/E) Load Ratio. The T/E Load Ratio is defined as the quantity of heat energy that is coincident with a quantity of electrical energy. In making these calculations, you will attempt to find an optimum match between your facility's T/E Load where

$$\text{Thermal/Electric Load Ratio} = \frac{\text{Thermal Demand}}{\text{kW Demand}}$$

If the thermal and electric load factors calculated earlier are high, there is a good possibility that your facility's demand for thermal energy occurs at about the same time as your demand for electric energy. In that case, your facility's Annual Average Thermal/Electric Load Ratio is approximately equal to the Annual Average Btu/hour Demand divided by the Annual Average kW Demand. For convenience, we will use the term "mbh" to represent 1,000 Btu/hr.

$$\frac{\text{Annual Average Thermal/}}{\text{Electric Load Ratio}} = \frac{\text{Annual Average mbh Demand}}{\text{Annual Average kW Demand}}$$

If either the thermal or the electric load factor is small, then a worst-case assumption should be made. This assumption is called the Minimum Demand Thermal/Electric Load Ratio and is calculated as follows:

$$\frac{\text{Mimimum Demand Thermal/}}{\text{Electric Load Ratio}} = \frac{\text{Annual Minimum mbh Demand}}{\text{Annual Minimum kW Demand}}$$

The use of either of the Thermal/Electric Load Ratios above must be tempered with the knowledge that these are only approximations of the load correlations and are useful only for a preliminary evaluation. Most facilities will require a more detailed examination of thermal and electric load profiles to obtain the number of hours per year that loads overlap and can be served with a cogeneration plant.

Generally, cogeneration opportunities are good for facilities with T/E Load Ratios above 5 and are best for ratios above 10, with Average Annual Btu/hour Demand above 10,000,000 Btu/hour, or 10,000 mbh.

If your T/E Load Ratio is less than two, it is reasonable to assume that you are not a good candidate for cogeneration. This would certainly be the case if your thermal demand is extremely small during the summer months and is only significant during a few winter months. However, if your Annual Electric Load Factor is small, you may consider using electric power generation only, in a strategy known as peak shaving. This is discussed in more detail under Operating Strategies in this chapter.

Equipment Sizing Considerations

For efficient cogeneration, the T/E Load Ratio in your facility must correspond to the T/E Output Ratio of the cogeneration systems. T/E Output Ratios vary for different prime movers and cogeneration plant configurations.

In an analysis of cogeneration options at your facility, you must decide whether to size the cogeneration plant to match your peak electric load, which will produce some waste heat and lower overall plant efficiency, or to size the cogeneration plant to match your heat load and supplement your electric needs with more expensive purchased power. If your load factor is small for either thermal or electric demand, you should size toward the minimum value of that demand.

Operating Strategies

Electric Dispatch

One method of operating a cogeneration plant is to supply your facility's total electrical requirements as a first priority and generate steam as a second priority. This mode of operation is referred to as "electric dispatch."

Under the electric dispatch mode, the cogeneration plant is sized for Annual Peak kW Demand, with some additional capacity for growth. A cogeneration plant sized in this manner can operate totally independently of the electrical utility company.

Standby electric service from your local utility is required for scheduled and unscheduled equipment shutdown. The cost of this standby power can be very expensive, especially if demand is high or if the utility company must provide additional generation or transmission equipment to serve your facility. Total on-site backup using standby emergency power generating equipment is rarely economical.

Electric dispatch requires that you operate your cogeneration plant to meet your electric load demand requirement. Operation of the major electric loads of your facility should be coordinated with the cogeneration plant to prevent sudden load spikes or irregular, repetitive load spikes that could exceed the capacity of your generating plant.

Also, if your annual electric load factor is low, your cogeneration equipment will be oversized for your facility load most of the time. This will reduce efficiency, increase maintenance costs and lower the return on investment.

Peak Shaving

Peak shaving is the practice of selectively dropping electric loads or generating on-site electricity during periods of peak electric demand. This procedure is not cogeneration; normally, no heat recovery equipment is installed and no heat is recovered from the generating process. Peak shaving is most commonly used to reduce annual electric utility costs. Peak shaving with an electric generator has a lower initial cost than a true cogeneration system since no heat exchangers or associated recovery equipment is installed.

Thermal Dispatch

Thermal dispatch operation is the complement to electric dispatch. In this mode of operation, the cogeneration plant is sized to meet your facility's Annual Peak Btu/hour (or mbh) Demand. You will then purchase a part of your electric power from the local utility and cogenerate the rest. For this mode of operation to be successful, thermal load requirements should parallel each other in the same way required for electric dispatch.

For maximum savings on electric energy, electric generation should be near its maximum capacity during the summer peak electric demand period. This means that the ideal candidate for thermal dispatch operation would have maximum thermal loads occurring during summertime electric peak periods. If this is not the case, you will save less on electric energy.

Hybrid Strategy

A "hybrid strategy" utilizes the best features of the electrical and thermal dispatch strategies. As the name suggests, this operating strategy is a hybrid of the electric dispatch and thermal dispatch operating strategies periodically adjusted to minimize operating costs and maximize return on investment.

With the hybrid strategy, the cogeneration plant is sized smaller than for electrical or thermal dispatch. The hybrid strategy calls for the plant to be operated to achieve maximum savings during electric peak demand periods. At other times, it would be operated in a thermal load following mode. To satisfy the total thermal load requirements, it may be necessary to maintain existing boilers or install additional ones. As in the thermal dispatch strategy, your facility still must purchase a portion of its electric energy from the utility.

Prime Mover Selection

Choosing the most appropriate prime mover for a cogeneration project involves evaluation of many different criteria, including

- Hours of operation
- Maintenance requirements
- Fuel requirements
- Capacity limits

Following are some general guidelines for evaluating each of the different prime movers with respect to these criteria.

Hours of Operation

Most cogeneration plants are designed for continuous operation, with the exception of some small reciprocating engine packages. Large plants are not economical if they cycle on and off on an hourly or even daily basis.

Because cogeneration equipment is relatively expensive as compared with boilers of similar thermal capacity, the equipment must be operated as much as possible to achieve an acceptable return on investment.

If an analysis of your facility shows a low electric load factor, indicating relatively few hours of high electric demand, cogeneration is probably not a cost-effective option. Instead, you may wish to examine the potential of peak shaving with reciprocating engine generators.

Maintenance Requirements

All cogeneration equipment requires some type of periodic maintenance. The frequency and amount of maintenance varies considerably between prime movers.

Reciprocating Engine Generators have the highest maintenance requirements. Like automotive engines, diesel engines require routine maintenance such as oil and filter changes as often as once a week. However, natural gas fired reciprocating engines require more overhauls of the head and block, and they have a shorter life expectancy.

Gas Turbine Engines are mechanically simpler and require less frequent maintenance. They can operate for longer periods between major maintenance intervals, and the major maintenance is usually simpler, such as bearing inspection and replacement.

Steam Turbines require even less mechanical maintenance than gas turbines since they have no combustion equipment attached. Occasional bearing inspection and replacement is generally the extent of steam turbine maintenance. The cost is a function of the size and number of turbines.

Fuel Requirements

Reciprocating Engines are not flexible with respect to their fuel requirements. Generally, you buy either a diesel engine or a natural gas fired engine. Special design adaptations can produce engines that burn other fuels, such as low Btu gas or heavier oils. Occasionally, natural gas engines can be adapted to switch between natural gas and propane as an alternate fuel. Diesel engines are more efficient in their fuel-to-energy conversion. However, air pollution regulations may impose greater constraints on these engines than gas turbines, usually requiring catalytic converters and carburetion limits.

Gas Turbines can be switched from natural gas to diesel fuel to take advantage of price fluctuations and backup fuel requirements. In addition, they can be adjusted for other fuel oils, low Btu gas, and, occasionally, organic by-products of industrial processes such as "black liquor" from pulp and paper mills.

Steam Turbines are limited only by the fuel for their steam source. In addition to the above fuels, coal, wood, waste, peanut shells, any suitable biomass, and incinerated municipal waste can be used to generate steam for steam turbines.

Capacity Limits

The three prime movers discussed here have separate and distinct capacity ranges. Reciprocating engine generators range in size from about 40 kW to over 3,000 kW. Generally, small electric demand plants with still smaller heat requirements can be satisfied with reciprocating engines. The quality of the heat recovered from these engines can be a limitation. Only about 35% of the recoverable heat is available as 125 psig steam; the rest of it is available only as 180°F hot water.

Almost all of the gas turbine heat is recoverable as 125 psig steam. The lower limit of gas turbine equipment size is about 480 kW, and the upper limit is over 30,000 kW. Combined-cycle gas and steam turbine plants can produce over 100,000 kW.

Steam turbines are the most limited with respect to power generation. Their practical lower limit is about 1,000 kW. Their conversion efficiency for power generation is below 15% if practical upper limits of 200 psig of superheated inlet steam and 100 psig of saturated outlet steam are maintained. This efficiency can be improved by increasing the inlet pressure and temperature, which dramatically increases the cost of the steam production plant. Another alternative for improving electric generating efficiency is to use a condensing turbine, but this lowers the temperature of the outlet steam to the point that it cannot be used for much more than low-temperature hot water production.

Figure 9-10 on the following page shows the thermal and electric energy output typically available from several different cogeneration system configurations.

Figure 9-10. Energy Production in Cogeneration Systems

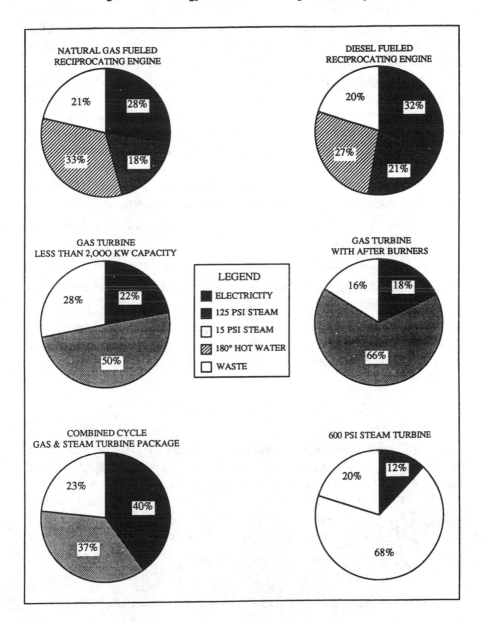

Energy Savings Analysis

To arrive at an estimate of the energy savings potential of cogeneration at your facility, you will need to make two simplifying assumptions. For this analysis we will assume that the proposed cogeneration system is sized such that 100 percent of its thermal and electric output is utilized. We will also assume that the plant will operate at or near full output capacity at all times. These assumptions represent a "best case" operating scenario for your cogeneration system. In reality, your plant probably will not operate in this manner; however, these assumptions are necessary here for the purpose of this evaluation.

In addition to the above assumptions, you will need to obtain the following information on your proposed cogeneration plant:

- The thermal and electric output capacities of the cogeneration equipment

- The fuel consumption rate of the cogeneration equipment

- The number of hours per year that you plan to operate the cogeneration system

This information can be obtained from equipment data available from the manufacturer. You will need to determine the annual operating hours based on knowledge of your facility operating characteristics.

Energy Production

The thermal and electric energy produced by your cogeneration plant will be used at your facility to offset thermal energy produced by your boiler plant and electricity purchased from your local utility. The next step is to determine the annual thermal and electric energy production of the proposed cogeneration equipment.

The annual electric energy production is calculated as follows:

$$\frac{\text{Annual Electric}}{\text{Energy Production}} = \frac{\text{Electric Output}}{\text{Capacity}} \times \frac{\text{Annual}}{\text{Operating Hours}}$$

The annual thermal energy production is calculated as follows:

$$\frac{\text{Annual Thermal}}{\text{Energy Production}} = \frac{\text{Thermal Output}}{\text{Capacity}} \times \frac{\text{Annual}}{\text{Operating Hours}}$$

Fuel Consumption

To determine the fuel use requirements of your proposed cogeneration plant, you will need to know the rate of fuel consumption for the equipment being evaluated here. This information can be obtained from the equipment manufacturer. Your cogeneration plant fuel consumption is calculated as follows:

$$\frac{\text{Cogeneration Fuel}}{\text{Consumption}} = \frac{\text{Equipment Fuel}}{\text{Consumption Rate}} \times \frac{\text{Annual}}{\text{Operating Hours}}$$

Since you have assumed that all of the output will be used by your facility, you will now calculate the value of the cogenerated energy based on your current utility costs.

Current Utility Costs

To determine your annual average cost per kWh, first tabulate electric energy costs from the bills that you used for gathering the electric demand and consumption data. (Generally, a more detailed analysis of your electric rate structure is needed to develop an accurate unit cost of the purchased power you will offset with cogeneration. For this evaluation, we will assume the unit cost of electricity is the same as the annual average unit cost of electricity.) Next, divide your annual electric cost by your annual electric consumption.

$$\text{Annual Average Cost/kWh} = \frac{\text{Annual Electric Cost}}{\text{Annual Electric Consumption}}$$

You can use the following method to determine your thermal energy cost:

- Tabulate monthly fuel costs from the bills used to obtain the fuel consumption data earlier.

- Determine your average Fuel Unit Cost by dividing the Annual Fuel Cost by your Annual Fuel Consumption.

$$\text{Fuel Unit Cost} = \frac{\text{Annual Fuel Cost}}{\text{Annual Fuel Consumption}}$$

- Determine your current cost per thermal MMBtu of steam or hot water. To do this, first multiply the Fuel Unit Cost by the Btu/unit value from Figure 9-9, then divide by your boiler's Fuel-to-Steam Efficiency. Divide the result by 1,000,000 to convert to MMBtu units.

$$\frac{\text{Current Cost}}{\text{Thermal MMBtu}} = \frac{\text{Fuel Unit Cost X Btu/Fuel Unit}}{\text{(Fuel-to-Steam Efficiency) (1,000,000)}}$$

Annual Savings

The annual energy savings of your cogeneration plant is calculated as the savings from electric and thermal energy production less the cost of operating the plant. The annual electric energy savings can be determined from the following equation:

$$\frac{\text{Annual Electric}}{\text{Energy Savings}} = \frac{\text{Annual Electric}}{\text{Energy Production}} \times \frac{\text{Annual Average}}{\text{Cost/kWh}}$$

The annual thermal energy savings is given by the following:

$$\frac{\text{Annual Thermal}}{\text{Energy Savings}} = \frac{\text{Annual Thermal}}{\text{Energy Production}} \times \frac{\text{Current Cost/}}{\text{Thermal MMBtu}}$$

Operating and Maintenance Costs

Operating and maintenance costs are, to a large degree, dependent on plant operating hours and, therefore, proportional to fuel consumption. For the purpose of this evaluation, maintenance costs can be approximated as 15% of fuel costs for reciprocating engines and 7% of fuel costs for gas or steam turbines. Use a lower figure for steam turbine maintenance costs if some boiler maintenance is already included in your operating expenses.

The cost of the fuel to operate your cogeneration plant can be determined using the following equation:

$$\frac{\text{Cogeneration}}{\text{Fuel Cost}} = \frac{\text{Cogeneration Fuel}}{\text{Consumption}} \times \frac{\text{Fuel}}{\text{Unit Cost}}$$

You now have all the information necessary to determine your annual savings from the operation of a cogeneration plant. This is calculated as follows:

$$\frac{\text{Annual}}{\text{Savings}} = \frac{\text{Annual Electric}}{\text{Energy Savings}} + \frac{\text{Annual Thermal}}{\text{Energy Savings}} -$$

$$\frac{\text{Cogeneration}}{\text{Fuel Cost}} - \frac{\text{Operating and}}{\text{Maintenance Cost}}$$

At this point, it is important to note that the annual savings derived from this analysis is only a preliminary estimate of the savings potential at your facility and is based on simplifying assumptions you have made.

Economic Analysis

Initial Cost
To develop the total initial cost of the cogeneration plant you are evaluating, first refer to Figure 9-11 below for approximate equipment cost ranges per kW of installed cogeneration electric capacity.

Figure 9-11.
Approximate Cost Per kW of Cogeneration Plant Capacity

Reciprocating Engine Packages (High Speed)

Large 900 to 3,000 kW packages . . . .$600 to $400/kW
Medium 400 to 800 kW packages . . . .$500 to $700/kW
Small 45 to 300 kW packages. . . . . $700 to $1200/kW

Gas Fired Turbine Engine Packages

Large 4,000 to 10,000 kW. . . . .$800 to $1,000/kW
Medium 500 to 4,000 kW . . . . $1,200 to $1,800/kW

Steam Powered Turbine Engines*

Less than 125 psig inlet turbines. . . . . . . $100 to $130/kW
Less than 250 psig inlet turbines. . . . .$90 to $120/kW

*Note initial costs for steam powered turbines are highly dependent on the entering and leaving pressures of the turbine. The best efficiencies are for 1,000 to 3,000 psi superheated steam. The cost of small-scale steam boilers in this range is prohibitive, and the cost of condensing steam turbines is not listed since considerable extra equipment is required. The costs as given do not include the cost of heat recovery equipment.

The cost figures in Figure 9-11 do not include the cost of new buildings to house the equipment, nor do they include allowances for the electrical wiring necessary to connect your facility to your utilities. If additional space must be constructed or major work is required to connect the utilities, these expenses must be added to the equipment cost estimates to determine your total initial cost. Also, these costs were assembled at an earlier time and may need to be revalidated for your current analysis.

Investment Analysis

The final decision to build a cogeneration plant is usually based on investment analysis. Broadly defined, this is an evaluation of costs versus savings. The costs for cogeneration include the initial capital cost for the equipment or the cost of operating and maintaining that equipment, costs; fuel costs, finance charges, tax liabilities, and other system costs, that may be specific to your application. Savings include offset electrical power costs; offset fuel costs; revenues from excess power sales, if any; tax benefits; and any other applicable savings that are offset by the cogeneration plant, such as planned replacement of equipment.

A number of economic analysis techniques are available for comparing investment alternatives. The Simple Payback Period is the most commonly used and the least complex, and is usually adequate for a go, no-go decision at this stage of project development. It is the period of time required to recover the initial investment cost through savings associated with the project. Simple Payback Period is calculated as follows:

$$\text{Simple Payback Period} = \frac{\text{Initial Cost, \$}}{\text{Annual Savings, \$/Yr}}$$

The Simple Payback Period will give you an indication of the attractiveness of cogeneration at your facility and will help you decide whether a full-scale feasibility study is warranted.

Figure 9-12 is a summary of the information used in this evaluation of cogeneration at your facility.

It is important to note that many of the variables used in this evaluation are assumed values based on your knowledge of the operating characteristics of your facility and the simplifying assumptions made for this preliminary analysis. In a full-scale cogeneration feasibility study, you will want to evaluate several equipment types and sizes in conjunction with different operating strategies for each system configuration. This type of evaluation should be performed for you by an experienced consulting engineer.

Figure 9-12.
Cogeneration Feasibility Evaluation Summary

UTILITY DATA ANALYSIS

Electric Energy Consumption	———————— KWH/year
Annual Peak Electric Demand	———————— KW
Electric Energy Cost	$———————— /year
Thermal Energy Consumption	———————— BTU/year
Annual Peak Thermal Demand	———————— BTU/hour
Annual Fuel Cost	$———————— /year
Annual Electric Load Factor	———————— %
Annual Thermal Load Factor	———————— %
Thermal/Electric Load Ratio	————————

ENERGY ANALYSIS

Cogeneration Plant Capacity	
Electric Output	———————— KW
Thermal Output	———————— BTU/hour
Annual Operating Hours	———————— hours/year
Annual Electric Energy Production	———————— KWH/year
Annual Thermal Energy Production	———————— BTU/year
Cogeneration Plant Fuel Consumption	———————— BTU/year
Current Utility Costs	
Average Annual Electric Cost	$———————— /KWH
Average Thermal Energy Cost	$———————— /MMBTU
Annual Savings	
Electric Energy Savings	$———————— /year
Thermal Energy Savings	$———————— /year
Cogeneration Fuel Cost	$———————— /year
Operating and Maintenance Costs	$———————— /year

ECONOMIC ANALYSIS

Annual Savings	$———————— /year
Initial Cogeneration Equipment Cost	$————————
Simple Payback Period	———————— years

10

Control Systems

This chapter will focus on the following subjects:

INTRODUCTION

Energy engineers have been and will continue to make significant contributions for the accomplishment of energy-efficient processes, buildings and transportation systems. In the hands of the energy engineer, control systems are an important tool to make these systems energy efficient; the majority of the presently existing systems were designed primarily as labor- and materials-saving (low first costs) rather than as energy-saving (low operating costs) equipment.

For example, the practice in the design and selection of heating, ventilating and air-conditioning (HVAC) systems has been based on maximum load requirements. Usually, HVAC systems operate at less than design load conditions and require controllers to regulate operations during partial load conditions. Design optimization of the configuration and selection of HVAC systems and controllers is essential to meet the required environmental conditions with minimum energy consumption. Energy-related characteristics of the HVAC equipment and processes have been discussed in Chapters 7 and 8. The purpose of the present chapter is to describe the controls needed for an energy-efficient design and operation of building systems.

BASIC CONCEPTS

A *control system* is a means by which some quantity of interest in a machine, mechanism or other equipment is maintained or altered, in accordance with a desired manner. A control system consists essentially of five parts:

1. Process (system)
2. Measuring means (sensor)
3. Error detector
4. Controller
5. Final control element

These five elements can be interconnected either in an *open-loop* format or in a *closed-loop* format.

An open-loop system is a system in which the output has no effect upon the input signal. A clothes dryer may be considered as an open-loop control system in which input is the status of the electric switch (on/off) and the output is the dryness (moisture content) of the clothes. When the dryer is turned on, it runs for a pre-set time and then automatically turns off. The actual moisture content of the clothes has no effect on the on-time.

The dryer could be put in a closed-loop system by installing a moisture-sensing probe which will continuously measure the moisture content of the clothes being dried and compare this value with a desired value. The difference between the two values could be used by a controller to control the on/off switch to the dryer.

Similarly, if a home thermostat were installed outside the house, the output (furnace heat) would have no effect on the input to the thermostat (in this case, outdoor temperature). If we closed this loop by bringing the thermostat indoors, the output (furnace heat) would now affect the input (indoor temperature). This closed-loop thermostat would continue to operate until the indoor temperature reached the desired value (setpoint). In this closed-loop control system, the output is fed back and compared with the desired value, and the difference between the two is used as an actuating signal to actuate the control device (gas valve) for a gas furnace.

Thus a closed-loop (feedback) control system is a system in which the output has an effect upon the input quantity to maintain the desired output value. A block diagram of a feedback control system showing all basic elements is shown in Figure 10-1.

If Figure 10-1 were to represent a residential heating system using a natural gas furnace, different elements would be as shown in Table 10-1.

CONTROL MODES

Figure 10-1 may be condensed to Figure 10-2. In this figure, for the residential heating system, the manipulated variable m(t), gas flow rate at any instant in time, could have one of the two possible values, namely zero or one, depending upon the thermostat status e(t) which could be on or off. Such a control action is called on/off or two-position control mode.

Thus, the two-position control is a type of control action in which the manipulated variable is quickly changed to either a maximum or minimum value depending upon whether the controlled variable is greater or less than the setpoint. The minimum value of the manipulated variable is usually zero (off).

The equations for two-position control are

$m = M_1$ when $e > 0$ *Formula (10-1)*

$m = M_0$ when $e < 0$

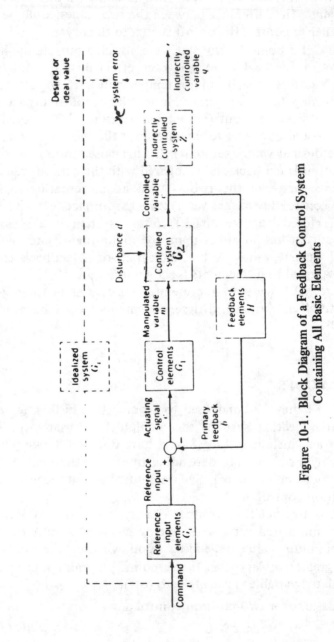

Figure 10-1. Block Diagram of a Feedback Control System
Containing All Basic Elements

Table 10-1. Basic Elements of a Residential Heating System

Element	Example in a residential heating system
Command v	Desire to set the indoor temperature
Reference input element G_v	Thermostat knob
Reference input r	Position of thermostatic switch
Primary feedback b	Position of bimetallic strip
Actuating signal e	On/off status of switch
Control elements G_1	Relay-controlled gas valve
Manipulated variable m	Gas flow rate (on or off)
Controlled system G_2	Furnace burner, ducts, house, etc.
Controlled variable c	Indoor air dry-bulb temperature
Disturbance d	Heat gains, losses, infiltration, etc.
Feedback element H	Bimetallic strip in the thermostat
Indirectly controlled system z	Occupant
Indirectly controlled variable q	Thermal comfort level
Idealized system G_i	Imaginary system which would result in an indoor environment where level of occupant's satisfaction is 100 percent.
System error	Thermal discomfort level

Where

M_1 = maximum value of manipulated variable.
M_0 = minimum value of manipulated variable.

A differential or dead-band in two-position control causes the manipulated variable to maintain its previous value until the controlled variable has moved beyond the setpoint by a predetermined amount. In actual operation, this action may be compared to hysteresis as shown in Figure 10-3.

A differential may be intentional, as is common in domestic thermostats when employed for the purpose of preventing rapid operation of switches and solenoid valves and to enhance the life of the system.

Two-position control is simple and inexpensive, but it suffers from inherent drifts. Rapid changes of the controlled variable are

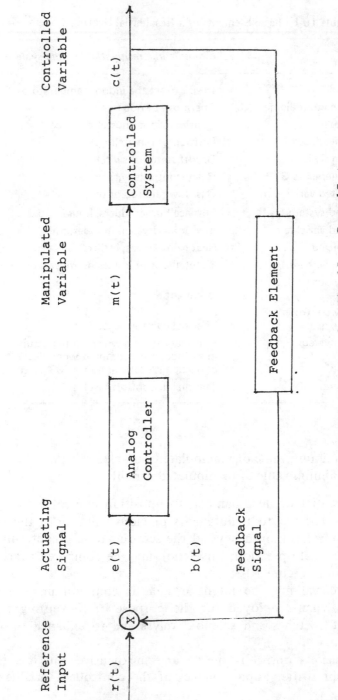

Figure 10-2. Generalized Block Diagram for a Closed-Loop Control System

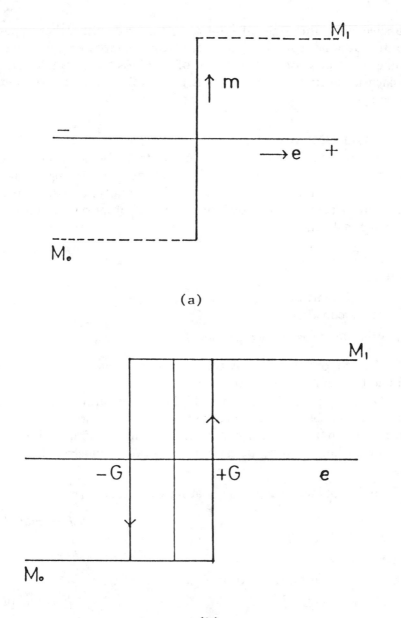

(a)

(b)

**Figure 10-3. (a) Two-Position Control Action;
(b) Two-Position Control Action with Differential Gap**

possible with this type of control. Compared with other types of control actions, two-position control can be more energy-intensive. Some examples of applications of two-position control include residential heating/cooling systems and rooftop units in commercial buildings.

Proportional Control

Proportional control is a type of control in which there is a continuous linear relation between values of the actuating signal and the manipulated variable. For purposes of flexibility, an adjustment of the control action is provided and is termed proportional sensitivity. Proportional control may be described as

$$m = K_c e + M \qquad\qquad\qquad \textit{Formula (10-2)}$$

Where

K_c = proportional sensitivity
M = a constant

and other terms as defined previously.

The proportional sensitivity, K_c, is the change of output variable caused by a unit change of input variable.

The constant M in Formula (10-2) may be termed as the calibration constant because the selection of a value for M determines the normal (zero actuating signal) value of the manipulated variable. The operation of proportional control action is illustrated in Figure 10-4.

For a unit step change in actuating signal

$$e = 0 \qquad t < 0 \qquad\qquad\qquad \textit{Formula (10-3)}$$
$$e = E \qquad t \geqslant 0$$

where E is a constant; substituting in (10-2)

$$m - M = K_c E \qquad\qquad\qquad \textit{Formula (10-4)}$$

The change in manipulated variable corresponds exactly to the change in deviation with a degree of amplification depending upon the setting of proportional sensitivity K_c. Thus, a proportional controller is simply an amplifier with adjustable gain.

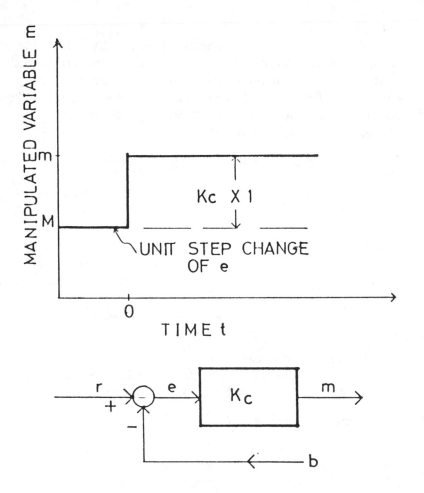

Figure 10-4. Proportional Control Action

Proportional control is more sensitive to the error signals than is two-position control. It is the least expensive of continuous-type controls but is more expensive compared to the two-position control. Calibration procedures are more difficult for proportional control than for the two-position control. Proportional control is used in air distribution systems. hydronic systems and in central systems for commercial buildings.

Integral Control

Integral action is a type of control action in which the value of the manipulated variable m is changed at a rate proportional to the actuating signal. Thus, if the actuating signal is doubled over a previous value, the final control element is moved twice as fast. When the controlled variable is at the set point (zero actuating signal), the final control element remains stationary.

Mathematically, integral control may be expressed as

$$\dot{m} = \frac{1}{t_{int}} e \qquad\qquad\qquad Formula\ (10\text{-}5)$$

or, in integrated form

$$m = \frac{1}{t_{int}} \int e\,dt + M \qquad\qquad Formula\ (10\text{-}6)$$

Where

    M = constant of integration

    t_{int} = integral time (defined as the time of change of manipulated variable caused by a unit step change of e)

The operational form of the equation is

$$m(s) = \frac{1}{t_{int}\,s} e(s) \qquad\qquad Formula\ (10\text{-}7)$$

and is shown in Figure 10-5.

For a step change of actuating signal

$$e = 0 \qquad t < 0$$
$$e = E \qquad t \geqslant 0 \qquad\qquad Formula\ (10\text{-}8)$$

where E is a constant. Substituting in Formula (10-6) and integrating

$$m - M = \frac{1}{t_{int}} Et \qquad\qquad Formula\ (10\text{-}9)$$

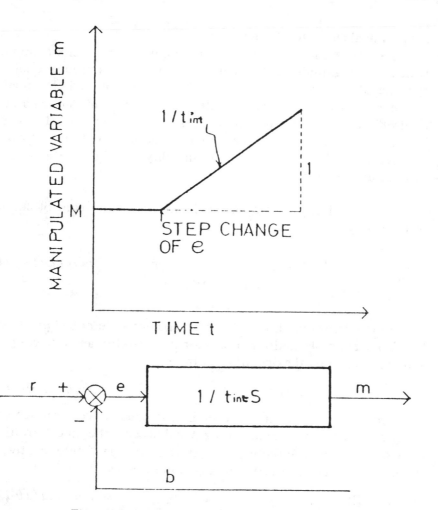

Figure 10-5. Operation of Integral Control Action

Thus, the manipulated variable changes linearly with time and "integrates" the area under the actuating signal function. For a unit step change, if actuating signal (E = 1.0), the slope of the line is inverse of integral time (Figure 10-5).

Integral control has the advantage over proportional control in that it tends to zero the offset, but it requires more expensive calibration procedures. Its maintenance is more difficult. Some applications of integral control include control of boilers, solar storage systems and meat-processing plants.

Proportional-Derivative Control

Derivative control action may be defined as a control action in which the magnitude of the manipulated variable is proportional to the rate of change of actuating signal. This control mode has many synonyms, such as "pre-set," "rate," "booster" and "anticipatory control" action. Derivative control response is always used in conjunction with the proportional mode. It is not satisfactory to use this response alone because of its inability to recognize a steady-state actuating signal.

Mathematically, a proportional-derivative (PD) control action is defined by

$$m = K_c e \mid + K_c t_d \dot{e} \mid + \qquad M \qquad \text{Formula (10-10)}$$
$$\text{[proportional]} \quad \text{[derivative]}$$

where t_d = derivative time and other variables as described previously. PD is the simple addition of proportional control and rate control action as shown by the operation Formula

$$m(s) = K_c (1 + t_d s) e(s) \qquad \text{Formula (10-11)}$$

Proportional-derivative action is not adequately described by applying a step change of actuating signal because the time derivative of a step change is infinite at the time of change. Consequently, a linear (ramp) change of actuating signal must be used:

$$e = Et \qquad \text{Formula (10-12)}$$

Where

E = a constant

t = time

Substituting Formula (10-12) and its first time derivative into Formula (10-10)

$$m - M = K_c E (t + t_d) \qquad \text{Formula (10-13)}$$

The actuating signal is defined at time t, whereas the manipulated variable is defined at $(t + t_d)$. The net effect is to shift the

manipulated variable ahead by time t_d, the derivative time. As shown
in Figure 10-6, the controller response leads the time change of
actuating signal. Derivative time is defined as the amount of lead,
expressed in units of time, that the control action is given. In other
words, derivative time is the time interval by which the rate action
advances the effect of proportional control action.

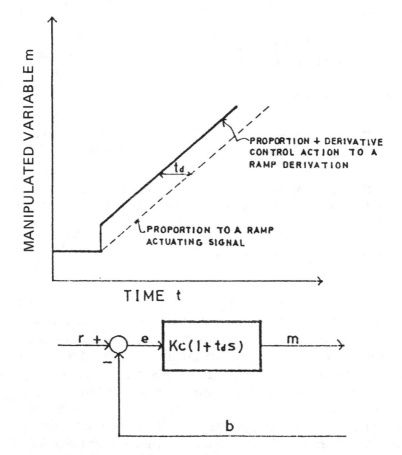

Figure 10-6. Proportional-Derivative Control Action

Proportional-derivative control has the advantage of a rapid
response to the magnitude and to the rate of change in loads. However, PD control can become unstable easily because it has no zeroing
capability. This control is useful for controlling environments in
buildings with large variations in occupancy.

Proportional-Integral Control

Integral control action is often combined additively with proportional control action. The combination is termed proportional-integral or proportional plus reset control and is used to obtain advantages of both control actions.

Proportional-integral control action is defined by the following differential formula:

$$\frac{dm}{dt} = \frac{K_c}{t_{int}} \ e \ |_{[integral]} + K_c \ \frac{de}{dt} \ |_{[proportional]} \qquad \textit{Formula (10-14)}$$

or, in integrated form

$$m = [\frac{K_c}{t_{int}} \ \int e \ dt]_{[integral]} + [K_c \ e + M]_{[proportional]} \qquad \textit{Formula (10-15)}$$

where terms are as previously described. These formulas illustrate the simple addition of proportional and integral control actions. In operational form:

$$m(s) = K_c \ (\frac{1}{t_{int}s} + 1) \ e(s) \qquad \textit{Formula (10-16)}$$

where the system function $K_c/(t_{int}s)$ identifies the integral action and the system function K_c identifies the proportional action.

Proportional-integral (PI) control action has two adjustment parameters, the proportional sensitivity K_c and integral time t_{int}. The propotional sensitivity is defined the same way as for the proportional control action. With the integral response turned off ($t_{int} \rightarrow \infty$), the proportional sensitivity is the number of units change in manipulated variable in per-unit change of actuating signal e. As clear from Formula (10-16), the proportional sensitivity K_c affects both the proportional and integral parts of the action.

The integral action adjustment is achieved through integral time. For a step change of actuating signal e, the integral time, t_{int}, is the time required to add an increment of response equal to original step change of response as shown in Figure 10-7. Another term used with this type of control is reset rate, defined as the number of times per minute that the proportional part of response is replicated. Reset rate is therefore called "repeats per minute" and is the inverse of integral time.

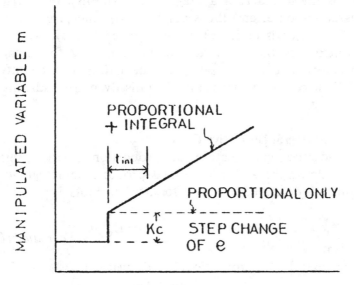

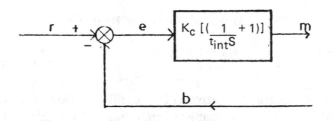

Figure 10-7. Operation of Proportional-Integral Control

For a step change of deviation

e = 0   t < 0

e = E   t > 0   *Formula (10-17)*

Substituting in Formula (10-14),

$$m - M = K_c E \left(\frac{t}{t_{int}} + 1 \right) \qquad \text{\textit{Formula (10-18)}}$$

This is the formula for a straight line. The first term, t/t_{int}, is the integral response, and the second term is the proportional response. The latter is indicated by the dotted line of Figure 10-7.

PI control has the advantage of compensating for changes in input in addition to compensating the deviations in the controlled variable. High cost of maintenance is a disadvantage with this type of control.

Proportional-Integral-Derivative Control

The additive combination of proportional action, integral action and derivative action is termed proportional-integral-derivative action. It is described by the differential formula

$$\dot{m} = \frac{K_c \, e}{t_{int}} \qquad + K_c \, \dot{e} \qquad + K_c t_d \dot{e} \qquad \text{\textit{Formula (10-19)}}$$

$$\text{[integral]} \qquad \text{[proportional]} \quad \text{[derivative]}$$

or

$$m = \frac{K_c}{t_{int}} \int e \, dt \; + \; K_c \, e + M \qquad + K_c t_d \dot{e} \qquad \text{\textit{Formula (10-20)}}$$

$$\text{[integral]} \qquad \text{[proportional]} \quad \text{[derivative]}$$

$$m(s) = K_c \left(\frac{1}{s t_{int}} + 1 + t_d s \right) e(s) \qquad \text{\textit{Formula (10-21)}}$$

Proportional-integral-derivative (PID) control action is illustrated in Figure 10-8 in which the change of manipulated variable is shown for a ramp function of the actuating signal.

$$e = Et \qquad \text{\textit{Formula (10-22)}}$$

Substituting this ramp function and its time derivative into Formula (10-20):

$$m - M = K_c E \left(\frac{1}{t_{int}} \int t \, dt + t + t_d \right) \qquad \text{\textit{Formula (10-23)}}$$

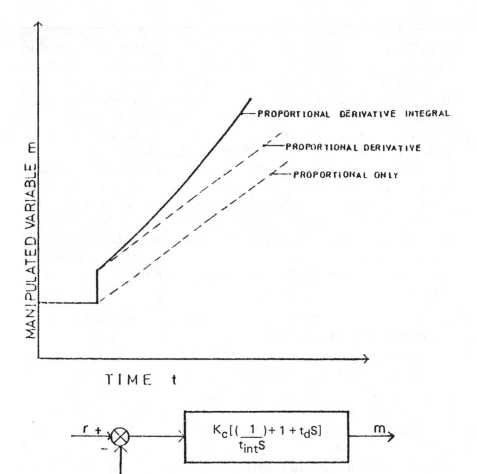

Figure 10-8. Operation of Proportional-Integral-Derivative Control

Integrating the first term

$$m - M = K_c E \left(\frac{t^2}{2t_{int}} + t + t_d \right)$$

                                                                                        Formula (10-24)

The proportional part of the control action repeats the change of actuating signal (lower straight line) in Figure 10-8. The derivative part of the control action adds an increment of manipulated variable proportional to the area under the deviation line and, as Figure 10-8 shows, the increment increases because the area increases at an increasing rate. The combination of proportional, integral and derivative actions may be made in any sequence, because these actions are described by linear differential equations.

PID control action has many advantages. It can compensate for magnitude and rate of change in input. It zeroes the deviation in controlled variable and is the most energy-efficient. However, it is the most expensive of all the controls and is difficult to calibrate. It is very hard to keep PID control stable; it requires frequent calibration checks. PID control is applied in environmental control chambers for scientific research and in variable air volume HVAC systems.

TYPES OF CONTROLLERS

Controllers for an energy (conversion and/or utilization) system may be one of the following types:

1. Electric or electronic controllers
2. Pneumatic controllers
3. Computer-based controllers

Electric or Electronic Controllers

Electric switches operated manually, or with relays and timers, are an important class of controllers for implementing several control strategies like start-stop of equipment, duty cycling, night setback, control of lighting, etc. Single-pole, double-throw (SPDT) switching circuits are used to control three-wire uni-directional motors. SPDT circuits are also used for heating-cooling applications.

Manually controlled electric switches used to "turn it off when it's not needed" offer great opportunities for energy savings. For example, a manufacturing company happened to hire an energy engineer whose first assignment was to identify and implement two energy-saving strategies within the first month of his starting the job. The condition was that the first cost of these projects should be zero and the savings should at least be equal to his salary. Manually operated electric switches were the answer for the engineer. He identified that the company had a policy of leaving all its office lights on for the janitor crew at night. He implemented a policy that all the employees were required to turn off their office lights manually, while leaving at 5:00 p.m. Further, the janitors would turn on the lights in an office only when they are cleaning that office. It turned out that 10,000 fluorescent tubes (40 watts each) could be turned off for 4 hours/day—5 days/week—52 weeks/year. The company was paying $0.07/kWh and $8.00/kW. Thus, savings from this strategy alone were

$$\frac{10,000 \times 40 \times 4}{1000} \text{ (kWh/day)} \times 5 \text{ days/week} \times$$
$$52 \text{ weeks/year} \times \$0.07/\text{kWh} = \$29,120/\text{year}.$$

Secondly, he identified that an exhaust fan in the grinding shop was kept running all the time, even though the shop operated for 16 hours/day and 5 days/week. He implemented a policy that the workers at the end of the second shift would *manually turn off* this fan and the workers at the start of the first shift would manually start this fan. The specifications of the fan and heating equipment were as follows:

Horsepower of the fan = 10 hp
Air exhausted = 20,000 ft^3/minute
Cost of electricity = $0.07/kWh
Efficiency of the gas heater = 0.70
Cost of gas = $0.60/therm
Heating season = 24 weeks/year
Setpoint = 70°F
Avg. outdoor temperature = 35°F

Calculations:
a. *Gas savings*

$$\frac{(20,000 \text{ ft}^3)}{\text{minute}} \times \frac{(60 \text{ minutes})}{\text{hour}} \times \left[\frac{(5 \text{ days})}{\text{week}} \frac{(8 \text{ hours})}{\text{day}} + \frac{(2 \text{ days})}{\text{week}} \frac{(24 \text{ hours})}{\text{day}} \right] \times$$

$$\frac{(24 \text{ hours})]}{\text{day}} \times \frac{24 \text{ weeks}}{\text{year}} \times \frac{(0.077 \text{ lbm})}{\text{ft}^3} \frac{(0.248 \text{ Btu})}{\text{lbm}°\text{F}} \times$$

$$(70°\text{F} - 35°\text{F}) \times \frac{(1 \text{ Therm})}{10^5 \text{ Btu}} \left(\frac{1}{0.70}\right) \left(\frac{\$0.60}{\text{Therm}}\right) = \$14,518/\text{year}.$$

b. *Electricity savings*

$$10 \text{ hp} \times \frac{0.746 \text{ kW}}{\text{hp}} \left[\frac{5 \text{ days}}{\text{week}} \times \frac{8 \text{ hours}}{\text{day}} + \frac{2 \text{ days}}{\text{week}} \times \frac{24 \text{ hours}}{\text{day}}\right] \times$$

$$\frac{(52 \text{ weeks})}{\text{year}} \times \frac{\$0.07}{\text{kWh}} = \$2,389/\text{year}.$$

Total savings from manually operated switches were \$46,027/ year, which covered the starting salary of the energy engineer, which was \$40,000/year.

The example illustrates the usefulness of manually controlled switches.

TIMERS

Either single-pole, single-throw (SPST) or SPDT circuits can be interfaced to timers. These timers can range from very simple clocks to sophisticated central-time clocks with multiple-channel capability for controlling equipment on different time schedules.

Timers can be used to implement control strategies like night setback, duty cycling and automatic start-stop of equipment.

Night Setback

Energy expended to heat unoccupied buildings up to comfort conditions is wasted, and most buildings are indeed unoccupied at night-time. Energy is saved by setting back the temperature levels at these times.

Night setback offers opportunities for significant energy savings with little or no capital expenditure in most of the residential, institutional, commercial and industrial buildings. However, one should be careful that the night setback does not:

(i) Effect process conditions, e.g., tolerances on the manufactured goods

(ii) Cause the air conditioners to turn on to meet the setback temperatures

Duty Cycling

In duty cycling, loads are cycled on and off by a timing device. Duty cycling has been used to duty cycle the operation of exhaust fans and space heating/cooling equipment with some possible savings. Savings are only a possibility because some experts believe that duty cycling is not an economical way of achieving savings in the long run. The idea that duty cycling can cause detrimental damage to equipment must be considered. If, for example, a company can save $10,000 per year on its electricity bill but has to spend about the same amount in maintenance and repair, then the use of duty cycling is not feasible.

In some instances, the breakdown of equipment is not acceptable at all. This is the case in a hospital. If an unforeseen problem occurs in the operating room, no financial savings could pay for the problems that might result.

Duty cycling has been with us for years. Ever since the energy crisis, duty cycling has become a broader area. Since air-conditioning systems are usually based on the most extreme operating conditions, the basic theory behind duty cycling is that the mechanical equipment is oversized during a large part of the operating life.

The standard thermostat has been used to cycle air conditioners for decades. Thermostats essentially cycle the air-conditioning equipment on and off to maintain comfortable environmental conditions. Consequently, mechanical equipment controlled by thermostats is designed to tolerate cycling for normal operation. Problems may be caused by the addition of the cycle-timing devices in addition to the thermostat controls. Therefore, the problem of accelerated deterioration of mechanical equipment has been attributed to duty cyclers.

The idea that the age of the equipment may be altered by duty cycling has to be considered first. There have been ideas on both sides. Some feel that the life will be extended and some feel otherwise.

In one case, it was reported that duty cycling control towers caused the breaking up of gear boxes. When discussed at a forum on duty cycling, it was found that, when the fan changed speed from high to low, the gears acted as a brake for slowing the fan. The added stress on the gears was not taken into consideration when the towers were designed. This problem was easily fixed by a) shutting off the fan for approximately 45 seconds so the fan speed decreased enough to avoid any more damage to the gears or b) a more sophisticated way to use decelerating relays.

A serviceman mentioned trouble he had had with air-handling equipment in a television studio. The cycling pattern was 27 minutes off and 21 minutes on. Within one year, 25% of the fan drives had failed. The problem was either a motor-drive end bearing had failed or the fan-drive end bearing had failed. The solution to this problem was simple. There was over-tension in the belts. This was initially done to keep the air-conditioning equipment from squeaking. After the belts were loosened to the proper specification, there were no problems reported within the next year.

The same serviceman also duty cycled air-handling equipment at a hospital. The off time was never less than 5 minutes. There were no problems reported there.

Another concern has been the decrease in compressor or starter motor life. It is estimated that the starter motor life would be reduced by one half for every 10-degree Celsius increase in winding temperature. For example, a motor with a 15-year normal life that is cycled once a day with an increase in winding temperature of 20°C will last only about 4 years. In other words, if the same motor with a 30°C increase in temperature were cycled four to five times a day, it would last only 2 years.

In addition, the bearings in the compressor and starter motor may get damaged by duty cycling. Moreover, the contacts and control equipment could also have a shorter life. This decrease in life would be dependent upon the number of cycles.

The problem with the contacts' wearing out prematurely occurs because the original equipment is not made for the excessive use that occurs during duty cycling. All that has to be done is to put in heavy-duty contacts.

When it comes to the problem with bearings, an HVAC dealer in Peoria stated that, if the roller bearings are greased properly, there will be no problem. However, he did not recommend duty cycling sleeve bearings. Heavy-duty or over-specified sleeve bearings must be used. For collar-mounted bearings (either screw-locking or eccentric), the collars may work loose if duty cycled; so these must be watched very carefully.

As for the life of the compressor and compressor motor, this is a more complex problem. Some motors cannot be duty cycled. This will depend upon several parameters. Problems could occur because the motor was not properly maintained during its years of operation. Such a motor should not be cycled. Even if the older motor was properly maintained, it should not be aggressively cycled (cycled too often). Improper cycling can inhibit oil return to the compressor and cause it to cease. Most older motors were not manufactured for duty cycling; that is why special care should be taken.

Currently, larger motors are not used as often. A 100-hp motor will consist of four 25-hp motors. The amperage of the in-rush at start is much less for these smaller motors, and the chance of damage actually occurring is significantly less. Duty cycling of these smaller motors is much more efficient also. They are either on or off and run at optimum efficiency at all times.

Generally, heat is the worst enemy of motors. If a motor is duty cycled, it runs cooler. Even though the highest temperature occurs during start-up and there may be very many start-ups when duty cycling is used, the manufacturer takes this into consideration. This is done when the minimum on and minimum off times are specified. The average temperatures at which the starter motors and compressors run are determined. The average duration without duty cycling will not be exceeded, when duty cycling, if the minimum on and off times are followed. Thus, if a compressor is cycled off 10% of the time, the number of run hours will not be decreased. In this case, the number of days or years of the compressor life will be increased by 10%.

Publicity concerning the problems with duty cyclers led mechanical equipment manufacturers to issue stipulations in the warranties of air-conditioning equipment in the past. One such stipulation was that, if the air conditioner was duty cycled at all, the warranty was voided. This was true on some units until approximately 3 years ago.

Now, the warranties are straightforward. For example, the warranty for a compressor is for 5 years. This is straight across the board (no stipulations against duty cycling). For a top-of-the-line compressor, the warranty could be for 15 years because of the additional safety equipment that is included. The only guidelines specify that the minimum on and off times stated by the manufacturer must be followed to protect the equipment against short cycling.

While some people consider duty cycling is bad for the mechanical equipment, many manufacturers have introduced duty-cycle timers to their lines and will give the buyer information on cycling on and off times. Why so? Can duty cycling actually reduce energy usage without the added expense of premature equipment failure? The answer is yes. Duty cycling can reduce energy costs and, if the cycle rates are not excessive, duty cycling will not cause the premature failure of well-designed mechanical equipment.

Determining On Time and Off Time

The National Electrical Manufacturers Association (NEMA) has published a standard (NEMA Standard MG 10) in which the problems of duty cycling of electrical motors are examined. The standard was prepared as a guide for selecting the optimal on and off times of a given electrical motor, taking into account motor type, horsepower rating, speed, starting frequency, restrictions on in-rush current, power demand charges and the extra winding stress imposed by repeated accelerations. A portion of the table outlining NEMA recommendations is shown in Table 10-2.

**Table 10-2.* Allowable Number of Starts and Minimum Time
Between Starts for Design A and Design B Motors**

	2-Pole			4-Pole			6-Pole		
HP	A	B	C	A	B	C	A	B	C
2.0	11.5	2.4	77	23	11	39	26.1	30	35
5.0	8.1	5.7	83	16.3	27	42	18.4	71	37
7.5	7.0	8.3	88	13.9	39	44	15.8	104	39
15.0	5.4	16	100	10.7	75	50	12.1	200	44
20.0	4.8	21	110	9.6	99	55	10.9	262	48
25.0	4.4	26	115	8.8	122	58	10.0	324	51
30.0	4.1	31	120	8.2	144	60	9.3	382	53
40.0	3.7	40	130	7.4	189	65	8.4	503	57
50.0	3.4	49	145	6.8	232	72	7.7	620	64

*Adapted from NEMA Standard MG10.

The parameters above represent the following:

hp = NEMA horsepower rating of the motor
A = Maximum number of starts per hour
B = Maximum product of A times load inertia (lb x ft^2)
C = Minimum reset time in seconds (convert to minutes)

The table lists the minimum off time and maximum number of starts permitted per hour. What is actually needed is the information for determining the duty period (on time plus off time). Using the numbers in the table, Item A (maximum starts per hour), the shortest duty period permissible for a 2-pole, 25-hp motor, taking into consideration that the motor and load are sized correctly, can be calculated as follows:

$$\frac{60 \text{ min/hr}}{4.4 \text{ starts/hr}} = 60 \text{ min}/4.4 \text{ starts} = 13.53 \text{ min/start}$$

Therefore, the duty period = 14 minutes.
The minimum off time is 115 seconds (approx. 2 minutes).

On Time + Off Time = Duty Period
On Time + 2 Min. = 14 Minutes
On Time = 12 Minutes.

To verify the above data, exchange the 14-minute duty period for the 4.4 starts per hour in the above equation. The number of starts per hour is 4.28.

Since 4.28 is less than 4.4, which was specified as the maximum number of required starts per hour, the duty period of 12 minutes on and 2 minutes off will meet the requirements of the NEMA Standard MG 10. Increasing either the off time and/or the on time from the minimums listed above will decrease the number of starts per hour and save wear and tear on the mechanical equipment.

The NEMA standards recommend the minimum on and off time for cycling motor loads. It is also imperative to contact the compressor manufacturer for their recommendation as a minimum off time to allow the compressor to pump down and not start against a high-pressure head. Remember that duty cycling must be done properly in order to get the maximum savings.

Energy-Saving Cycle Strategy

Now that the minimum on and off times have been calculated
to protect the equipment from short cycling and premature failures,
an energy-saving duty cycling strategy must be determined. The
measure of the effectiveness of a duty cycle program is to determine
that the loads are cycled in a manner to save electrical demand (to
be discussed in programmable controllers) as well as consumption
(kWh). Straight-time duty cycling is accomplished by starting the
duty cycler at one point in time (for example, at 8:00 a.m.). Monday
through Friday for a cycle of 15 minutes on time and 15 minutes off
time, then stopping the duty cycle (say at 5:00 p.m.).

Whenever more than one load is to be duty cycled, there are
two methods to provide the cycling program. The first method is
to start all the loads at the same time and have them cycle on and
off at the same time. This method is called parallel duty cycling.
The second method to provide duty cycling to more than one
load is to stagger the on and off times to avoid the in-rush current
while evenly spreading the distribution time to each load. The decision now is which method to choose and what times to use.

The demand period is the time interval that the utility company
checks the power consumption of a user to determine the maximum
number of kW used during this interval. In most cases this period
is 15 minutes.

The duty period is equal to the on time plus the off time. So,
if the duty period is less than the demand period interval, the parallel
duty cycling can be used. During the demand interval, the utility can
detect if all of the loads are running at the same time. If the duty cycle
period exceeds the demand time, then the staggered duty cycling
is necessary if savings in kW demand are to be achieved.

Mechanical equipment manufacturers have specified that
minimum duty cycle periods must be, in most cases, greater than the
demand periods to ensure their equipment's life span. So, staggering
load starts are usually a must to save demand charges. The mechanical equipment manufacturers are primarily concerned about short-
cycle protection for the air-conditioner compressors. This allows
the compressor to pump down before trying to start against a high-
pressure head and thus burn out the motor.

Stagger Cycling and Savings

To show an example of the stagger method of duty cycling, assume that the manufacturer was contacted and he recommends a duty cycle of 20 minutes on and 10 minutes off as the minimum duty cycle. Anything greater in the on or off time would be acceptable. The duty period is 30 minutes. Say three compressors, each rated at 10 kW (approx. 3 tons), have to be cycled and the demand interval is 15 minutes.

$$\underline{\text{ON PERIOD} + \text{OFF PERIOD} = (20 + 10)\text{ min.}}$$

number of loads = 3 loads

The stagger rate equals 10 minutes per load.

Figure 10-9 shows the relationship between the stagger rate of 10 minutes per load, demand interval and the kW savings possible using the stagger function. Figure 10-10 has been generated from Figure 10-9. The demand for the three loads would be the highest demand recorded over any 15-minute interval. Since the highest demand recorded was 19.99 kW and the maximum potential demand would have been 30 kW, the savings are:

$$\frac{30\text{ kW} - 19.90\text{ kW}}{30\text{ kW}} = 0.2933 = 2933\%$$

If duty cycling were done using the parallel method, there would be no savings of demand because all three loads would have been on for 20 minutes. This peak would show up during the first demand period and the second demand period.

Regardless of the cycling method, the actual energy consumption would be the same because the on time would, of course, be the same.

Percent consumption savings:
OFF TIME X 100/DUTY PERIOD
10 MIN OFF X 100/30 MIN = 33.33%

Duty cycling can save electrical expenses when tied to air-conditioning equipment, provided that the proper installation is done and proper maintenance followed. (For example, the mechanical equipment is not short-sycled—the correct on and off times are calculated for the equipment). Also, maintenance routines including proper lubrication, tightened belts, and clean filters should be regularly performed.

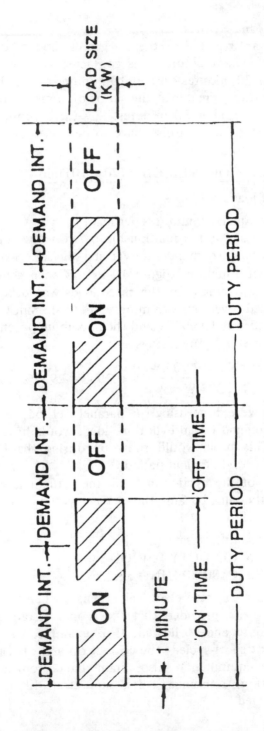

Figure 10-9. Stagger Rate

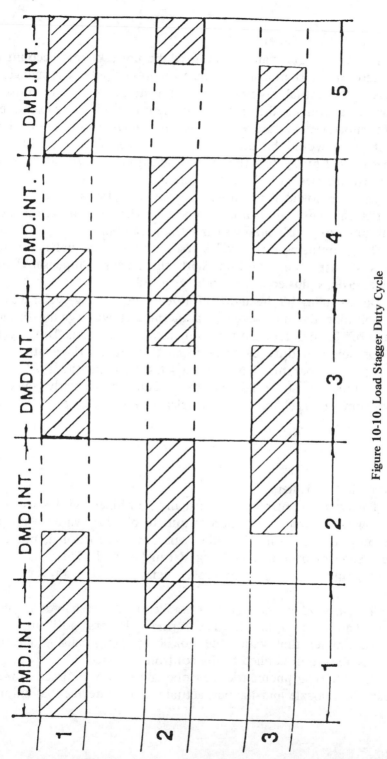

Figure 10-10. Load Stagger Duty Cycle

Thermostats Considered

The duty cycle example discussed did not take into account the cycle rate of the thermostat in the estimates of potential savings. To calculate the actual savings, the thermostatic cycle rate must be known for a given condition of cooling requirements. For example, if the thermostat cycle rate were 30%, the savings in both demand and consumption would be divided by .30 to calculate the additional savings caused by cycling the load in addition to having the thermostat wired in series with the duty cycler.

The thermostat is a temperature-compensating duty cycler. It is essential in many applications where environmental temperature conditions along with energy savings are paramount.

Designers using duty cycles traditionally must balance energy savings against space comfort. With the newer compensated duty cycling, savings plus comfort can be attained.

Duty cycling has been done for years, but a duty cycler was not accomplishing the duty cycling—a thermostat was cycling the load on and off to maintain the set temperature conditions. Duty cyclers and compensated duty cycling programs based on time rather than set specific temperature have replaced the standard thermostat. The time element must be taken into account, particularly since the utility company is charging for electricity based on time.

Pneumatic Controllers

Pneumatic controllers are usually combined with sensing elements with a force or position output to obtain a variable air pressure output. The control mode is usually proportional, but other modes such as proportional-integral can be used. These controllers are generally classified as non-relay, relay, direct- or reverse-acting type.

The *non-relay* pneumatic controller uses a restrictor in the air supply and a bleed nozzle. The sensing element positions an air exhaust flapper that varies the nozzle opening, causing a variable air-pressure output applied to the controlled device.

A *relay-type* pneumatic controller, directly or indirectly through a restrictor, nozzle and flapper, actuates a relay device that amplifies the air volume available for control.

Direct-acting controllers increase the output signal as the controlled variable increases. For example, a direct-acting pneumatic thermostat increases output pressure when the sensing element detects a temperature rise.

Reverse-acting controllers increase the output signal as the controlled variable decreases. A reverse-acting pneumatic thermostat increases output pressure when the temperature drops.

As pointed out earlier, pneumatic controllers can be operated in either proportional or proportional-integral mode. System characteristics in this control mode having influence on energy efficiency are proportional band, reset time, over-capacity of the heating/cooling equipment and part-load operation. Proportional band is the ratio of the percent change in the actuating signal to the percent change in the manipulated variable. Reset time is defined as the time required by the reset to repeat the proportional action.

Programmable Controllers

A programmable controller is a control device that has logic potential but is not powerful enough to be called a computer-interfaced controller. This type of control fills a need for applications requiring more than a timer but not a computer. A programmable controller can do all that timers can do and much more but at a cost considerably less than that of a computer.

Because of the logic capability, programmable controllers are very useful in implementing control strategies like

a) Demand control
b) Economizer control
c) Enthalpy control
d) Combustion air control
e) Temperature reset

Demand Control

Demand charge is that portion of the electric bill which reflects the utility's capital requirements in generating plant, transmission lines, transformers, etc., to provide power to a facility. The demand load is the maximum concurrent load to occur during any 15- or 30-minute interval during the billing month, and it is measured in kilowatts. It is in the interest of the customer and the utility to keep the peak demand as low as possible. While this does not directly save energy, it definitely saves money. The following methods can be used to reduce peak demand:

(i)   Off-peak schedule
(ii) Thermal storage (see Chapter 12)
(iii) Demand-limiting controls
(iv) Programmable controllers

In demand control by a programmable controller, when the controller senses that the electrical demand is approaching a critical (preset) level, it shuts off equipment on a programmed priority basis to restrict the demand from passing the critical level. Since demand charge generally amounts to approximately 25%-35% of the total electric bill, savings from controlling demand can be significant.

Economizer Cycle

An economizer cycle is the adaptation to the fresh-air intake which permits the use of outside air for cooling when temperatures are sufficiently low. When nighttime outdoor temperatures are below indoor setpoint by 5°F or more, the controller shuts off the refrigeration system and the return air dampers and opens the outdoor air dampers fully. Using outdoor air for night cooling will save energy in most areas.

Enthalpy Control

During occupied periods, the opportunities to use outdoor air for cooling will depend on the outdoor wet-bulb (WB) temperature as well as dry-bulb (DB) temperature. If the outdoor air is brought into the building above 60°F WB, the cooling load is increased. The wet-bulb temperature is a measure of the total energy content (enthalpy) of the air.

In enthalpy control, the controller is interfaced to DB sensors and to WB sensors for both the outdoor as well as the indoor air.

The controller is programmed to calculate the enthalpies; if the enthalpy of the outdoor air is less than the enthalpy of the indoor air, 100 percent of the outdoor air is used for cooling, resulting in energy savings. (See Chapter 7 for Enthalpy Control Savings Calculations.)

Combustion Air Control

The programmable controllers can be used to control the excess air in a combustion process. By sensing CO_2 or O_2 or CO levels in exhaust, the controller can adjust the combustion air intake for optimal combustion efficiency. Continuous control through the use of a programmable controller allows the air intake to be adjusted in accordance with the demand on the unit.

Temperature Reset

The logic capability of the programmable controllers can be used to reset the setpoints on the chillers and the boilers. The controller can be interfaced to DB, WB, wind velocity and solar radiation sensors. The arithmetic logic of the controller can be used to calculate the optimal setpoints and reset accomplished which can result in significant energy savings and can enhance the comfort levels in the buildings.

COMPUTER CONTROL

The control of physical systems with a digital computer is becoming more and more common. Aircraft autopilots, mass transit vehicles, oil refineries, paper-making machines and countless electromechanical servomechanisms are among the many existing examples. Furthermore, many new digital control applications are being simulated by microprocessor technology, including on-line computer controllers in automobiles and household appliances. Among the advantages of digital logic for control are the increased flexibility of the control programs and the decision-making or logic capability of digital systems.

One desirable characteristic of a control system is to have a satisfactory response. "Satisfactory response" means that the output, c(t), is to be forced to follow or track the reference input, r(t), (See Figure 10-2) despite the presence of disturbance inputs to the plant (system) and despite errors on the sensor. It is also essential that the tracking succeeds even if the dynamics of the plant should change somewhat during the operation. The process of holding c(t) to r(t) is called *regulation*. A system which has good regulation in the face of changes in the system dynamics is said to have low *sensitivity* to system's parameters. A system which has both good disturbance rejection and low sensitivity is called *robust*. Use of digital computers as controllers can help create robust systems.

Figure 10-11 shows the use of a digital computer as a control element. Note that the notation on this figure has been changed compared with Figure 10-2 to incorporate digital computers as controllers.

In Figure 10-11, consider first the action of the analog-to-digital (A/D) converter on a signal. This device acts on a physical variable, most commonly an electrical voltage, and converts it into a stream of numbers (pulses). In Figure 10-11, the A/D converter acts on the indicated error signal, ê, and applies the numbers to the digital computer. It is also common for the sensor output, y, to be

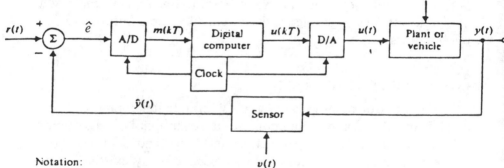

Notation:
r = reference or command inputs
u = control or actuator input signal
y = controlled or output signal
$\hat{y}$ = instrument or sensor output, usually an approximation to or estimate
    of y. (For any variable, say θ, the notation $\hat{\theta}$ is now commonly taken
    from statistics to mean an estimate of θ).
$\hat{e}$ = $r - \hat{y}$ = indicated error
e = $r - y$ = system error
w = disturbance input to the plant
v = disturbance or noise in the sensor
A/D = analog-to-digital converter
D/A = digital-to-analog converter

Figure 10-11. Digital Control System Block

sampled and have the error formed in the computer. We need to know the times at which these numbers arrive if we are to analyze the dynamics of the system.

We may assume that all the numbers (pulses) arrive with the same fixed period T, called the *sampling period.* In practice, digital control systems sometimes have varying sample periods and/or different periods in different feedback paths. Typically, there is a clock as part of the computer logic which supplies a pulse every T seconds (See Figure 10-12), and the A/D converter sends a pulse (number) to the computer each time the pulse from the clock arrives. Thus, in Figure 10-11 we identify the sequence of numbers as m(kT). We conclude from the periodic sampling action of the A/D converter that some of the signals in the digital control system, like m(kT), are variable only at discrete times. We call these variables *discrete signals* to distinguish them from variables like ê and y which change continuously in time. A system containing only discrete variables is called a *discrete-time* system. A system having both discrete and continuous signals is called a *sampled-data* system.

In addition to generating a discrete signal, however, the A/D converter also provides a *quantized* signal. By this we mean that the output of the A/D converter must be stored in digital logic composed of a finite number of digits. Most commonly, of course, the logic is based on binary digits (i.e., bits) composed of 0's and 1's, but the essential feature is that the representation has a finite number of digits. A common situation is that the conversion of ê to m is done so that m may be thought of as a number with a fixed number of places of accuracy. If we plot ê versus m, we may get a plot as shown in Figure 10-13.

In Figure 10-13, we would say that m has been truncated to one decimal place, or that m is *quantized* with q of 0.1 since m changes only in fixed quanta of, in this case, 0.1 units. A signal which is both discrete and quantized is called a *digital signal.*

In a real sense, the problems of analysis and design of *digital controls* are concerned with accounting for the effects of the sampling period T and the quantization size q. If both T and q are small, digital signals are nearly continuous, and continuous methods of analysis can be used.

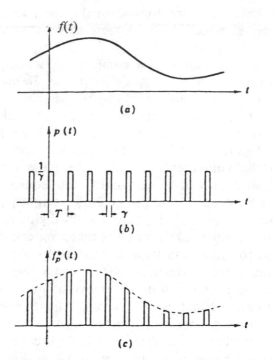

Figure 10-12. (*a*) Continuous Function $f(t)$:
(*b*) Sampling Pulse Train $p(t)$; (*c*) Sampled Function $f_p^*(t)$.

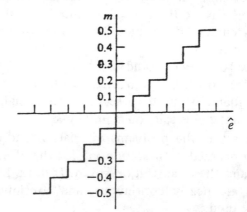

Figure 10-13. Plot of Output vs Input Characteristics of A/D Device

Design of Digital Systems

As mentioned previously, we can place the systems of interest in three categories according to the nature of the signals present. These are:

1. Discrete systems
2. Sampled-data systems
3. Digital systems

In *discrete systems,* all signals vary at discrete times only. For the analysis of discrete systems, one must learn z-transforms of discrete signals and "pulse"-transfer functions for linear constant discrete systems. Of special interest is the characterization of the dynamic response of discrete systems. As a special case of discrete systems, we consider discrete equivalents of continuous systems, which is one aspect of the current field of *digital filters* (Kalman filter).

If quantization effects are ignored, digital filters are discrete systems which are designed to process discrete signals in such a fashion that the digital device (a digital computer program, for example), can be used to replace a continuous filter.

A *sampled-data* system has both discrete and continuous signals. In these systems, we are concerned with the question of data extrapolation to convert discrete signals (as they emerge from a digital computer) into the continuous signals necessary for providing the input to one of the plants. This action typically occurs in conjunction with the D/A conversion.

Digital controllers for the building systems may be one of the following types:

a. Direct digital control (DDC) or distributed control or remote control or on-site control
b. Supervising control or central control
c. Hierarchy control (DDC plus supervisory control)

A central control system provides the energy engineer with a means of constant surveillance of the building and helps in making efficient and effective use of physical plant systems and personnel. Proprietary central computerized control systems are marketed by each of the major temperature control manufacturers. These systems have common features and can accomplish a similar range of tasks, but each manufacturer uses coding and computer languages which are unique to the system.

Each manufacturer's system is made up of standard hardware, but the application is always tailored to the specific project. Basically, any system is composed of the following four major components:

- Interface panels which are located at strategic points throughout the building.

- The transmission system between the central console and all interface panels. This system can be single-core cable for digital transmission or multi-core cable for multiple transmissions.

- A central control console and associated computer hardware located in a control room. The console, computer and associated hardware form the point at which the operator enters all instructions and retrieves all data.

- Software programs. Programs for common applications such as start-stop action, load shedding, reset, alarm, optimization, etc. are available from the major control system suppliers.

Experiences with central control systems reported thus far in literature have not been very positive. Some of the reasons cited for ineffective use of the central control systems are inadequate training facilities for the operators, inadequate buyer commitment, poor vendor assessment and ineffective interface with the building system. In spite of these difficulties, if the energy engineer is involved with larger, more complex buildings (or groups of buildings), possible use of central computer-based control systems should be fully explored.

References
1. D'Azzo, J.J.; and Houpis, C.H. *Feedback Control Systems: Analysis and Synthesis.* New York. McGraw Hill Book Co., 1983.
2. Mehta, D. Paul. Dynamic Thermal Responses of Buildings and Systems. Ph.D. Dissertation. Iowa State University, Ames, IA, 1979.
3. _____ *HVAC Systems and Applications Handbook.* ASHRAE, Atlanta, Ga., 1987.
4. Smith, C.B. *Energy Management Principles.* Pergamon Press, 1981.
5. Kennedy W.J.; and Turner, W.C. *Energy Management.* Prentice-Hall Inc.; Englewood Cliffs, N.J., 1984.
6. Dubin, F.S.; and Long, C.G. *Energy Conservation Standards.* New York. McGraw Hill Book Co., 1978.

11

Computer Applications

This chapter will focus on the following subjects:

INTRODUCTION

Digital computers became available in the 1940's. In the following four decades, the computational power of computers has increased more than a million times, while the cost, power requirements and programming difficulties have reduced dramatically. This has led to a rapid increase in the use of computers by engineers in general and energy engineers in particular during the last decade. Several areas can be identified where computers have been incorporated by the energy engineers in their practice:

a) Information and communication
b) Computation
c) Design
d) Instrumentation
e) Control

The details of the use of computers as control elements were discussed in Chapter 10. The objective of this chapter is to describe the application of a computer in other areas, especially as it impacts the engineering analysis and design. For analysis, computer applications basically involve the use of computers to do calculations. Computers find their greatest advantage in making large quantities of simple calculations, a task where energy engineers typically find them very helpful.

BASIC HARDWARE

The main components of a computer are shown in Figure 11-1. All computers, no matter how simple or complex they are, contain these components.

The heart of a computer is the CPU (Central Processing Unit), which is simply a group of electronic devices which can exist in one of the two formats or states. These states could be interpreted as "yes" or "no"; "1" or "0"; "on" or "off," etc. The computer is sensitive to the status of these states and makes calculations using binary arithmetic. The operation is simple but sophisticated, incorporating speed and accuracy with which these calculations can be made. The main functions of the various components are described as follows:

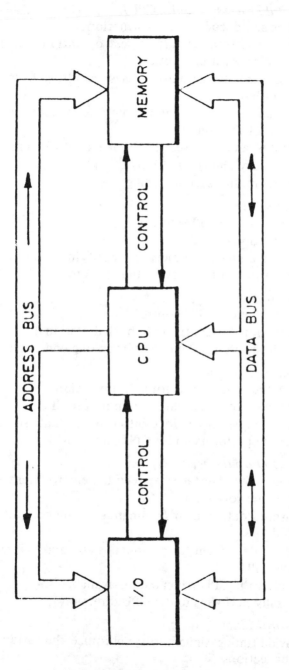

Figure 11-1. Computer Components

1. *The Central Processing Unit (CPU)*
    a. fetches and decodes each instruction
    b. performs arithmetic and logical operations on data as required by each instruction
    c. determines internal status (flags) resulting from operations with data
    d. can transfer data to and from the memory and I/O sections of the computer
    e. generates and receives handshake control signals from the memory and I/O sections
    f. provides master system timing

2. *The Address Bus*
    a. generally under CPU control
    b. uni-directional
    c. provides a unique numerical identifying code for selecting a particular memory or I/O element

3. *The Data Bus*
    a. generally under CPU control
    b. bi-directional (data travel in both directions)
    c. provides route for data transfer among computer parts

4. *The Memory Section*
    a. stores the instructions for CPU operation
    b. stores data to be operated upon by the CPU
    c. may be physically divided into program/data sections
    d. generally performs under CPU direction

5. *The Input/Output Section (I/O)*
    a. provides the interface between the environment and the rest of the computer
    b. transmits data and/or commands to the computer peripherals
    c. receives data from the peripherals and makes it available to the CPU
    d. can alert the CPU when data is available
    e. generally performs under CPU direction

6. *The Control Lines*
    a. provide timing signals to synchronize the various computer sections

    b. provide path for handshake signals

    c. provide a means of interrupting CPU operations

    d. individual lines generally uni-directional

BASIC SOFTWARE

As a result of the rapid increase in the use of computers by the energy engineers, a large number of computer programs have appeared These programs, in most cases, are available to users other than the developer. Generally, these programs are available in one of the following ways:

a. The program is not proprietary and the source code can be purchased for in-house implementation.

b. The program is considered proprietary and can be used on a time-sharing basis or by leased-object language codes.

c. The program is proprietary and can be used only by submit-timing input data to the program developer.

d. The program is developed to meet the specific needs of a specific customer. This could be accomplished either by using the outside services or by developing in-house facilities.

CLASSIFICATION OF COMPUTER PROGRAMS

The computer programs which are available to an energy engineer for computation and design could be divided into the following classes:

1. *Heating and/or Cooling Load Calculation Programs*

The programs within this class are meant to calculate heating and/or cooling loads for a building or zone. Load calculations are well suited to computer analysis due to the large number of relatively simple calculations involved, the amount of data manipulation needed and the fact that these calculations are routinely performed on nearly all building projects. It was, therefore, one of the earliest applications of the computer to building design. A large number of programs currently exist which vary in the methods used. In general, these programs can be classified as either design condition load calculations (usually a single design condition or design day, e.g.,

'Bless'*) or as hour-by-hour load calculations, which are usually more suited for energy analysis programs, as is the case with the program "NBSLD."

2. Energy Analysis Programs

Programs meant to estimate the annual energy consumption of buildings are contained in this class. Energy analysis programs can be classified according to the method by which they treat the annual weather data necessary to perform the estimation. The degree-day method proportions a design heating load over a heating season based on the number of degree days in the heating season. The bin method is based on the number of hours per year that the outside temperature falls within specified temperature ranges. Hour-by-hour calculations consider each hour of a typical year's weather data. They generally consist of a load subroutine which calculates the heating and cooling loads and a system simulation portion, as is the case with the program "AXCESS." Several programs, e.g., "RUND1620" and "ECUBE," also contain an economic analysis section which calculates the cost of the energy required and makes various economic calculations.

3. Duct Design Programs

Duct design programs are used to assist in the areas of duct sizing, cost estimation and sound analysis. Examples of such programs are DUCT, HVDUCT, DUCTAN, EXHAUST, DUCTS, DADDS and so on. Input to these programs normally consists of required CFM for each terminal unit and a nodal network which is developed based on a line diagram of the ductwork system. Some programs are designed to also calculate heat losses or gains from ductwork and to recalculate CFM's accordingly. The analysis of fan noise attenuation is part of several of these programs.

4. Piping Design Programs

The piping design programs can generally be classified as: (1) pipe-sizing programs, which calculate the size of pipe required based on specified GPM and a description of the piping system, e.g., "MOIP" or (2) pipe-flexibility programs, which are utilized to estimate the stress and deflections of piping systems, as is the case with the program "AUTOFLEX." The sizing programs will normally size

*See Table 11-1 for details on these programs.

piping and estimate pump head for systems based on velocity and pressure drop limits, and some will calculate heat gain or loss from piping sections. A number of the programs produce bills of material or cost estimates for the piping system. The piping flexibility programs are intended to assist in a stress analysis of piping systems. These programs can generally handle thermal, uniform or constant loadings and will calculate deflections, stresses, forces and moments.

5. *Equipment Selection Programs*

The programs contained in this class have to do with the selection of HVACR equipment, as is the case in "MFOURTEEN," "RH" and "CONDX." Areas covered include coil selection (hot water, chilled water, steam or direction expansion), refrigeration units, air-handling units and ceiling diffusers.

6. *Solar Programs*

The utilization of solar energy for the heating and cooling of buildings has been receiving considerable attention, and a number of programs for simulating the performance of these systems have appeared, e.g., "SUN," "TRANSYS" and "SIMSHAC." Solar programs can be considered to be specialized energy analysis programs. These programs will take into account collector area and performance characteristics, storage capacity, building load and weather data to predict the performance of the solar heating and/or cooling system.

7. *CAD for HVAC Systems*

Computer-aided design and drafting programs (CAD) have been developed that not only provide designers with computerized methods for the basic drafting of HVAC systems and shop drawings but that also simplify HVAC analytical and design tasks. Further, CAD programs enhance the interdisciplinary design of buildings, their mechanical systems and other systems and ease the task of modifying designs. Communications among the various design groups increase through the use of CAD programs.

All systems display a drawing or part of a drawing on a screen and provide a means of constructing straight lines, circles, polygons, arcs, section crosshatching, components, dimensioning and lettering. Pre-defined elements can be added at any scale and in any orientation. Elements can be added or deleted on the screen. Close-up views of areas of the drawing are available on the screen. Options such as a background grid, forcing of orthogonality of lines, snap-to-grid, snap-to-tangent and filleting can speed up the drafter's work.

CAD programs usually have layering capabilities, so different parts of a drawing can be displayed or modified separately or together. For example, in a duct drawing, supply ducts may be put on one layer and return ducts on another so either or both can be seen, changed or removed. Layering can also help an HVAC designer coordinate designs with other disciplines. For example, an HVAC designer working on a duct drawing can have the CAD system display the building's architecture, structure, piping and electrical systems in the background to check for space conflicts. Errors and obstructions are easily detected and corrected with the use of CAD programs.

Scale drawings express not only physical size but also how different parts are connected and where they are located within an area. As an example, an architectural designer can draw a building's walls, partitions and windows and have the computer automatically extract the number and size of windows, the areas and lengths of walls and the areas and volumes of rooms and zones.

Some CAD programs can link the graphic elements to the non-graphic characteristics of those elements. With these programs, an operator can store information from a drawing for later use in reports, schedules, design procedures or notes. A pump, for example, can be displayed accurately on a drawing, and its flow rate, voltage, weight, manufacturer, model number, cost and other data can be stored in a file or database. The link between graphics and attributes makes it possible to review the characteristics of an item simply by pointing to the graphics on the drawing or to generate schedules of items located in a certain area on a drawing. Conversely, the graphics-to-data links allow the designer to see the graphic items on a drawing that have a certain characteristic by searching for those characteristics in the data files and then making the associated graphics stand out as brighter, flashing, bolder or different-color displays.

Computer-aided design also makes building design drawings and information more useful than manually generated drawings. CAD HVAC drawings and their associated data can be used by building owners and maintenance personnel for ongoing facilities' management, strategic planning and maintenance. They are useful for computer-aided manufacturing such as duct construction, where duct design drawings serve as the basis for flat patterning, nesting and automatic cutting of ductwork and fittings. Two-dimensional CAD drawings can be used to create a three-dimensional model of the building and its components. This 3-D model can then help to visualize all or part of the building; to check for interferences between building parts; and to perform lighting, shading, daylighting, acoustic or energy analysis of the building and its spaces. The 3-D building model can also aid in developing sections and details for building drawings.

Further, CAD drawings and data can aid in specification preparation by allowing automated cross-referencing of drawings, drawing notation and building documents. CAD drawings of HVAC systems also serve as a basis for extracting quantity and size information of HVAC components that can be used for drawing equipment schedules and cost estimating.

SELECTION OF COMPUTER PROGRAMS

Selection of a suitable computer program for a specific application by an energy engineer requires a careful approach for evaluating the available programs. Generally, answers to the following questions are needed:

a) Are there any programs available which would satisfy the specified need?

b) Which of the available programs would be best for the situation?

c) How can those programs be obtained?

ASHRAE-sponsored research has produced a document titled *"A Bibliography of Available Computer Programs in the General Area of Heating, Refrigerating, Air Conditioning and Ventilating."* This document is one of the most comprehensive listings of HVAC-related software available today. The reader is urged to refer to this document to find answers to the above-listed questions.

Once a group of programs has been identified in the general area of specified application, each should be subjected to detailed scrutiny. Searching questions should be asked to make sure that a program of proper size is selected. This phase of program selection is very important. CIBS (Chartered Institute of Building Services) has stressed this point very well by developing a questionnaire to aid in the selection of a program for estimating the building energy use. CIBS suggests that the following information should be available:

1. *General Information*

The program name, giving a brief description of the program, and the name of the company or organization owning the program; also the company or organization responsible for the development of the program should be stated if different from owner. If the program can be made available for general use, can this be on another company's computer equipment or on a bureau basis? On what computer system has the program run successfully and in which modes is the program used? Can changes be made to individual items of input data, without repeating the entire input? In what computer language is the program written? Is there available documentation for the program user such as user manuals, pro forma data sheets, details of theoretical bases of programs, detailed specifications and flowcharts and program listings? Finally, what input data verification is performed by the program, and does the program output include a printout of the input data in a checkable form?

2. *Plant Capacity Estimated by Program*

If heating and/or cooling plants are sized by this program, is it necessary to specify the means by which calculations are made (for example, by peak demand in a given year)? Is it further necessary to specify how the start-up demand (after the weekend, say) is controlled in such circumstances or by any other peak design criteria? What assumptions are made for partial load efficiencies of plant chosen automatically?

3. *Plant Capacity Estimated by Another Program*

If plant capacity is calculated using another program, give name and details. Is a minimum information list available for this program?

4. *Weather Data*

What standard weather data is available to the engineer and how is it constructed (for example, by hourly values all year, or by mean weekly or monthly)? If the weather data is based on a design condition, the source of this data should be given, e.g., the CIBS Guide, Average of Meteorological Office data naming the year, etc. If weather tapes are used, details of source and year should be given, e.g., Meteorological Office, 1967. If the user wishes to supply his own weather data, what details are required of solar radiation, either direct or diffuse, temperature, humidity, wind speed and direction for the approximate time interval (days, weeks or months)?

5. *Load Profile Calculations*

a) Calculation of Time Interval. The frequency at which calculations are performed should be given (for example, hour-by-hour calculations of every day in the year, representative days from every week or representative days from every month). Other combinations available should be stated.

b) Internal Load Profiles. How are load profiles for commodity use, hot water, equipment, lighting and occupancy considered?

c) Internal Comfort. What criteria are used for maintaining internal comfort in terms of temperature, humidity, etc. for both heating and cooling seasons?

d) Infiltration. State on what basis infiltration and air change are handled during heating season (infiltration) and cooling season (natural ventilation).

e) External Heat Gains/Losses. Are calculations performed identically to an existing A/C load program? If so, the name and availability of information lists for that program should be stated.

6. *Building Structure*

Consideration should be given to building thermal storage, the method of defining internal environment, comfort conditions and gains.

7. *Radiation Transmission Through Glass and Blinds*
The availability of the program to cope with:
a. Varying glass thickness
b. Double- or triple-glazing
c. Heat-absorbing/heat-reflecting glass

    d. Internal/external blinds
    e. Combinations of special glazing and blinds
    f. Facade shading
    g. Range and intervals or orientation possible
    h. Air conduction through glazing
    i. Solar blind operation times

   8. *Gains Through Opaque Walls and Roofs*

Consideration should be given to the method by which solar radiation on external opaque surfaces is considered and the conduction calculation procedure for walls (lag and decrement).

    a. Plant Operation. Is the plant operational 7 days a week or 5 days a week, and can the plant running hours be varied for intermittent operation?

    b. Use of Load Profiles. These load profiles are space loads, i.e., the heat which must be added to or removed from any given space to maintain the specified design condition. Summation of these loads over the year can provide a means of comparison of performance of various building envelopes, although such summations are by no means a measure of energy used by heating or cooling equipment. Power requirements for lights, etc., cannot meaningfully be added at this stage. Does the program thus compare space load requirements and, if this is so, then are heating and cooling requirements shown separately? By use of COP/boiler efficiency, can these loads be converted to provide equipment loads, and, further, is the conversion done hourly by using variable COP/boiler efficiency or alternatively using constant COP/boiler efficiency? Finally, are the power loads from lighting, lifts, etc., then added?

   9. *Simulation of Building Services Systems*

    a. Heating Systems. Are heating controls possible by using time clock, or clocks, night setback, optimizers or by any other method? Can frost and condensation protection be included? Is heating plant response time taken into account, and, if so, in what way? Is heating plant efficiency considered as constant?

    b. Air-Conditioning Systems. Can only heating and cooling periods be specified? Is heating by A/C equipment or by direct heating? Is it possible for both to operate if required? Can each zone have different systems? Which individual systems can be modeled (for example, single duct, dual duct, V.A.V., 2-pipe induction, 4-pipe induction or others)? Are the individual performance and part-load efficiencies of plant items considered (for example, fans, pumps, chillers, cooling towers/air-cooled condensers, cooling coils, heat recovery, controls—proportional band, etc., and thermostats)? Can overall controls be investigated, such as economizer and enthalpy control? Also, is plant response time taken into account and, if this is so, how?

c. Operating Conditions. Can variations in building operation be considered, such as weekend, holidays, plant operating continuously or intermittently? Can a characteristic part-load efficiency curve be input for multiple unit installations, such as multiple boilers and multiple refrigeration units?

10. *Consumed Energy Profiles*

How are profiles presented: by overall totals, by type of energy or by type of use? Give a brief description of output, specifying options where appropriate.

FUTURE TRENDS

Future applications of computers are likely to develop "intelligent buildings" based on the use of "expert systems." Expert systems are the systems which make decisions based on "knowledge base" accumulated from one or more human experts. Some powerful expert systems have already been developed for medical diagnosis, chemical structure predictions, aircraft design and computer system configuration. Buildings seem to be the next potential candidate for using expert systems in their design and operation.

References
1. Howell, R.H., and Sauer, H.J., *Bibliography on Available Computer Programs in the General Area of Heating, Refrigerating, Air Conditioning and Ventilating.* ASHRAE, Atlanta, GA, 1981
2. Spielvogel, L.G., *Computer Energy Analysis for Existing Buildings.* ASHRAE Journal, August, 1975.
3. Smith, C.B., *Energy Management Principles.* Pergamon Press, 1981.
4. _____ *HVAC Systems and Applications Handbook.* ASHRAE, Atlanta, GA, 1987.
5. _____ *CIBS Computer Applications Panel/Task Group Program Evaluation/Building Energy Use Estimation Programs.* BSE, Vol. 45, Feb. 1978.

Table 11-1. Details of Commercial Programs

Program	Availability	Purpose	Input Data Needed
BLESS	V.C. Thomas; Daverman Associates; 200 N.W. Monroe; Grand Rapids, MI 49502 616/451-3525	To calculate heating/cooling loads.	Summer and winter design conditions; building design parameters; room exposure data; glass, wall, and roof information.
NBSLD	James P. Barnett, Thermal Engr-Sec; Center for Building Technology, NBS Washington, DC 20234	To calculate heating/cooling loads and to evaluate room temperature fluctuations.	Building parameters like wall, ceiling and floor characteristics; window area and location, operating schedules. Weather data from U.S. Weather Bureau tapes.
AXCESS	E.S. Douglas, Edison Electric Institute 90 Park Ave. New York, NY 10016	To estimate the energy requirements of various alternatives.	Weather data and control information; base load profiles; base design heating/cooling load; space type data; zone data; HVAC system details.
RUN01620	Duke Power Co. 422 S. Church St. Charlotte, NC 28242 704/373-4304	To estimate building energy usage comparing input fuel requirements.	Base load items and durations of operation, heating/cooling loads; details of HVAC system.
ECUBE	American Gas Assoc. 1515 Wilson Blvd. Arlington, VA 22209 703/524-2000	To calculate annual energy requirements in buildings.	Details of internal heat gains; electrical load and process load; solar load; hourly weather data; part load equipment performance data; fuel type, economic input data.

(Continued)

Program	Availability	Purpose	Input Data Needed
DUCT	APEC Executive Office Grant Deneau Tower Suite M-15 Dayton, OH 45402 513/228-2606	To design, size and analyze ducts.	Duct specifications; design criteria; insulation; cost data.
HVDUCT	ACTS Computing Corp. 29200 Southfield Rd. Southfield, MI 48076 313/557-6800	To design high velocity air duct systems.	Pressure losses in different ducts; line lengths; fitting types; duct diameters.
DUCTAN	Deere and Company John Deere Road Moline, IL 61265	To size and design ducts.	Duct dimensions; angle in degrees of grille deflection; static regain efficiency factor; air flow.
EXHAUST	Deere and Company John Deere Road Moline, IL 61265	To design industrial dust collection and air systems.	Hood entry loss factors; duct geometry and dimensions; hood air flow rates.
DUCTS	General Electric Co. 401 N. Washington St. Rockville, MD 20850 301/340-4000	To size and design ducts.	Duct size, length, air flow and fittings code.
DADDS	Daverman Associates 200 Monroe, N.W. Grand Rapids, MI 49502 616/451-3525	To design ducts.	Length and fitting size; outlet air quantities; fan and noise criteria.

(Continued)

Program	Availability	Purpose	Input Data Needed
MOIP	APEC Executive Office Grant-Deneau Towers Suite M-15 Dayton, OH 45402 513/228-2602	To size piping.	Pipe specifications, design criteria, material and labor costs; Nodal description of piping.
AUTOFLEX	Auton Computing Corp. 10 Columbus Circle New York, NY 10019 212/581-5250	To determine piping flexibility and to analyze stresses.	Piping code; system description; geometry; external forces, moments, restraints, and anchors.
MFOURTEEN	Giffels Assoc. Inc. Marquette Building 243 W. Congress Detroit, MI 48226 313/961-2084	To calculate hot, cold water demand, size supply piping and hot water heater.	Fixture count; continuous demands; and water heater information.
RH	Hankins & Anderson, Inc. 2117 N. Hamilton St. P.O. Box 6872 Richmond, VA 23230 804/353-1221	To select reheat coils for variable volume and induction reheat boxes.	Heating load; box size; water and air temperatures.
CONDX	ACTS Computing Corp. 29200 Southfield Rd. Southfield, MI 48076 313/557-6800	To match air cooled condensing units and direct expansion cooling coils.	Cooling load; entering wet-bulb and dry-temperatures; CFM of air; face velocity.

(Continued)

Program	Availability	Purpose	Input Data Needed
SUN	Berkeley Solar Group 1815 Francisco Street Berkeley, CA 94703	To estimate the performance of solar systems.	Average weather data; system parameters; loads.
TRANSYS	Solar Energy Lab University of Wisconsin Madison, WI 53706	To simulate solar systems for dynamic performance.	Size, quality and configuration of components; hourly weather data.
SIMSHAC	Gearold R. Johnson Mechanical Engineering Colorado State Univ. Fort Collins, CO 80523	To simulate solar systems for the dynamic performance.	Component description, parameters and geometry; building characteristics; and weather data.

12

Thermal Storage

This chapter discusses the following:

INTRODUCTION

Thermal storage can be defined as the temporary storage of energy at high or low temperature for use when it is needed. Most of the energy waste occurs because, in many situations, there is a time gap between energy availability and energy use. For example, in large buildings with interior zones that require year-round cooling, the energy typically exhausted to the atmosphere through cooling towers in the winter could be stored in water for heating during the nighttime and unoccupied periods. Similarly, solar energy could be stored during the daytime for night heating.

Unfortunately, the words "heat storage" for storage of energy at high temperature and "cool storage" for storage at low temperature have become common usage. This practice is contradictory to the concepts stressed in thermodynamics, where "heat" is used for energy in transition and not in storage. Similarly, "hot/cold" terms refer to temperature and not to heat.

Finally, the "energy availability" term used above refers to the time of availability and not to the quality of energy (useful work) as used in thermodynamics. For our purpose in this chapter, the words "thermal storage" will be used for energy in the form of internal/kinetic/potential energy stored on a temporary basis at a time when it is not needed for later use when it is needed. Similarly, "cool" and "heat" storage will mean storage of energy in the form of internal energy of a medium at a temperature of 20°F-70°F and 70°F-160°F, respectively.

ADVANTAGES OF THERMAL STORAGE

The concept of energy storage is not new and has been used for centuries in the design and operation of occupied spaces. In fact, a study of the design of Taj Mahal, one of the most well-known buildings in the world, shows that the concepts of thermal storage were used in its design to make the building energy-efficient and comfortable. The building, which is a 17th-century specimen of Mughal architecture that was built by Emperor Shah Jahan, uses high values of thermal mass to bridge the time gap between the extreme day and night temperature swings in the local climate throughout the year.

Currently, thermal storage techniques are of great interest to an energy engineer because of the following possible advantages:

a) Savings in the first and operating costs of the equipment are possible, especially in the design of new facilities;

b) Use of alternative energy sources like solar and wind energy becomes feasible;

c) Off-Peak rates from the utility company can be utilized.

d) Energy use efficiency can be improved through the use of technologies like heat reclamation.

Significant savings in the first cost of the equipment are possible because thermal storage permits the installation of smaller-size heating or cooling equipment than would otherwise be required. Conditions favoring thermal storage are:

a) High loads of relatively short duration

b) High electric power demand charges

c) Low-cost electrical energy during off-peak hours

d) Building expansion (installation of storage may eliminate the need for heating/cooling equipment addition);

e) Need to provide cooling for small-duration heavy loads such as cooling of restaurants and computer rooms;

f) Need to supplement a limited-capacity cogeneration plant,

and so on.

Example 12-1

As an example of the possible benefits from thermal storage, let us consider a hypothetical case of a building having a cooling requirement of 500 tons for 8 hours a day (9 a.m.-5 p.m.) and a second cooling load of 1000 tons for 2 hours a day (5 p.m.-7 p.m.). The designer could have different choices for installing a chiller as shown in Table 12.1.

Table 12-1. Choices for Chiller

	Choice No. 1	Choice No. 2	Choice No. 3
Chiller capacity	1000 Tons	500 Tons	250 Tons
Thermal storage capacity	—	1000 Ton hrs	3500 Ton hrs
First cost of chiller	$400,000	$250,000	$150,000
First cost of storage	—	$60,000	$100,000
Total cost	$400,000	$310,000	$250,000
Savings		$90,000	$150,000

The above example clearly demonstrates how the installation of thermal storage can reduce the first cost of the cooling equipment. It may be pointed out that, if the designer chooses a chiller of 500 or 250 tons, it will cut down the demand charges also. The actual savings from the reduction in the demand charges will depend on the location, type of building and the rate structure of the utility.

It may also be observed in the above example that, if the durations of the cooling requirements for either load are likely to increase in the future, a 250-ton chiller will not serve the purpose even with thermal storage. In that situation, Choice No. 2 may be the only alternative available, as long as a 500-ton chiller can meet the storage cooling requirements during the non-occupied periods.

BASIC DEFINITIONS

Thermal storage could be accomplished through *sensible heat storage* or as *latent heat storage.* In sensible heat storage, storage is accomplished by raising or lowering the temperature of the storage medium like water, sand, oil, rock beds, ceramic bricks, concrete slabs, etc. In latent heat storage, storage is accomplished by a change in the physical state of the storage medium, with or without a change in its temperature. Change of state may be from solid to liquid or vice versa. Examples of the phase change materials (PCM) used for energy storage are ice (water), salt hydrates and some organic materials. Phase change materials can store larger amounts of energy per unit mass compared to the sensible heat storage media and hence result in smaller and lighter storage devices with lower storage losses and higher storage efficiencies.

Storage efficiency is defined as the energy delivered by the storage system divided by the energy delivered to the storage system. Storage efficiencies on the order of 90 percent have been reported for well-stratified water tanks which are having daily duty cycling.

Thermal storage systems may be classified as *full storage* systems or *partial storage* systems. In full storage systems, the equivalent of the entire cooling or heating load energy for the design day is stored in the storage off-peak and used during the following peak period. In partial storage systems, only a portion of the daily load is generated during the preceding off-peak period and stored. In partial storage, the load is satisfied by simultaneous operation of the cooling or heating equipment and withdrawal from storage during the peak period.

STATUS OF ENERGY STORAGE TECHNOLOGY

Currently, the following methods are used for accomplishing thermal storage:

    a) Water tanks

b) Ice storage
c) Phase change materials
d) Rock beds
e) Ground-coupled storage

A brief description of each one of the methods is given below.

a) Water tanks

Water tanks are the most commonly used method of thermal storage, and ASHRAE Standard 94.3-1986 has been developed which describes the procedures for measuring performance of water tanks. Water has the highest specific heat of all common materials, namely 1 Btu/lbm°F (4.18 KJ/Kgm°K), which makes it a very suitable medium for sensible heat thermal storage. Thermal storage in water has the following advantages:

(i) Interfacing the thermal storage with the HVAC equipment is easy.

(ii) With proper stratification, uniform discharge temperature is possible.

(iii)Water tanks may be located above ground or underground.

(iv) Simple control techniques can be used.

(v) Operating cost is very reasonable.

Water tanks may be made of steel for above-ground installations or concrete for burial in the ground. Geometrical shape for the tanks may be cylindrical or rectangular.

Stratification of the stored water has attracted considerable attention by researchers because it has a significant effect on the thermal performance of water tanks, especially when they are used for storing chilled water. Some methods used to reduce mixing of stored and returned water are as follows:

(i) *Temperature Stratification,* in which water stored in tanks is stratified thermally with the lighter, warmer water on top. However, in chilled water storage where water in most cases swings between 41°F (5°C) and 59°F (15°C), the density difference is too small to use this method.

(ii) *Empty Tank Method* uses two tanks, and the return water is pumped into a second tank. If more than two tanks are used, the equivalent of one tank should be empty all the time. In situations where hot and chilled water are to be stored simultaneously, the equivalent of two tanks should be empty all the time.

(iii) *Labyrinth Method*, developed by Japanese designers, moves water through interconnecting cubicles, as shown in Figure 12-1.

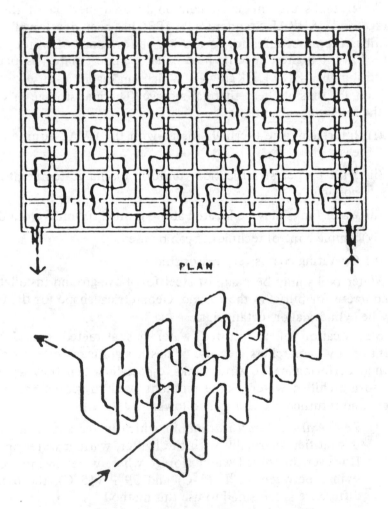

PLAN

Figure 12-1. Labyrinth Concept
(Source - HVAC Systems and Applications. ASHRAE Handbook 1987)

(iv) *Flexible Diaphragm* is still another method of temperature stratification, which is available under a license from Engineering Interface of Toronto. In this method, a sheet of coated fabric is attached at the middle of the tank and floats up and down as the volume of stored and return water in the tank varies.

Storage Capacity of Water Tanks

Storage capacity C of water tanks in Btu's can be calculated using the relation

$$C = M \, Cp \, \Delta T \qquad\qquad Formula \ (12\text{-}1)$$

Where M = Mass of stored water (lbm)
      Cp = Specific heat of water (1 Btu/lbm° R)
      ΔT = Temperature swing (° R)

Example Problem 12-2

Calculate the volume of the water tank required for thermal storage for Choices No. 2 and No. 3 of Example 12.1.

Choice No. 2

Required thermal storage = 1000 Ton hours
                        = 12,000 X 1000 Btu's
                        = 12×10^6 Btu's

Assuming $\Delta T = 20°$ R and taking Cp = 1 Btu/lbm ° R in Formula (12-1)

$$M = C/c_p \Delta T = \frac{12 \times 10^6 \ (Btu)}{1 \ (Btu/lbm°F) \ 20 \ (°R)} = 0.6 \times 10^6 \ lbm \ of \ water$$

$$\text{Volume of water stored} = \frac{\text{Mass}}{\text{Density}}$$

$$= \frac{1.2 \times 10^6 \ (lbm)}{62.5 \ (lbm/ft^3)}$$

$$= 9,600 \ ft^3$$

Choice No. 3

Required thermal storage = 3500 Ton hours
                        = 42×10^6 Btu's

Assuming $\Delta T = 20°R$ and taking $c_p = 1$ Btu/lbm°R in Formula (12.1)

$$M = C/C_p \Delta T \qquad = \frac{42 \times 10^6 \text{ (Btu)}}{1 \text{ (Btu/lbm°F) 20 (°R)}}$$

$$= 2.1 \times 10^6 \text{ lbm of water}$$

$$\text{Volume of water stored} \quad = \frac{\text{Mass}}{\text{Density}}$$

$$= \frac{2.1 \times 10^6 \text{ lbm}}{62.5 \text{ (lbm/ft}^3)}$$

$$= 33,600 \text{ ft}^3$$

b) Ice storage

Ice storage is based on the use of latent heat storage compared to the use of sensible heat storage in water tanks. Water also has one of the highest latent heats of fusion, namely 144 Btu/lbm. Ice storeage has the following advantages:

(i) Ice storage has smaller standby losses.

(ii) Ice storage needs smaller volume.

(iii) Ice storage can be easily interfaced with the existing refrigeration units.

Ice storage has lower refrigeration efficiencies compared with water storage because lower evaporator temperatures are needed to make ice than to chill water. In a direct expansion coil making ice, evaporator temperature may go as low as 10°F. This severely impacts the coefficient of performance of the system.

Ice storage could be accomplished using the following techniques:

(i) *Direct Expansion Coils.* In this method, a direct expansion coil is fitted inside a storage tank filled with water. Ice layers about 3 inches thick are formed around the coil. Stirrers could be used to aid uniform buildup of ice. Cool storage is utilized by circulating return water in the tank, which melts the ice and uses its latent heat of fusion for cooling. Commercial units having a capacity as large as 100,000/lbm of ice are available.

(ii) *Brine Coils*. In this method, a direct expansion coil is used to cool a brine solution (for example, a solution of ethylene glycol and water), which in turn is circulated through a coil in the storage tank. Brine has the advantage of lower refrigerant quantity requirements. Thickness of ice formed is generally lower in brine coils compared with the direct expansion coils.

(iii) *Plate Ice Makers*. In this method, the evaporators are vertical plates which are fitted above a water/ice storage tank. Water from the tank is pumped over these plates. When the water has been cooled to 32°F, it starts making thin films of ice on the plates. These layers of ice are discharged to the storage tank by circulating hot gas in the plates.

Capacity of Ice Storage

For operation at 32°F, the capacity of ice storage is equal to 144 Btu/lbm. For operation at higher temperatures, say 45°F, the storage capacity is increased by the sensible heat between 45°F and 32°F; i.e., capacity is $(144 + 1.0 \times [45-32]) = 157$ Btu/lbm. If a fraction of the tank is ice, a correction must be applied accordingly.

c) Phase Change Materials

Water is not the only medium which is used for latent heat storage. Several other materials, known as phase change materials (PCM), have been used where heat of fusion between solid and liquid phases has been used for thermal storage with the same advantages which were mentioned for ice storage. PCM are more appropriate compared with ice when the storage temperature desired is other than 32°F. For example, solar energy storage systems operate in the range of 80°F -124°F, and storage systems for off-peak electricity use can operate at temperatures well above 1000°F. Salt hydrates, organic materials and clathrates are the most frequently used PCM. Table 12.2 lists some important properties of such materials.

Salt hydrates are compounds of salt and water. Some salt hydrates have heats of fusion which are comparable to ice. Generally, these materials are non-toxic, non-flammable and not very expensive.

Table 12-2. Properties of Commercially Available Phase Change Materials (Lane 1985)*

Material	Type	Melting Point, F°	C°	Heat of Fusion Btu/lb	kJ/kg	Latent Heat Btu/ft	MN/m³	Cost
MgCl₂.6H₂O	Quasi Congruent	243	117	72.5	168.6	6500	242	medium
Na₄P₂O₇.10H₂O	Incongruent	158	70	–	–	–	–	medium
Sodium Acetate .3H₂O	Incongruent	136	58	97.0	225.6	7700	287	medium
MgCl₂.6H₂O/Mg(NO₃)₂.6H₂O	Eutectic	136	58	56.9	132.2	5400	201	medium
Paraffin Wax	Congruent	122	50	114.3	265.9	5500	205	low
Na₂S₂O₃.5H₂O(Hypo)	Semi-congruent	118	48	86.4	201.0	9000	335	medium
Neopentyl Glycol	Congruentᵃ	110	43	56.3	131.0	–	–	very high
CaBr₂.6H₂O	Congruent	93	34	49.7	115.6	6000	224	high
Na₂SO₄.10H₂O	Incongruent	89	32	108.0	251.2	9000	335	low
Na₂SO₄.10H₂O (with admixtures)	Incongruent	89	32	89.0	207.0	7300	272	low
CaCl₂.6H₂O	Semi-congruent	82	28	74.0	172.1	7000	261	low
CaCl₂.6H₂O (modified)	Congruent	81	27	82.0	190.7	7800	291	low
PE Olycol	Congruent	74	23	64.7	150.5	4500	177	high
Glauber's salt with various percentages of NaCl. NH₄Cl and KCl	Incongruent	70 to 38	21 to 3	70 to 41	163 to 95	7000 to 4000	260 to 150	low
CaBr₂.6H₂O/CaCl₂.6H₂O	Congruentᵇ	57	14	61.0	141.0	6800	253	high
Water/Ice	Congruent	32	0	144.0	355.0	9000	335	very high

ᵃPhase change is solid/solid   ᵇIsomorphous
*Source — ASHRAE. HVAC Systems and Applications Handbook, 1987.

Salt hydrates have one major disadvantage, namely they melt incongruently or semi-congruantly. They melt to a saturated aqueous phase and a solid phase, which is the anhydrous salt or a lower hydrate of the same salt.

Mixtures of two or more materials, mixed in a controlled ratio so that a minimum melting point is obtained, are called Eutectics. These materials melt completely at the melting point, and they have the same composition in the liquid and solid phases.

Organic materials, such as paraffin, have characteristics like low thermal conductivity, low density and contraction of volume on freezing. Lower densities result in larger storage volumes. However, they have few problems of thermal stratification, incongruent melting or subcooling.

Clathrates materials are those compounds in which one chemical entity is bonded inside the structure of another. For example, some commonly used refrigerants (R-11, R-12, R-22) can use water molecules for bonding to form clathrates. Some clathrates melt in the range of 40°F to 56°F.

ASHRAE Standard 94.1-1985 may be referenced for the procedures for testing the performance of PCM storage systems.

d) Rock beds

Thermal capacity of rocks or pebbles can be used to store energy. These materials are packed in a container which contains some wire-screen to support the bed of the material. The container is provided with air plenums for inlet and outlet of the fluid, which is generally air. Air flows through the bed in one direction during the transfer of heat to the bed and in the opposite direction during transfer of heat from the bed. Major use of rock beds is for the storage of solar energy.

e) Ground-coupled storage

Earth was the first natural thermal storage. Earth can be coupled to an HVAC system to function as a heat source or sink. There are two types of earth-coupled systems, namely direct heating/cooling systems and heat pump systems. In direct heating/cooling system coupling to the earth, energy is stored in a localized volume of earth and transferred back when needed for heating or cooling. In earth-coupled heat pump systems, heat is transferred to the earth in summer and transferred from the earth in winter for heating.

Coupling is accomplished by burying a large, closed-loop heat exchanger, either in a vertical or in a horizontal format. The medium used to transfer energy to or from the earth is a solution of ethylene glycol or calcium chloride in water.

REFERENCES

1 _____ *HVAC Systems and Applications Handbook*. ASHRAE. Atlanta, GA, 1987.

2 Smith, C.B. *Energy Management Principles*. Pergamon Press, 1981.

3 Tran, N., et al. "Field Measurement of Chilled Water Storage Thermal Performance." *ASHRAE Trans.* 95(1),1989

4 Sohn, C.W.; and Tomlinson, J.J. "Diurnal Ice Storage Cooling Systems in Army Facilities." *ASHRAE Trans.* 95(1), 1989

5 Wildin, M.W.; and Truman, C.R. "Performance of Stratified Vertical Cylindrical Thermal Storage Tanks, Part 1: Scale Model Tank."*ASHRAE Trans.* 95(1), 1989

6 Lane, G.A. "Congruent-Melting Phase-Change Heat Storage Materials." *ASHRAE Trans.* 88(2), 1982.

7 Lane, G.A. "PCM Science and Technology: The Essential Connection." *ASHRAE Trans.* 91(2), 1985.

8 Lorsch, Harold G. "Thermal Energy Storage for Solar Heating." *ASHRAE Journal,* Nov. 1975.

13

Passive Solar
Energy Systems*

INTRODUCTION

Passive solar energy systems use natural, nonmechanical forces—including sunshine, shading, and breezes—to help heat and cool homes with little or no use of electrical controls, pumps, or fans.

It should not be assumed that passive solar building heating measures are less effective or that their application is less widespread than the active measures. In fact there are few, if any, buildings that would not benefit in comfort and energy cost savings by the use of one or more passive measures.

Passive heating systems often depend on "thermal storage mass" to absorb and store the sun's energy for use within a building. Storage mass includes materials such as rock, stone, concrete or water, which store large quantities of heat as they absorb the sun's energy. This storage is accompanied by a moderate rise in temperature—the larger the temperature rise, the greater the energy stored. Later, as the thermal storage mass cools, the stored heat is released to heat the living space. "Phase change" materials may also be used for storage.

New passive buildings are generally integrated into their sites. Their design takes advantage of the natural environment to heat and cool the building by natural means. For example, evergreen trees are planted on the north side for winter protection, and deciduous trees on the south side to heighten winter sun access. Hillsides are used as earth berms that insulate the building. Indigenous materials, such as

*Source: Residential Conservation Service Auditor Training Manual.

adobe or rocks are used for thermal mass and insulation. Southerly orientation and glazing contribute direct solar gain for heating. Thermal mass on or near south glazing contributes to heating as well. Shading devices and overhangs are used to keep out the high summer sun when cooling is needed, and the thermal mass absorbs heat, effectively cooling the structure. Many of these passive features can be retrofitted to appropriately oriented existing structures.

Passive systems may require some attention from building occupants to perform at their peak. For example, someone may have to slide insulation into place to close off windows at night or during periods of extended cloud cover to reduce heat losses. Also, vents may have to be opened to control heat flow.

Well-designed passive buildings incorporate energy-saving features beyond standard improvements (insulation, weatherstripping and caulking). Depending on the specific location and design of the building, a variety of methods are suitable for additional energy savings in retrofit situations. For example, vestibules, now called airlock entries, are reappearing as a popular way to keep incoming blasts of cold or hot outside air to a minimum. Effective placement of vegetation such as trees, shrubs and vines is a natural, simple way to protect a building from temperature extremes.

This chapter discusses five passive solar systems.

- Thermosyphon domestic hot water systems
- Direct gain glazing systems
- Indirect gain systems
- Sunspaces
- Window heat gain retardant devices

THERMOSYPHON DOMESTIC HOT WATER SYSTEMS

Thermosyphon systems are generally considered to be passive measures most suitable for warm climates.

DIRECT GAIN GLAZING SYSTEMS

In direct gain systems, the building itself is a solar collector. Direct gain in its simplest form is probably the easiest way to apply solar energy to any building. It simply involves letting the sun shine in through added south-facing windows, thereby heating space directly. Nearly all of the sunlight entering a room is immediately converted to heat.

Thermal mass for storing excess heat (such as a concrete floor) may be located with direct exposure to the sunlight or in some other part of the building. All materials can store heat, but some store more than others. Table 13-1 lists the heat capacities (ability to store heat) of some common materials. Of the materials listed, air holds the least amount of heat and water the most.

Table 13-1. Heat Capacities of Common Materials

Material	Heat Capacity (Btu/ft^3—$^\circ$F)
Air (75°F)	0.018
Clay	13.9
Sand	18.1
Gypsum	20.3
Limestone	22.4
Wood, oak	26.8
Glass	27.7
Brick	28
Concrete	28
Asphalt	29
Aluminum	36.6
Marble	38
Copper	51.2
Iron	55
Water	62.5

To reduce heat loss and thus increase overall thermal performance, insulation may be placed next to the glass at night, either inside or outside. During the heating season, south-facing glass takes advantage of the sun's low position in the sky; in the summer when the sun is high in the sky, the glass may be shaded by overhangs or foliage. Both east and west glass, even in cold climates, can admit somewhat more solar energy than they lose if nighttime insulation is used. North glazing should be kept at a minimum.

Vertical glass admits almost as much heat during the winter as tilted glass (skylights) and is much easier to insulate and keep clean;

also, vertical glass does not break as easily. Many codes require tilted glass to be tempered, which is more expensive than regular window glass.

South-facing windows that are designed to distribute heat to as much of the building as possible are preferred. Clerestory windows with adequate overhangs should be considered before skylights.

If possible, south-facing windows that are exposed to the sun should be located so that the sunlight falls directly on thermal mass (heavy masonry fireplaces, masonry floors, water walls, etc.). Structures built with concrete floors (e.g., slab on grade) are frequently carpeted, eliminating the exposure of thermal mass. Replacing the carpeting with heavy ceramic tiles could result in greatly increased solar storage, reduced building heat load and lower heating bills.

Thermal mass provides nighttime heating by reradiating stored heat to living space. Thermal mass also tempers temperature swings that can occur. Some people will not want nighttime heating if it means altering their space to provide storage; others will not mind the changes necessary to provide some nighttime heating, to prevent overheating and to moderate temperature variations.

Applications

Addition of south-facing glazing should be considered as a potential passive solar retrofit wherever south-facing walls are available and an outside view is desired. Direct gain systems help to heat buildings.

Advantages and Disadvantages

Glass is relatively inexpensive, widely available and thoroughly tested. The overall direct gain system can be one of the least expensive means of solar heating, the simplest solar energy system to conceptualize and the easiest to build. In many instances, it can be achieved by simply enlarging existing windows. Direct gain systems, besides heating the interior, provide natural lighting and a view. To meet a small fraction of the heating needs of a building, direct gain systems do not necessarily need thermal storage.

On the other hand, ultraviolet radiation in the sunlight can degrade fabrics and photographs. If the desire is to achieve large energy savings, then relatively large glazing areas and correspondingly large amounts of thermal mass are required to decrease temperature swings. Thermal mass can be expensive, unless it serves a structural purpose. Interior daily temperature swings of 15 to 20 degrees are common even with thermal mass.

Construction Terms

Clerestory—Vertical window placed high in wall near eaves; used for light, heat gain, and ventilation.

Double Glazed—A frame with two panes of transparent glazing with space between the panes.

Fenestration—The arrangement, proportioning, and design of windows or doors in a building.

Glazing—Transparent or translucent material, generally glass or plastic, used to cover a window opening in a building.

Header—A horizontal structural member over an opening used to support the load above the opening.

Lite—A single pane of glazing.

Movable Insulation—Insulation (such as shutters, panels, curtains, or reflective foil draperies) that can be moved manually or by mechanical means.

Shading Coefficient—The ratio of the solar heat gain through a specific glazing system to the total solar heat gain through a single layer of clear double-strength glass.

Tempered Glazing—Glazing that has been specially treated to resist breakage.

Thermal Mass—Any material used to store the sun's heat or the night's coolness. Water, concrete, and rock are common choices for thermal mass. In winter, thermal mass stores solar energy collected during the day and releases it during sunless periods (nights or cloudy days). In summer, thermal mass absorbs excess daytime heat, and ventilation allows it to be discharged to the outdoors at night.

Thermal Storage Floors, Ceilings, and Interior Walls—Floors, ceilings, and interior walls that contain thermal mass and are used to collect and/or store heat in solar energy systems. They can be exposed to sun directly or receive only indirect solar heat to be effective.

INDIRECT GAIN GLAZING SYSTEMS

Any passive heating system that uses some intermediary material for heat collection and/or storage before passing that heat on to its desired place of use can be called an indirect passive system. The term "indirect gain systems" means the use of panels of insulated glass, fiberglass or other transparent substances that direct the sun's rays onto specially constructed thermal walls, ceilings, rockbeds, or containers of water or other fluids where heat is stored and radiated.

Trombe Wall

The Trombe wall uses a heat storage mass placed between glass and the space to be heated. Per unit of thermal storage mass used, the Trombe wall makes the best use of the material. While the temperature swing in the material is great, the temperature variation in the heated space is small. See Figure 13-1 for some features of a Trombe wall.

Water Wall

The water wall uses the same principle as the Trombe wall and involves replacing the existing wall, or parts of it, with containers that hold water. The water mass then stores heat during the day and releases the heat as needed.

Thermosyphon Air Panels (TAP)

TAP systems are often called "day heaters" because of their effective use during the day, since they have no storage. They are similar in appearance to active flat-plate collectors and are often mounted vertically. A TAP system has one or more glazings of glass or plastic, an air space, an absorber, another air space and (often) an insulated backing. Air flows naturally up in front of, behind or through the absorber and re-enters the building through a vent at the top. Figure 13-2 displays structural features of TAP systems.

Applications

Trombe walls, water walls and thermosyphon air panel systems all provide heat. Each system requires south-facing opaque walls with solar access.

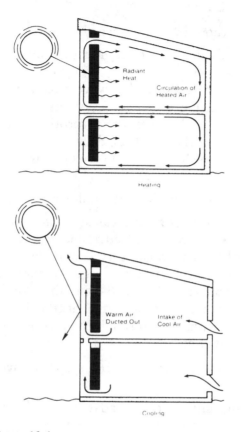

Figure 13-1. Trombe Wall: Heating and Cooling

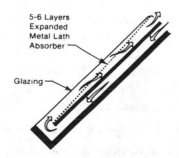

Figure 13-2. Airflow Through a TAP System

368 Handbook of Energy Engineering

Advantages and Disadvantages

Trombe and water walls have several advantages:

- Glare and ultraviolet degradation of fabrics are not problems.
- Temperature swings in the living space are lower than with direct gain or convective loop systems.
- The time delay between the absorption of solar radiant energy by the surface and the delivery of the resulting heat to the space provides warmth in the evening.

Trombe and water walls have the following disadvantages:

- Two south walls (a glazed wall and a mass wall) are needed.
- Massive walls are not often found in residential construction (although thermal storage walls may be the least expensive way to achieve the required thermal storage since they are compactly located behind the glass).
- In cold climates, considerable heat is lost to the outside from the warm wall through the glazing unless the glazing is insulated at night.

Thermosyphon Air Panels (TAP) systems have the following advantages:

- Glare and ultraviolet degradation of fabrics are not problems.
- TAPs provide one of the least expensive ways to collect solar heat.
- To provide a small fraction of the heating needs of a building, thermal storage is not needed.
- TAPs are easily incorporated onto south facades.
- TAPs are readily adaptable to existing buildings and require less skill for resident installation.
- Because the collector can be thermally isolated from the building interior, night heat losses are lower than for any other uninsulated passive system.

TAP systems have the following disadvantages:

- The collector is an obvious add-on device.
- Both careful engineering and construction are required to

ensure proper airflows, air seals and adequate thermal iso-
lation at night.

- The thermal energy is delivered as warmed air—it is difficult
  to store this heat for later retrieval because air transfers heat
  poorly to other mass.
- It's often impractical to add much collector area.
- Occupant use patterns, and natural airflows within the structure will strongly affect the usability of energy delivered;
  siting should thus be thoughtful and creative.

Construction Terms

Absorber—The surface in a collector that absorbs solar radiation
and converts it to heat energy. Generally, matte black surfaces are
good absorbers and emitters of thermal radiation, while white and
metallic or shiny surfaces are not.

Backdraft Damper—A damper designed to allow air flow in only
one direction.

Damper—A device used to vary the volume of air passing through
an air outlet, inlet or duct.

Drum Wall—A type of thermal storage wall in which the thermal mass is large metal drums filled with a storage medium, usually
water.

Masonry—Stone, brick rammed earth, adobe, ceramic, hollow
tile, concrete block, gypsum block or other similar building units or
materials, or a combination thereof, bonded together with mortar to
form a wall, pier, floor, roof or similar form.

Solid Masonry—Masonry in which there are no voids; for instance, concrete block with filled cores.

Thermal Lag—The ability of materials to delay the transmission
of heat; can be used interchangeably with time lag.

Trombe Wall (or Solar Mass Wall)—A massive wall that absorbs
collected solar heat and holds it until it is needed to heat the interior.

SOLARIA/SUNSPACE SYSTEMS

The terms "sunspaces" and "greenhouses" are both commonly
used to refer to this measure; they are used interchangeably in this
discussion. Sunspaces could be considered a form of direct gain sys-

tem. Sunlight is absorbed in the sunspace by thermal storage mass (such as bricks or water in containers), which then radiates heat. Existing doors and windows are frequently used to allow heated greenhouse air to flow into interior living spaces. The sunspace can be sealed off from the rest when too much heat is gathered or when there is insufficient sunlight to contribute to heating. Closing doors and windows between interior space and the sunspace at night or on particularly cloudy days may also be necessary.

Although single glazing for sunspaces will result in a maximum light transmission for plant growth and solar gain, single glass will permit a large amount of heat loss at night. In northern climates, double glazing will retard this heat loss. Additionally, movable insulation can be applied to prevent nighttime heat loss.

Thermal mass can be expensive. Therefore, if it can be reduced or eliminated and wider temperature fluctuations allowed, the cost of a sunspace will be substantially less. However, system efficiency will be reduced.

A greenhouse can be thought of as a buffer zone between the outside environment and the inside of the building. In this buffer zone, direct solar gain causes wide temperature fluctuations and a higher temperature indoors than outdoors at virtually all times of the day. These higher indoor temperatures buffer the adjacent living area, reduce building heat loss and can be used immediately to help heat living area.

The sunspace is a versatile passive solar measure. It can be added to many different architectural designs with pleasing results. Sunspaces are equally compatible with expensive and inexpensive homes. Figure 13-3 illustrates attached retrofit sunspaces. Figure 13-4 illustrates a retrofit greenhouse for a mobile home.

The four basic methods for transferring thermal energy from the greenhouse into interior living space are

- Direct solar transmission
- Direct air exchange
- Conduction through common walls
- Storage in and transfer from gravel beds or other thermal mass

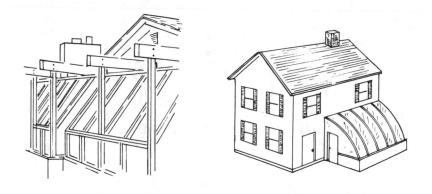

Attached
Pit Greenhouse

Figure 13-3. Attached Retrofit Greenhouses

Figure 13-5 illustrates an example of thermal mass in the form of water drums. These storage methods are often used in combination. For example, in addition to a common heat storage wall to conduct heat from the greenhouse to the building, forced or natural air flow (direct air exchange) can also be used.

Applications

A sunspace can provide heat and additional living space, as well as the opportunity to grow vegetables during most of the year, in nearly all climates. Building a sunspace is a good do-it-yourself project, especially for small groups of neighbors and friends.

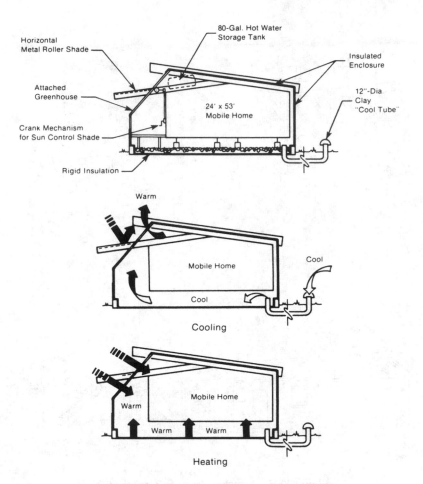

Figure 13-4. Attached Greenhouse on a Mobile Home

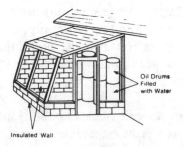

Figure 13-5. Example of Storage in an Attached Greenhouse

Advantages and Disadvantages

Greenhouses have several advantages:

- Temperature swings in adjacent living spaces are moderated.
- They provide space for growing food and other plants.
- They reduce heat loss from buildings by acting as a buffer zone.
- They are readily adaptable to most existing residential buildings.
- Since the greenhouse serves more than one function, it can be a natural and integrated part of the building design.

Because sunspaces are a form of direct gain system, their greatest disadvantage is that a poor design can lead to wide indoor temperature fluctuations. To combat this problem, indirect gain elements are often combined with direct gain, e.g., Trombe walls or water walls, to provide the tempering effect of thermal mass. In general, greenhouses are most economical when they have purposes in addition to providing heat and are built to a standard of quality that will enhance the functional and appraised value of the home.

A second consideration is having maximum sunlight penetration in the winter (for heat) and less penetration in the summer (to prevent overheating). To achieve this, a rule of thumb of latitude $\pm 15°$ is applicable for optimal greenhouse glazing tilt, with appropriate shading.

Aesthetics can also be a concern with sunspaces. The architectural lines of a house need to be aesthetically matched. Sometimes this leads to an extreme tilt and sometimes no tilt. In each case, interior design (addition or subtraction of thermal mass, outside venting, movable insulation, etc.) has to compensate for poor design from a passive solar energy point of view.

A fourth concern is movable insulation, which is necessary in most cases for retention of winter heat gain.

WINDOW HEAT GAIN RETARDANTS

Shading is the most effective method of reducing heat gain through transparent materials, and, ideally, a shading device should allow sunlight in during the winter. The term "window heat gain retardants" means mechanisms that significantly reduce summer heat gain through south-facing windows ($\pm 45°$ of true south) by use of devices such as awnings, insulated rollup shades, metal or plastic solar screens, or movable rigid insulation. Included as window heat gain retardants (WHGR) are any devices that provide shading. Other examples of these devices are sunscreens, exterior roll blinds, exterior shutters, venetian blinds, film shades, opaque rolls and insulating shutters and blinds.

Internal shading can reduce the amount of heat dispersed within a space. The most common internal shading devices are venetian blinds, vertical blinds, shades, draperies, reflective films and shutters. These devices can reject up to 65% of the solar radiation that strikes the glass directly.

External shading is more effective than internal shading; it can keep up to 95% of the solar radiation from entering the building (100% if opaque shutters are used). Many devices are available for exterior shading. Horizontal overhangs, with fixed and movable elements, are very effective on south windows, because the sun is highest as it approaches due south during midday. The sun also is higher in summer than in winter, and the overhang can be positioned to screen the sun in summer but admit it in winter. On east or west elevations, however, the sun's angle is too low to be blocked out by horizontal overhangs, and properly oriented vertical louvers have proved more beneficial. If the louvers are movable, the user can control them to provide a better view or greater admission of light when the sun is located on the opposite face of the building. Trees, vines, and adjacent buildings may also provide shade, depending on their proximity, height and orientation. Many summer shading devices can also provide window *insulation* in the winter. Care should be taken in their application so that beneficial winter solar heat gain is not lost. As an example, heat reflecting material may work best year-round if it is mounted on a retractable shade.

Applications

Window heat gain retardants help keep buildings cool in summer. They should be considered wherever summer solar heat gain through windows causes overheating. More particularly, substantial dollar (and resource) savings are possible when retardants are used to diminish expenses for air conditioning.

Advantages and Disadvantages

Any heat gain retardant device has advantages and disadvantages and to list all of them for the wide range of shading devices would be impractical. Instead, here are a few key ones for each device.

For sunscreens, visibility can remain as high as 86% while providing daytime privacy. Sunscreens block solar radiation reflected from nearby objects more effectively than other measures. However, drapes or blinds must be used for privacy at night.

With an exterior roll blind, energy-conserving effects can be achieved year-round. In summer, shading is provided by day and ventilation by night (with the blind raised). In winter, insulation is provided by night with the blind lowered, and solar gain by day with the blind raised. However, the blind cannot be tilted to provide a view.

For exterior shutters and blinds, protection of windows is provided from storm damage, vandalism, or intrusion. However, operation and maintenance of exterior devices can be time-consuming.

For awnings, ease of installation and attractiveness are advantages. However, they are subject to wind damage and they must be maintained. New fabric may be necessary every four to eight years if fabric awnings are used instead of more permanent wood or metal awnings.

Construction Terms

Awning—A shading device, usually movable, used over the exterior of a window.

Heat Reflective and Heat Absorbing Window or Door Materials —Glazing, films or coating applied to existing windows or doors; they have exceptional heat absorbing or heat reflecting properties.

Overhang—A horizontal or vertical projection over or beside a window used to selectively shade the window or door on a yearly basis.

Shading Device—A covering that blocks the passage of solar radiation; common shading devices consist of awnings, overhangs, or trees.

Shutter—Movable cover or screen for a window or door.

14

Energy Management

Energy management is the judicious and effective use of energy to maximize profits (minimize costs) and to enhance competitive positions. A successful energy management program is more than conservation. It is a total program that involves every area of a business. A comprehensive energy management program is not purely technical. It takes into account planning and communication as well as salesmanship and marketing. It affects the bottom line profits of every business; thus the individual who is assigned the role of "energy manager" has high visibility within the organization.

Energy management includes energy productivity, which is defined as reducing the amount of energy needed to produce one unit of output. Energy management includes energy awareness, which is essential in motivating employees to save energy. Probably the highest initial rate of return will occur through the establishment of a good maintenance management program.

This chapter reviews the basics of maintenance management and energy management organization.

MAINTENANCE MANAGEMENT

There are obvious losses from poor maintenance such as steam and air leaks and uninsulated steam lines. There are also losses from less obvious areas. For example Tables 14-1 and 14-2 illustrate the hidden effect of dirty evaporators and condensers on equipment performance. These losses go undetected and result in decreased capacity of equipment and an increase in energy usage. In Table 14-1, for a reciprocating compressor under the dirty condenser and evaporator conditions, capacity is reduced 25.4% and the increase in brake horsepower per ton is 39%. In Table 14-2, for an absorption chiller under similar conditions, capacity is reduced 23.8% and power requirements are increased 7.5%.

**Table 14-1. The Effects of Poor Maintenance on the Efficiency of a
Reciprocating Compressor, Nominal 15-Ton Capacity**

Conditions	(1) °F	(2) °F	(3) Tons	(4) %	(5) HP	(6) HP/T	(7) %
Normal	45	105	17.0	—	15.9	0.93	—
Dirty Condenser	45	115	15.6	8.2	17.5	1.12	20
Dirty Evaporator	35	105	13.8	18.9	15.3	1.10	18
Dirty Condenser and Evaporator	35	115	12.7	25.4	16.4	1.29	39

(1) Suction Temp, °F
(2) Condensing Temp, °F
(3) Tons of refrigerant
(4) Reduction in capacity %
(5) Brake horsepower
(6) Brake horsepower per Ton
(7) Percent increase in compressor bh per/ton

**Table 14-2. The Effects of Poor Maintenance on the Efficiency of
an Absorption Chiller, 520-Ton Capacity**

Conditions	Chilled Water °F	Tower Water °F	Tons	Reduction in Capacity %	Steam lb/ton/H	Per-cent
Normal	44	85	520	—	18.7	—
Dirty Condenser	44	90	457	12	19.3	3
Dirty Evaporator	40	85	468	10	19.2	2.5
Dirty Condenser and Evaporator	40	90	396	23.8	20.1	7.5

A third major loss is in missing the opportunity to upgrade the
facilities within the spare parts program. For example, when a motor
burns out it is usually replaced with the same model or sent to a
shop for rewinding. If the motor is sent to a rewind shop, it will
usually have poorer efficiency and power factor characteristics than
before. Thus the energy manager should consider upgrading the replacement with either a high-efficiency motor or requiring a higher
specification from the shop rewinding the motor.

The following summarizes key elements of the maintenance
management program.

Preventive Maintenance Survey

This survey is made to establish a list of all equipment on the property that requires periodic maintenance and the maintenance that is required. The survey should list all items of equipment according to physical location. The survey sheet should list the following columns:

1. Item
2. Location of item
3. Frequency of maintenance
4. Estimated time required for maintenance
5. Time of day maintenance should be done
6. Brief description of maintenance to be done

Preventive Maintenance Schedule

The preventive maintenance schedule is prepared from the information gathered during the survey. Items are to be arranged on schedule sheets according to physical location. The schedule sheet should list the following columns:

1. Item
2. Location of item
3. Time of day maintenance should be done
4. Weekly schedule with double columns for each day of the week (one column for "scheduled" and one for "completed")
5. Brief description of maintenance to be done
6. Maintenance mechanic assigned to do the work

Use of Preventive Maintenance (PM) Schedule—At some time before the beginning of the week, the supervisor of maintenance will take a copy of the schedule. The copy the supervisor prepares should be available in a three-ring notebook. He will go over the assignments in person with each mechanic.

After completing the work, the mechanic will note this on the schedule by placing a check under the "completed" column for that day and the index card system for cross-checking the PM program.

The supervisor of maintenance or the mechanic will check the schedule daily to determine that all work is being completed according to the plan. At the end of the week, the schedule will be removed from the book and checked to be sure that all work was completed. It will then be filed.

Preventive Maintenance Training

The supervisor of maintenance or a mechanic is responsible for assisting department heads in the training of employees in handling, daily care and the use of equipment. When equipment is mishandled, he must take an active part in correcting this through training.

Spare Parts

All too often equipment is replaced with the exact model as presently installed. Excellent energy conservation opportunities exist in upgrading a plant by installing more efficient replacement parts. Consideration should be given to the following:

- Efficient line motor to replace standard motors
- Efficient model burners to replace obsolete burners
- Upgrading lighting systems

Leaks—Steam, Water and Air

The importance of leakage cannot be understated. If a plant has many leaks, this may be indicative of a low standard of operation involving the loss not only of steam but also of water, condensate, compressed air, etc.

If, for example, a valve spindle is worn, or badly packed, giving a clearance of 0.010 inch between the spindle, for a spindle of ¾-inch diameter the area of leakage will be equal to a 3/32-inch diameter hole. Table 14-3 illustrates fluid loss through small holes.

Although the plant may not be in full production for every hour of the entire year (i.e., 8760 hours), the boiler plant water systems and compressed air could be operable. Losses through leakage are usually, therefore, of a continuous nature.

Thermal Insulation

Whatever the pipework system, there is one fundamental: It should be adequately insulated. Table 14-4 gives a guide to the de-

Table 14-3. Fluid Loss Through Small Holes

Diameter of Hole	Steam—lb/hour		Water—gals/hour		Air SCFM
	100 psig	300 psig	20 psig	100 psig	80 psig
1/16"	14	33	20	45	4
1/8"	56	132	80	180	26
3/16"	126	297	180	405	36
1/4"	224	528	320	720	64

Table 14-4. Pipe Heat Losses

Pipe Dia Inches	Surface Temp °F	Insulation Thickness Inches	Heat Loss (Btu/Ft/Hr)		Insulation Efficiency
			Uninsulated	Insulated	
4	200	1½	300	70	76.7
	300	2	800	120	85.0
	400	2½	1500	150	90.0
6	200	1½	425	95	78.7
	300	2	1300	180	85.8
	400	2½	2000	195	90.25
8	200	1½	550	115	79.1
	300	2	1500	200	86.7
	400	2½	2750	250	91.0

gree of insulation required. Obviously, there are a number of types of insulating materials with different properties and at different costs, each one of which will give a variance return on capital. Table 14-4 is based on magnesia insulation, but most manufacturers have cataloged data indicating various benefits and savings that can be achieved with their particular product.

Steam Traps

The method of removing condensate is through steam trapping equipment. Most plants will have effective trapping systems. Others may have problems with both the type of traps and the effectiveness of the system.

The problems can vary from the wrong type of trap being installed to air locking or steam locking. A well-maintained trap system can be a great steam saver. A bad system can be a notorious steam waster, particularly where traps have to be bypassed or are leaking.

Therefore, the key to efficient trapping of most systems is good installation and maintenance. To facilitate the condensate removal, the pipes should slope in the direction of steam flow. This has two obvious advantages in relationship to the removal of condensate: One is the action of gravity, and the other the pushing action of the steam flow. Under these circumstances the strategic siting of the traps and drainage points is greatly simplified.

One common fault that often occurs at the outset is installing the wrong size traps. Traps are very often ordered by the size of the pipe connection. Unfortunately, the pipe connection size has nothing whatsoever to do with the capacity of the trap. The discharge capacity of the trap depends upon the area of the valve, the pressure drop across it and the temperature of the condensate.

It is therefore worth recapping exactly what a steam trap is. It is a device that distinguishes between steam and water and automatically opens a valve to allow the water to pass through but not the steam. There are numerous types of traps with various characteristics. Even within the same category of traps, e.g., ball floats or thermo-expansion traps, there are numerous designs, and the following guide is given for selection purposes:

1. Where a small amount of condensate is to be removed, an expansion or thermostatic trap is preferred.
2. Where intermittent discharge is acceptable and air is not a large problem, inverted bucket traps will adequately suffice.
3. Where condensate must be continuously removed at steam temperatures, float traps must be used.
4. When large amounts of condensate have to be removed, relay traps must be used. However, this type of steam trap is unlikely to be required for use in the food industry.

To insure that a steam trap is not stuck open, a weekly inspection should be made and corrective action taken. Steam trap testing can utilize several methods to insure proper operation:

- Install heat sensing tape on trap discharge. The color indicates proper operation.
- Place a screwdriver to the ear lobe with the other end on the trap. If the trap is a bucket-type, listen for the click of the trap operating.
- Use acoustical or infrared instruments to check operation.

ENERGY MANAGEMENT ORGANIZATION

A common problem facing energy managers is that they have too much responsibility and very little authority to get the job done. A second problem is the lack of definition of the job. Probably the biggest problem is the lack of true commitment from top management. As pointed out in the text, an overall energy utilization program requires an "investment" in order to get the return desired.

The first phase of the program should start with top management establishing the organization, defining the goals and providing the resources for doing an effective job. Tables 14-5 and 14-6 illustrate a checklist for top management and typical energy conservation goals.

Table 14-5. Checklist for Top Management

A. Inform line supervisors of:
  1. The economic reasons for the need to conserve energy
  2. Their responsibility for implementing energy saving actions in the areas of their accountability
B. Establish a team having the responsibility for formulating and conducting an energy conservation program and consisting of:
  1. Representatives from each department in the plant
  2. A coordinator appointed by and reporting to management
     NOTE: In smaller organizations, the manager and his staff may conduct energy conservation activities as part of their management duties.
C. Provide the team with guidelines as to what is expected of them:
  1. Plan and participate in energy saving surveys
  2. Develop uniform record keeping, reporting, and energy accounting
  3. Research and develop ideas on ways to save energy
  4. Communicate these ideas and suggestions
  5. Suggest tough, but achievable, goals for energy saving
  6. Develop ideas and plans for enlisting employee support and participation
  7. Plan and conduct a continuing program of activities to stimulate interest in energy conservation efforts
D. Set goals in energy saving:
  1. A preliminary goal at the start of the program
  2. Later, a revised goal based on savings potential estimates from results of surveys
E. Employ external assistance in surveying the plant and making recommendations, if necessary
F. Communicate periodically to employees regarding management's emphasis on energy conservation action and report on progress

Adapted from *NBS Handbook*, 115.

Table 14-6. Typical Energy Conservation Goals

1. Overall energy reduction goals
    - (a) Reduce yearly electrical bills by _____%.
    - (b) Reduce steam usage by _____%.
    - (c) Reduce natural gas usage by _____%.
    - (d) Reduce fuel oil usage by _____%.
    - (e) Reduce compressed air usage by _____%.
2. Return on investment goals for individual projects
    - (a) Minimum rate of return on investment before taxes is _____.
    - (b) Minimum payout period is _____.
    - (c) Minimum ratio of $\dfrac{\text{Btu/year savings}}{\text{capital cost}}$ is _____.
    - (d) Minimum rate of return on investment after taxes is _____.

A key element in any energy management program is "flexibility." For example, in one year natural gas prices may accelerate steeply. Unless an organization reacts quickly, it soon may be out of business. Every energy management program should have a contingency or backup plan. Periodically, the risk associated with each scenario should be reviewed. Risk assessment can help the energy manager determine the resources required to meet various emergency scenarios. A method of evaluating the cost to business for each scenario is illustrated in Formula 14-1.

$$C = P \times C_1 \qquad\qquad \textit{Formula (14-1)}$$

Where

C = cost as a result of emergency where no contingency plans are in effect.

P = the probability or likelihood the emergency will occur.

C_1 = the loss in dollars as a result of an emergency where no plan is in effect.

Implementing the Energy Management Program

The first phase of the energy management program involves the accumulation of data. Table 14-7 illustrates the minimum information required to evaluate the various energy utilization opportunities. The energy manager must also review government regulations which will affect the program.

Table 14-7. Information Required to Set Energy Projects on the Same Base

Fuel	Cost At Present	Estimated Cost Escalation Per Year	Energy Equivalent
1. Energy equivalents and costs for plant utilities.			
Natural gas	$_____/1000 ft^3	$_____/1000 ft^3	_____Btu/ft^3
Fuel oil	$_____/gal	$_____/gal	_____Btu/gal
Coal	$_____/ton	$_____/ton	_____Btu/lb
Electric power	$_____/Kwh	$_____/Kwh	_____Btu/Kwh
Steam			
_____psig	$_____/1000 lb	$_____/1000 lb	_____Btu/1000 lb
_____psig	$_____/1000 lb	$_____/1000 lb	_____Btu/1000 lb
_____psig	$_____/1000 lb	$_____/1000 lb	_____Btu/1000 lb
Compressed air	$_____/1000 ft^3	$_____/1000 ft^3	_____Btu/1000 ft^3
Water	$_____/1000 lb	$_____/1000 lb	_____Btu/1000 lb .
Boiler make-up water	$_____/1000 lb	$_____/1000 lb	_____Btu/1000 lb

2. Life Cycle Costing Equivalents

After tax computations required	_____
Depreciation method	_____
Income tax bracket	_____
Minimum rate of return	_____
Economic life	_____
Tax credit	_____
Method of life cycle costing	_____
(annual cost method, payout period, etc.)	

All too often the energy manager is the last to find out information after the fact. For example, the procurement of motors may be the electrical department's responsibility. The energy manager must get his or her input into the evaluation process prior to purchase. One way is to establish a review process where important documents are initialled by the energy manager before they are issued. Key documents include

- Bid summaries for major equipment

- One-line diagrams
- Process flow diagrams
- Heat and material balances
- Plot plan
- Piping and instrument diagrams

It is much easier to add local instrumentation to the piping and instrument diagram prior to construction. Another way of insuring that the energy manager gets timely input is to incorporate these activities into the overall planning document, such as the Critical Path or PERT Schedules. Figure 14-1 illustrates a planning schedule incorporating input from the energy manager. Figure 14-2 illustrates how energy management activities interact with various departments.

Richard L. Aspenson, director of Facilities Engineering and Real Estate at 3M, summarized* the areas a good energy manager must master as follows:

Technical Expertise

Energy management normally begins with a solid technical background —preferably in mechanical, electrical or plant engineering. Managers will need a good grasp of both design theory and the nuts-and-bolts details of conservation programs. This includes a thorough understanding of the company's processes, products, maintenance procedures and facilities.

Communication

In the course of a single week, energy managers might find themselves dealing with lawyers, engineers, accountants, financial planners, public relations specialists, government officials, and even journalists and legislators. A good energy manager has to be able to communicate clearly and persuasively with all of these people—in *their* language. Above all, energy managers must be able to sell the benefits of their programs to top management.

Financial Understanding

To enlist the support of top management, energy managers will have to develop and present their programs as investments, with predictable returns, instead of as unrecoverable costs. They will have to demonstrate what kind of returns—in both energy and cost savings—can be expected

---

*"The Skills of the Energy Manager," *Energy Economics, Policy and Management*, Vol. 2, No. 2, 1982.

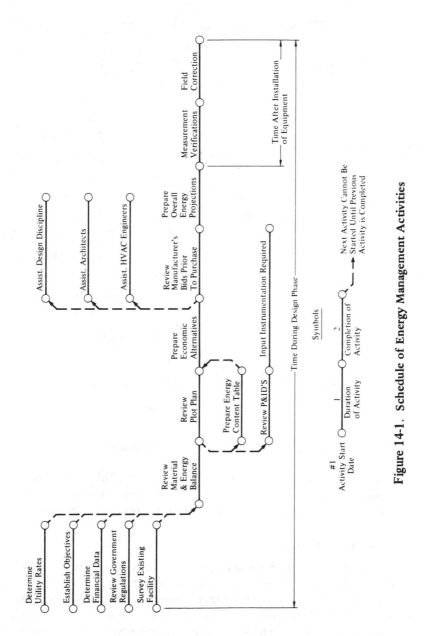

Figure 14-1. Schedule of Energy Management Activities

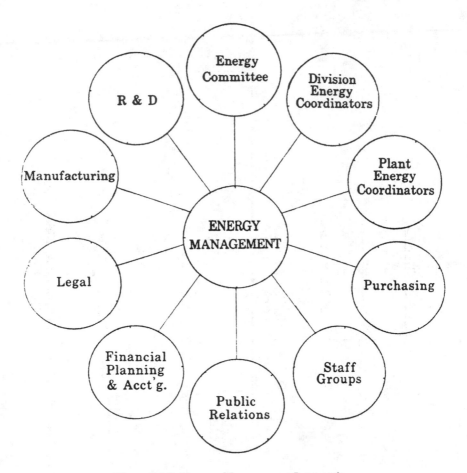

Figure 14-2. Energy Management Interactions

from each project, and over what period of time. This means first of all developing some credible way to measure returns—a method that will be understood and accepted by the financial officers of the company.

Planning and Strategy

A basic part of energy management is forecasting future energy supplies and costs with reasonable accuracy. This means coming to grips with the complexities of worldwide supply, market trends, demand projections, and the international political climate. There is no way, of course, to predict all of these things with certainty, but every business—especially energy-intensive ones—will need some kind of reliable forecasting from now on.

Community Relations and Public Policy

Energy managers have some responsibility to go outside their companies to share their ideas and experience with a variety of publics. Trade and professional associations can become clearing houses for new ideas. Legislators and government agencies need, and often welcome, expert help in setting standards and policies. And the public needs help in understanding what is at stake in learning to use limited supplies of energy wisely and efficiently.

Energy management can be an exciting, challenging and rewarding job if you put your skills to work.

We as Energy Managers are a special breed of people who have unlimited opportunities if we develop the skills and implement the programs that will improve the profitability of our respective businesses. We can and must influence public policy decision by using our technical knowledge in a constructive and objective way.

Pointers for the Energy Manager

1. Be aggressive. Learn how to say *no* in a diplomatic way.
2. Energy is not a soft sell. Stick your neck out.
3. Be sure you anticipate questions before you request approval on energy expenditures. Prepare factual data—think ahead.
4. Be creative. Identify needs. Prioritize your action plan tasks.
5. Be a positive person—one with perpetual motion—a catalyst for action.
6. Establish a free-thinking environment. Be a good listener.
7. Develop a five-year plan. Update yearly. Include a strategy for implementation—action plans.
8. Establish credibility through an accurate energy accountability system.
9. Efficient use of energy is an evolutionary process. Be patient but demanding. Follow through on commitments.
10. Give individual recognition for achievements.

Energy managers can turn the problems of a changing energy situation into opportunities. At the same time, we can grow to become better corporate citizens of our society.

15

Competitive Power[1]

COMPETITION VS REGULATION

For decades, electric utilities have been regulated monopolies. State regulators and their utilities operate under a "regulatory compact." The state grants the utility a monopoly franchise; within a certain geographic area, the utility has the exclusive right to sell electricity. In return for that franchise, the utility has an obligation to provide service to any qualified customer that wants service.

Two factors supported the monopoly status. First, the generation, transmission, and distribution of electricity was believed to be a natural monopoly, such that provision of service by two or more entities would increase costs significantly. Second, the supply of electricity was considered a necessity, "clothed with the public interest" (California Public Utilities Commission 1993).

During the past several years, more and more questions have been raised about both utility regulation and its monopoly status. Are utilities still natural monopolies? Is state regulation continuing to serve the broad public interest? Might customers be better served by companies that had to compete with each other? How can small electricity consumers, for whom competition may offer few benefits, be protected? How might a competitive electric-utility industry best meet environmental-protection goals?

Competition is increasingly discussed these days as a way to deal with a variety of problems. Several industries, including telecommunications, banking, airlines, and natural gas have been or are in the process of being deregulated and made more competitive. The pros and cons of meeting society's wants and needs by competition vs regulation vs managed competition are shaped largely by individual values. Those who believe that free markets work best tend to oppose government regulation because such regulations restrict individual freedom and impose inefficiencies on the economy. Those who believe that government programs can work favor government oversight because markets may yield outcomes that are inequitable and inconsistent with society's long-run objectives.

Although competition offers many benefits, it may also create some problems (Table 15-1). On the plus side, competition stimulates the introduction of new products and services and the demise of products and services that no longer meet customer needs. Although nothing inherent in competition favors the short-term over the long-term, competitive markets appear to price outputs to optimize short-term economic efficiency. On the other hand, societal values, such as environmental quality, long-term conservation of nonrenewable resources, research to develop new energy sources, and equity concerns (e.g., protection of low-income households), are largely ignored in a competitive environment.

Regulation, on the other hand, can protect these broader societal interests. However, it does so at the cost of timely and efficient decision making. Consider the complexity, expense, and amount of time required to reach decisions at state regulatory commissions. Also, whether regulatory decisions truly reflect customer interests is often unclear. If certain interests "capture" the regulators, their decisions may not benefit customers broadly.

Because of the limitations in competitive markets, all markets are bound by government or industry rules and procedures. As Reich (1991) noted:

> The idea of a "free market" apart from the laws and political decisions that create it is pure fantasy anyway. The market was not created by God on any of the first six days (at least, not directly), nor is it maintained by divine will. It is a human artifact, the shifting sum of a set of judgments about individual rights and responsibilities....
>
> ...In modern nations, government is the principal agency by which the society deliberates, defines, and enforces the norms that organize the market. Judges and legislators, as well as government executives and administrators, endlessly alter and adapt the rules of the game— usually tacitly, often unintentionally, always under the watchful eye and sometimes under the guiding hand of interests with clear stakes

Table 15-1. Characteristics of competitive vs regulated markets

Issue	Competition	Regulation
Time focus	Short-term	Long-term
Efficiency	Economic (low cost, high value)	Societal (economic, environment, equity)
Products/services	Many	Few
Price differentiation	Substantial	Limited to a few customer classes
Recognition of differences among customers	Yes	No
Decision making	Fast	Slow, litigated

in the outcomes of particular decisions. To the extent that rhetoric frames the issue as one grand choice between government and market, it befogs our view of the series of smaller choices about an endless set of alternative ways to structure the rules of ownership and exchange.

We will return to Reich's point about the role of government in establishing, maintaining, and conditioning markets later in the discussion of retail competition.

FORMS OF COMPETITION
IN THE ELECTRIC-UTILITY INDUSTRY

The second background point concerns the forms of competition taking place in the utility industry today (Fig. 15-1). The *Public Utility Regulatory Policies Act* (U.S. Congress 1978) helped to create new nonutility companies that build and operate power plants. The *Energy Policy Act of 1992* (U.S. Congress 1992) furthered that process by creating a class of generation companies exempt from the details of state and federal regulation. Thus, competition at the production end already exists.

If existing utilities are able to design, build, and operate power plants at costs below those of independent power producers (IPPs), then utilities may continue to be major players on the generation side. On the other hand, if the IPPs are more efficient and can build and operate power plants more cheaply and as reliably as today's utilities, then the utility role in generation will decline over time (assuming that utilities do not unfairly control access to transmission). Even here, the utility role is likely to be important in life-extension and repowering of existing power plants.

Transmission and distribution (T&D; the wires) are natural monopolies. However, tough questions remain about what parties should operate and control today's system and build new systems. The *Energy Policy Act* recognized the importance of transmission access to making generation truly competitive. It gave the Federal Energy Regulatory Commission (FERC) the authority to order a transmission-owning utility to provide access to another party wanting to move electricity from a power plant to a wholesale customer. But much is unresolved concerning access to, pricing of, and expansion of transmission services.

To illustrate, FERC can order a utility to construct new transmission facilities to meet the needs of another party. However, state siting and regulatory authorities determine whether and when those facilities are actually built. Resolving this potential state/federal jurisdictional issue may be difficult.

Utilities might continue to dominate transmission markets, perhaps through the creation of regional transmission groups. Alternatively, regional entities (analogous to natural-gas pipelines) might arise whose sole function is to build and operate the T&D systems. Hogan (1993) suggests splitting the transmission monopoly into two parts. The first, which he calls Gridco, would be responsible for the construction, operation, and maintenance of the transmission system. The second, which he calls Poolco, would manage the minute-by-minute dispatch of power plants and use of the transmission grid to transport power from generating units to load centers.

Small modular power plants (diesel generators, small combustion turbines, fuel cells, and photovoltaics) may become more cost-effective. If that happens, generating units can be located close to load centers, reducing the need for transmission of bulk power supplies. Although transmission would remain a monopoly, its importance would diminish in this scenario.

Although transmission is likely to be regulated primarily by FERC, state PUCs will retain regulatory control over distribution. Therefore, PUCs could have a substantial effect on the extent and nature of retail competition. Such competition, probably the most controversial element of the debates on the future of the utility industry, involves competition for retail customers among:

- Alternative fuels (e.g., switching from electricity to natural gas for certain end uses or industrial use of new electrotechnologies);

- Customer cogeneration or self-generation;

- Actions taken by customers, either on their own or with assistance from energy-service companies, to reduce peak demands and electricity use (thereby cutting electricity bills);

- Utilities seeking to encourage large customers to move from one local service area to another (utility to utility competition); and

- Alternative suppliers of electricity (i.e., retail wheeling).

The first four forms of retail competition already exist, although to varying degrees around the country. To the extent that customers believe they can meet their energy-service needs better and less expensively without the local utility, they can switch fuels, cogenerate or self-generate to meet their own electric needs, adopt energy-efficiency measures to reduce their electric bills, or move outside the local utility's service area. Whether customers take such actions depends, in part, on the level of retail electricity prices relative to the costs of these alternatives.

Retail wheeling (under which end users are free to choose their electricity suppliers) engenders the loudest and strongest debates, probably because of its effects on the traditional regulated monopoly franchise. If regulators allow retail customers to choose their electricity supplier, implicitly at least, they are taking away the local utility's franchise. In return, the local utility would (or at least should) no longer have any obligation to serve.

Proponents of retail wheeling believe it is crucial for customers, especially large industrials, to be free to choose their own electricity supplier (Hughes 1993). Absent such freedom, they will be unable to compete effectively in international markets, and local utilities will not be forced to improve efficiency.

Opponents believe that retail wheeling will lower costs only for large customers at the expense of utility shareholders and the remaining residential and commercial (core) customers (Cavanagh 1994). In addition to their worry about stranded investment, retail-wheeling opponents are concerned about the loss of the environmental benefits of DSM and renewables and the elimination of the utility's role as a portfolio manager on behalf of all retail customers.

Figure 15-1 shows several possible outcomes for competition at each of the three levels discussed here: generation, T&D, and retail. Different combinations from each column, in a mix-and-match mode, seem feasible.

To the extent that competition at the wholesale level lowers the costs of electricity production, the avoided costs against which DSM must compete will go down. Therefore, less DSM will be cost-effective. During much of the 1970s and 1980s, long-run marginal costs were above average costs. Thus, consumers faced price signals that, on average, encouraged them to use more electricity than was economically efficient. To the extent that long-run costs are now below average costs (and average rates), price signals encourage consumers to overinvest in energy efficiency. This change in the relationship between long-run marginal and average costs weakens the rationale for utility DSM programs. However, many important nonprice barriers to customer adoption of energy-efficiency measures remain, barriers that utilities can help overcome (Levine et al. 1994).

Under one scenario, at the retail level the franchise monopoly will hold, and today's utilities will continue to serve most of their existing customers, retaining the obligation to serve in return for which they retain their monopoly rights (top right box in Fig. 15-1). Only a few large industrials cogenerate, self-generate, or work out special deals that allow them to leave the system.

At the other end of the retail-competition spectrum, the franchise monopoly no longer exists (bottom right box in Fig. 15-1. All customers are free to choose their own supplier. Several utilities, as well as indepen-

dent brokers, vie for retail customers. Brokers, who may own only tele-
phones, computers, modems, and faxes, make "deals" in which they
aggregate retail customers and purchase power from a combination of
suppliers and arrange for T&D delivery of that power to customers (Bretz
1994).

If the first retail scenario prevails, the effects on utility DSM programs are likely to be minimal. On the other hand, if there is considerable
competition for end-use customers, utility DSM programs may change
dramatically. In particular, utilities will no longer be able to charge all
customers for the costs of these programs because prices will be set in a
competitive market and not in a regulatory proceeding. Utilities will (1)
either charge participating customers for their DSM services or (2) be able
to spread the costs of DSM programs across only the remaining core
customers. Therefore, utilities will run fewer programs that offer rebates
to customers that install energy-efficient equipment; such programs will
be aimed only at core customers. Instead, utilities will sell information,
contracting services, measures and their installation, and performance
guarantees to those customers choosing to buy these DSM services. And
the emerging brokers may also offer DSM services to retail customers
through energy-service companies.

The future of utility DSM programs is largely independent of the
extent and form of competition at either the generation or transmission
levels, although its success will depend on the costs of generation and
transmission. Utility DSM programs are likely to be affected primarily by
competition at the retail level.

STATE REGULATION OF RETAIL COMPETITION

It seems likely that FERC will play a stronger role than the state
PUCs in shaping wholesale competition. But at the retail level, state PUCs
are likely to be much more influential than FERC (Table 15-2), consistent
with the role of government espoused by Reich above. Commissions may
choose to shape the extent, pace, and beneficiaries of competition for retail
electricity customers for three reasons.

First, it may be economically inefficient to allow customers to arbi-
trarily select their electricity suppliers. This situation could occur if the
utility's retail prices do not fully reflect the costs of providing services to

Fig. 15-1. The forms that competition might take in generation, transmission and distribution, and retail services in the electric-utility industry.

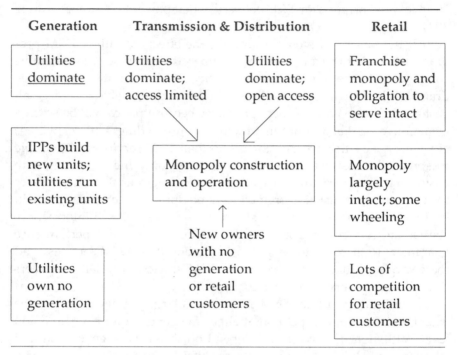

Table 15-2. Possible PUC rules to structure retail wheeling

Define core vs noncore customers

Set terms and conditions for noncore customers
- Exit fee
- Reentry fee
- Advance notice to exit or reenter
- T&D wheeling charge
- Prices for other services (e.g., backup, power quality)

some customers and these customers then choose other options (e.g., cogeneration) that reduce their costs but not the overall costs to society.

Second, it may be inequitable to allow such freedom. Customers that leave the utility system may create a "stranded investment" problem for the remaining customers and the utility's shareholders. Although this is primarily a transition problem, the amount of money at stake (as much as 10 to 20% of utility assets) is large enough to warrant careful attention.*

Third, in a fully competitive retail market, utilities may want to drop the societal functions they currently furnish, including provision of demand-side services to customers. For various reasons, including environmental quality, economic productivity, and the provision of capacity and energy resources, commissions may want to be sure that such services are still available to customers.

Discussing the steps that PUCs could take to shape retail competition is important because these actions could have major effects on future utility DSM programs. In addition, a recent survey found that PUCs are not yet ready to deal with increased retail competition, lacking policies that require, encourage, permit, discourage, or prohibit retail wheeling (R.J. Rudden Associates and Fitch Investors Service 1993). According to that survey, only 19% of the responding PUCs had policies in place to deal with recovery of stranded investment, and only 31% had policies to protect core customers from higher prices.

PUCs could decide which customers are eligible to become noncore. In some jurisdictions, PUCs have set minimum size limits for noncore customers of natural-gas utilities. PUC decisions on size might reflect their judgments on which customers are most able to negotiate well with suppliers and which customers can most benefit from that choice. On both scores, large industrial customers may be those most likely to be able to select their own electricity suppliers.

Although the California PUC began by setting minimum size limits for noncore customers of the local gas utilities, these limits soon dropped and were then replaced by unbundled rates. These rates covered separately the commodity, capacity, transportation, and storage costs of delivering gas to customers. On the electric side, Commissions could establish

---

*One way to address this problem would he to allow utilities to write up their T&D assets by the same amount that they write down their generating assets. This procedure would rationalize prices for both generation and T&D and keep utility shareholders and core customers whole.

separate rate schedules for each utility service, including energy, capacity, transmission service, distribution service, reserve margin, power-factor correction, power quality, and so on.

For potential noncore customers, PUCs could establish requirements for exit and/or reentry. To illustrate, the PUC could require exiting customers to reimburse the utility (i.e., remaining customers) for the undepreciated investments made by the utility on behalf of the departing customer. Or the Commission could allow the utility to charge the departing customer the difference between the revenues that customer would likely pay the utility over a set number of future years and the utility's year-to-year avoided costs (Fig. 15-2). If the utility has considerable excess capacity and therefore low avoided costs for several years, the exit fee could be substantial. On the other hand, if the utility faced rapid load growth and needed capacity soon, the exit fee could be much smaller or even zero. In such a case, the departing customer would lower long-term costs for the remaining customers.

Such a requirement would have two powerful effects. First, it would ensure that neither utility shareholders nor remaining customers would have to pay for "stranded" investment. Second, the charge might discourage customers from pursuing uneconomic retail-wheeling options.

Commissions could require noncore customers to provide advance notice if they wish to return to core (full service) status; TransAlta, Canada's largest investor-owned utility proposed just such a five-year advance-notice requirement (Huntingford and Nelson 1993).

Considerable work remains to determine the economic rationale for either exit or reentry fees. In particular, commissions need to establish the relationship between such fees and the utility's current retail tariff and its short- and long-run marginal costs.

Commissions might impose on noncore customers a distribution-system fee, the proceeds of which could be used to pay for the functions that the utility undertakes on behalf of society in general and certain core customers in particular. (FERC could impose a similar fee on transmission services.) The fee, set on a ¢/kWh or $/kW basis, could be justified in different ways. One approach would set the fee equal to the difference between the utility's current retail rate and its marginal supply cost (Moskovitz et al. 1993). Setting the fee in this way would ensure that retail customers do not make uneconomic purchase decisions.

The fees collected in this way, beyond the utility's direct costs for

Fig. 15-2. Schematic showing exit fees for large customers that leave the local utility's system in (1) 1995 or (2) in 1998. In each case, the exiting customer pays the utility the net present worth of the difference between avoided cost and retail rate for five years.

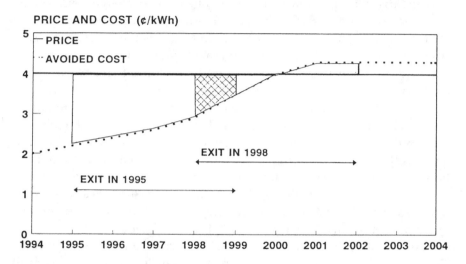

building and operating its distribution system, could be used for various social purposes. These purposes include:

- lifeline rates for low-income customers,

- low-income weatherization,

- acquisition of renewable resources that are not yet cost-effective,

- acquisition of DSM resources that provide broad economic and environmental benefits but that raise prices, and

- research and development on advanced energy-supply and DSM technologies.

These functions might be especially important to implement the carbon-dioxide reductions called for by the national *Climate Change Action Plan*.

Massachusetts Electric Company (MECO) filed a request along these lines with FERC (Corcoran 1994b). The case involves the Massachusetts Bay Transportation Authority (MBTA), which until recently was a MECO customer but now is requesting only wheeling services from MECO. MECO's request includes, in addition to the usual charges for transmission and distribution services, a stranded-investment charge based on its power-supply costs. MECO claims that it arranged for power supplies on the expectation that the MBTA would remain a customer. The proposed stranded-investment charge will automatically be reduced to reflect actual load growth that, presumably, will use the capacity made excess by the MBTA's departure from the MECO system.

Commissions may want to expand utility flexibility in negotiating prices with the larger customers that have choices. Commissions can require utilities to offer energy-efficiency services to such customers in negotiating lower prices and costs. For example, Southern California Edison's (1993) self-generation deferral rate is intended to prevent uneconomic bypass. It permits the utility to offer large customers a rate lower than its standard industrial tariff if the customer can "demonstrate serious intent and feasibility to self-generate." The PUC guidelines for this rate require the utility to offer a conservation alternative to the special rate. This alternative involves utility financing of an investment in energy-efficient equipment at the customer's facility.

To illustrate, Southern California Edison recently agreed to install various measures at the Eisenhower Medical Center under the self-generation deferral rate. The measures include efficient chillers, a hydronic economizer, and high-efficiency lamps and ballasts. These measures will cost about $600,000 and should reduce utility bills by about $900,000/year.

Finally, PUCs will likely define the utility's "obligation to serve" differently for core and noncore customers. The utility's obligations to provide reserve margin, power quality, and other services to noncore customers may be minimal or zero. On the other hand, PUCs may require utilities to offer backup services to such customers, the pricing of which would affect the attractiveness of retail wheeling.

Whether any of these mechanisms are legal and politically feasible will vary from state to state. Large industrial customers (those most likely to benefit from easy access to other suppliers) will argue strongly against such "obstacles" to competition. In addition, substantial practical problems may arise in designing equitable mechanisms. For example, what criteria would be used to decide which customers would pay an exit fee or a reentry fee? Would it be based on size and, if so, on annual revenue, sales, or peak demand? Could the exit fee be applied to industrial customers that physically leave the utility's service area or only to those customers that switch suppliers? Would the reentry fee be applied to new customers moving into the service area or only to customers previously served by the local utility? If the latter, would the charge be a function of the number of years that the customer had been off the system?

A 1993 New York Public Service Commission decision provides precedent for the kinds of requirements discussed above (Kelleher and Wood 1994). The Commission approved an industrial Subscription Option program that allows large commercial and industrial customers to "opt out" of the standard DSM programs offered by Niagara Mohawk. Customers that chose to opt out are no longer eligible for participation in these programs and no longer have to pay for part of those DSM-program costs. However, as a condition for opting out, the industrial customer has to pay for a detailed energy audit at its facility and provide that audit to Niagara Mohawk. (Initial results from these audits show average potential energy savings of 7% with a less-than-four-year payback.) The utility offers a shared-savings program to help these customers finance installation of measures recommended in the energy audit. This decision shows the feasibility of imposing DSM-related requirements on customers as a condition for certain treatment (e.g., access to retail wheeling).

The California PUC (1993), in its review of regulatory reforms, suggested four strategies for state regulation of electric utilities. Although much broader than this focus on retail competition, the PUC's strategies contain some of the same elements:

- Limited reform, in which current cost-of-service regulation is retained, but rate cases are held every year, and most balancing accounts are eliminated.

- Price-cap model, in which the regulatory compact remains, but the utilities gain considerable pricing flexibility, and the link between utility rates and costs is broken.

- Limited customer choice, in which a limited segment of customers is given access to competitive markets. The remaining (core) customers are served under traditional cost-of-service regulation, but the utility's obligation to serve noncore customers is substantially reduced.

- Restructured utility industry, in which the regulated utilities divest themselves of all generating assets and become common-carrier transmission and distribution companies. Utilities no longer enjoy a monopoly retail franchise, but in return they no longer have an obligation to serve those customers that choose other suppliers.

The role of utility DSM programs would likely differ substantially across these four strategies. Under limited reform, such programs could continue with little change from today. Moving down the list, the programs would be progressively smaller and shift more of the costs to participating customers.

RETAIL COMPETITION, PRICES, AND DSM

A key question concerning retail competition is the role of price. Will utilities and others compete for retail customers on the basis of unit price (¢/kWh) alone? Do consumers view electricity as a homogeneous product for which price is paramount? Or do they view electricity as a service with many attributes? Although we do not now know the answers to these questions, it seems likely that customers differ, with some emphasizing price and some emphasizing service.

Although much of the rhetoric suggests that price will dominate, it is helpful to consider the role that price plays in other competitive markets. Consider the extent to which people buy cars on the basis of unit price ($/pound), rather than on the basis of total cost, comfort, reliability, and other attributes. Similarly, do people go to movies on the basis of their unit cost in ¢/minute; do they buy books on the basis of ¢/page? These examples suggest that consumers will choose utilities on the basis of price, cost (the utility bill, which is the product of price and quantity), convenience, reliability, safety, and various consumer services, such as energy efficiency.

Regardless of whether price is the only factor on which suppliers will compete, price will surely be important. Industrial groups are aggressively comparing industrial electricity prices of their utility with those of surrounding utilities, and Wall Street financial advisers are doing the same. For example, the Tennessee Valley Industrial Committee, an association of large industrial customers of the Tennessee Valley Authority, commissioned a study to compare TVA industrial rates with those of other southeastern utilities (Drazen-Brubaker & Associates 1993). Rates varied by a factor of two, from 2.6 to 5.5¢/kWh (Utilities with high retail rates might use DSM as a way to compete with lower-cost providers.)

Regardless of whether customers focus on price, many utilities are acting as if they do. This means that utilities will look for more and more ways to control costs. Indeed, much of the recent downsizing at utilities is driven by management interest in lowering costs to become more competitive.

Just how important is DSM to a utility's prices? Answering this question requires information on the utility's DSM-program costs, the supply costs the utility would avoid because of the energy and demand reductions caused by the DSM programs, and the lost revenues caused by DSM-induced electricity savings. As explained by Hirst and Blank (1993), improvements in customer energy efficiency that lower sales also cut utility revenues. While utility costs go down also, the short-term cost reductions (primarily for fuel) are much less than the revenue reductions. In the long term, if DSM programs defer construction of new resources, costs may decline more than revenues do.

This difference between revenue and short-term cost reductions, called net lost revenues, is the fixed-cost component of electricity price, typically 50 to 75 % of the total cost of electricity. In many states, utilities recover these lost revenues from their customers. In other states, utility shareholders bear these losses until the next rate case.

In general, the effects of DSM on electricity price are small, on the order of a few percent (Faruqui and Chamberlin 1993). A forthcoming study, based on 66 data points, finds that the median price increase caused by utility DSM programs is 2.7% (Pye and Nadel 1994). Even in New York, which has aggressive utility DSM programs and low avoided costs, average electricity prices in the year 2000 will be only 6% higher than they would be without DSM programs (Gallagher 1994).

Cost overruns at nuclear plants and excess capacity are much more important factors in increasing electricity prices. On the other hand, the costs of these plants are sunk, whereas DSM program costs are at the

margin and therefore discretionary. However, tree trimming along distribution lines to reduce the frequency of outages, maintaining inventories
of transformers and cable to reduce the duration of outages, and providing telephone customer-assistance centers to improve customer service all
cost money and do not increase electricity sales. Thus, these activities, like
DSM, raise short-term prices. It is unclear whether these expenditures are
receiving the same scrutiny that DSM-program costs are.

If electricity prices were set equal to short-term variable costs, then
the lost-revenue problem would disappear. However, because short-term
costs are generally below long-term marginal costs, basing prices on
short-term costs would send misleading signals to customers, encourag-
ing them to ignore the long-term consequences of their energy-related
decisions. In particular, such pricing would lead them to underinvest in
energy-efficiency measures. On the other hand, if nonprice market barriers are the primary obstacle to customer adoption of cost-effective energy-
efficiency measures, setting prices equal to short-term marginal costs and
having utilities offer DSM programs may be the preferred route to achieving the benefits of improved energy efficiency. Also, pricing at short-term
marginal cost is what a competitive power market is likely to yield
(Hogan 1993).

To explore the effects of utility DSM programs on retail electricity
prices, consider a program that runs from 1995 through 1997. The utility
pays the full cost of installing DSM measures. The program cuts overall
demand by 1% as of 1998 and has a benefit/cost ratio (based on the total
resource cost test) of 1.6. If program costs are ratebased, rather than
expensed, net-lost revenues might account for about one-third of the
average price increase (Fig. 15-3).

If short-term variable costs are lower than in this example, the lost-
revenue component and total price impact would be greater. For example,
Public Service Company of Colorado (1993) proposed a set of DSM programs that provide substantial system benefits. Over the 20-year analysis
period, however, these programs are likely to raise electricity prices by
3%. Net-lost revenues contribute almost 60% to this price increase, with
program costs accounting for the other 40%. Thus, the rate impacts of
DSM programs depend both on program costs and on the differences over
time between avoided costs and retail prices.

This discussion suggests that utilities in the future are likely to
design and operate their DSM programs with greater attention to the rate-

Fig. 15-3. Effects of a hypothetical utility DSM program on electricity prices, showing the relative effects of program costs and lost revenues.

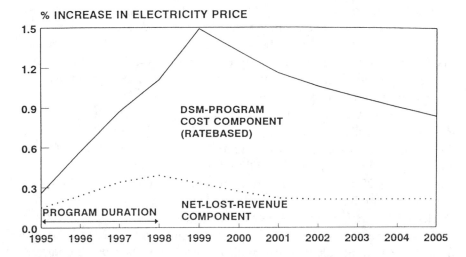

impact measure (RIM) and less attention to the total-resource-cost (TRC) test (California Commissions 1987). The two tests differ primarily in their treatment of lost revenues. The TRC test, which includes both participating and nonparticipating customers, ignores lost revenues. The RIM test, which focuses on nonparticipants, treats lost revenues as a cost.

DESIGNING DSM PROGRAMS TO LOWER RATE IMPACTS

The more market research utilities conduct (to better understand what customers want and what utilities can profitably provide), the more successful their DSM programs will be. Success includes increased participation rates and lower utility costs. If utilities can better understand their customers, they may be able to offer DSM programs that both improve customer service and provide the resource benefits of "traditional" DSM programs. As Stone (1993) notes, "Utilities that practice DSM aggressively have a sophisticated understanding of critical end-use technologies, of customers' decision processes and the factors that influence them, and of how their own cost [and price] structures position them competitively."

It will be interesting to see whether utilities use the opportunities provided by current regulation (i.e., mandates to run DSM programs that are paid for by all customers) to develop the marketing skills they need to deliver DSM in a competitive environment. In particular, will utilities be able to overcome their tradition of long-leadtime projects that required decision making based on a hierarchical structure and careful analysis? Or will they be able to shift to the rapid, flexible, decentralized decision making needed to market DSM programs to customers? An equally interesting issue is whether PUCs will allow utilities such flexibility in designing, running, and modifying their DSM programs.

A recent study suggests that some utilities, at least, are learning how to market skillfully to their customers (Keneipp et al. 1993). Pacific Gas & Electric has corporate account managers who work closely with customers that spend more than $0.5 million/year on PG&E utility bills. Each manager has 25 to 30 such accounts and has frequent personal contact with them. Similarly, Wisconsin Electric has an account executive who works with "chains" (e.g., supermarkets, department stores, discount stores, and fast-food stores) for which decisions may be made at the regional or national level. These skills are important because the largest potential savings are in the commercial/industrial (C/I) sector, and these customers are most at risk.

The keys to running successful DSM programs seem to be:

- Keep the program simple.

- Update the program to recognize changing market dynamics (e.g., that would allow use of a lower rebate), but do so only once a year.

- Work closely with trade allies (but remember that their interests may not match the utilities').

- Use financial incentives to motivate utility marketing staff.

- Train in-house and trade-ally staff on the program and its measures/technologies.

Learning how to market effectively will be increasingly important if utilities can no longer use rebates (paid for by customers in general) to entice customers to participate in DSM programs. On the contrary, utilities may have to sell DSM to customers and convince them that such services are worth their full cost.

Increasingly, utilities will unbundle their services so that each customer is offered (and purchases) an appropriate mix of kWh, kW, reliability, power quality, and DSM at an appropriate price. For those customers that focus on price, fewer services will likely be selected. Other customers may select DSM options if they are more concerned with bills than prices and if the value of these DSM services exceed their costs. Finally, utilities will increasingly use DSM as an alternative to price breaks.

Consider how a utility might respond to the "threat" of an industrial cogeneration project in a competitive environment. Traditionally, the utility might have (1) ignored the possibility and let the customer leave the system with no effort to retain that customer, (2) pointed out the complications and negative aspects of cogeneration in an effort to dissuade the customer from installing such a system, or (3) offered a discount rate to retain the customer.

In a more competitive environment, the utility might help analyze the economics of a cogeneration system. If the economics are attractive, the utility might offer to build and operate the system on behalf of the customer. The utility might benefit from owning and operating this system if it costs less than the net-revenue losses associated with the customer's departure from the utility system. Also, building and operating cogeneration systems might provide new business opportunities for utilities.

Alternatively, the utility might offer DSM as a way to cut the customer's utility bill. For example, Rochester Gas & Electric (RG&E) convinced the University of Rochester not to install a 10-MW cogeneration system (Demand-Side Report 1994). The key to this change of heart was the utility's offer of an $8-million program of energy-efficiency improvements. The university will pay about $6 million, the utility (through an energy-service company) will pay about $2 million, and the reduction in the university's utility bill should be about $2 million/year. In return, the university agreed to stay on the RG&E system for at least seven years. Southern California Edison's investment in efficiency improvements at the Eisenhower Medical Center, discussed earlier, is another example of this approach.

The likely key to success for DSM programs in the future will be increasing the participant contribution to program costs, thereby reducing the support for DSM from nonparticipating customers. Blank (1993) and Chamberlin, Herman, and Wikler (1993) suggest several possible ways to accomplish that objective.

Blank proposed a novel way to pay for DSM programs that he called the bonus-payment (BP) approach. The twin purposes of this approach are to reduce the rate impacts of DSM programs and to retain high participation levels. The utility, under this system, would pay all the up-front costs of the energy-efficiency measures plus a cash bonus. Thus, participants have no out-of-pocket costs when the measures are installed. In return, the utility recovers program costs and perhaps net lost revenues through the participants' monthly utility bills.

Blank's method is similar to shared-savings schemes tried in the past. However, these approaches may reduce participation in the program relative to what would occur if the utility offered rebates to participating customers. Blank's innovation is to have the utility provide an up-front bonus that, implicitly, compensates participants for the inconvenience associated with participation. This bonus is intended to be so attractive that it more than offsets the subsequent customer repayment.

From an analytical perspective, Blank's method is appealing if customers use discount rates in assessing energy-efficiency alternatives that are higher than the utility's discount rate (typically, the weighted average cost of capital).* If this condition applies, then customers gain the benefit of the bonus, which they value more highly than the (heavily discounted) monthly repayments. The utility, on the other hand, values the repayments more because of its lower discount rate.

Figures 15-4 and 15-5 compare the cash flows to participants and the utility, respectively, with traditional rebate programs and the BP approach. With a rebate, customers typically pay part of the cost of the DSM measures and their installation. On the other hand, customers keep all of the subsequent bill savings for themselves, as shown by the solid line in Fig. 15-4. With the BP approach, customers have a positive cash flow from the start; their subsequent monthly savings are much smaller because they are shared with the utility, as shown by the dotted line.

Figure 15-5 shows the cash flows from the two program designs for the utility. With the BP method, the utility's initial cost is higher, but its

---

*Considerable evidence exists that consumers *do* use (implicitly) higher discount rates in making energy-efficiency decisions than utilities use in making investment decisions (Ruderman, Levine, and McMahon 1987).

Figure 15-4. Annual cash flows seen by DSM-program participants under rebate and bonus-payment program designs.

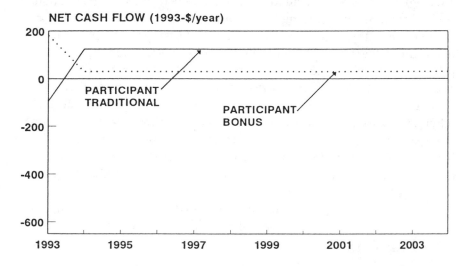

Figure 15-5. Annual cash flows seen by a utility rebate and bonus-payment program designs.

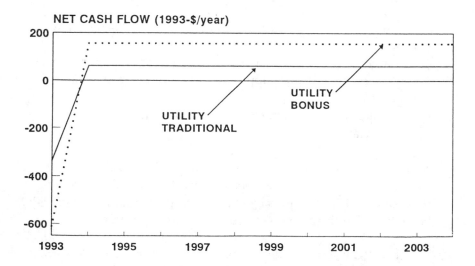

later revenues are also greater. Given the assumed differences in discount rates between participants and the utility (acting on behalf of customers as a whole), the bonus payment method should create a win-win situation.

Some utilities, including PacifiCorp and Southern California Edison, are moving to this model of DSM program. Although their programs do not include the up-front bonus that Blank recommends, both utilities provide turnkey services to participating customers in the commercial and industrial sectors. These services include analysis of energy-efficiency opportunities (including, in some cases, fuel-switching possibilities) and other productivity improvements; installation of measures; verification that the measures were installed and are operating properly (commissioning); and financing. In return, participating customers repay all the utility's costs, including interest payments, on their monthly utility bills (Hassan 1994).

Chamberlin, Herman, and Wikler (1993) suggest several ways to reduce the price increases normally associated with DSM programs. These possibilities include:

- Accept only those programs that pass the RIM test. This restriction will severely limit acceptance of programs that reduce electricity consumption, and focus DSM programs on reductions in peak demand.

- Accept programs primarily on the basis of the TRC test and
    - Ensure that all customers have access to at least some DSM programs,
    - Charge participants for energy services rather than for kWh and kW (similar to Blank's approach minus the up-front bonus payment), and
    - Allocate DSM-program costs and lost revenues within each customer class.

- Accept a package of DSM programs that passes the RIM test and has the highest TRC benefit. This approach involves selection of programs that pass the RIM test plus enough of the programs with high TRC benefits so that there is no net effect on electricity prices.

- Use the TRC test for residential programs and the RIM test for programs aimed at the larger commercial and industrial customers, with each class paying only for its own programs. This approach makes sense if small customers are responsive primarily to bills and large customers respond most to prices.

The timing of DSM programs can have substantial effects on the costs a utility avoids and therefore on the programs' cost-effectiveness (Richer 1993). DSM programs designed and timed to defer construction of new generating capacity can mitigate or eliminate rate increases.* Alternatively, running ambitious DSM programs when the utility's avoided costs are low relative to retail prices can exacerbate rate impacts. In addition, utilities could target DSM programs to those geographic areas where investments in T&D (as well as investments in generation) could be deferred, thus increasing the benefits of DSM programs.

On a more general level, utilities will need to understand, in much greater detail than they do today, what it costs to serve different groups of customers. The cost-of-service studies that utilities routinely conduct for rate cases will find new uses in utility marketing departments. Increasingly, these studies will include geographic disaggregation to reveal the large variations in avoided T&D costs across a utility's service area.

Whether the utility expenses DSM-program costs or treats these costs as capital investments that are part of the ratebase affects the timing of electricity-price increases (Hirst 1992). Expensing programs causes short-term price hikes, whereas ratebasing spreads the price increases over the amortization period. In the long run, if DSM programs are running at roughly a constant level from year to year, the differences between expensing and ratebasing in their effects on retail prices are minor. On the one hand, because customers likely have higher discount rates than utilities do, ratebasing is preferred. On the other hand, ratebasing is more expensive because of income-tax effects.

To reduce program costs, utilities may move upstream of their retail customers and undertake market-transformation programs. In such programs, utilities work with retailers, wholesalers, and manufacturers in an

---

*PUCs may still want utilities to acquire "lost-opportunity" resources even if the utility's short-term avoided costs do not call for ambitious DSM programs. Lost opportunities are actions that are time sensitive, including the chance to improve energy efficiency in a new factory or during extensive remodeling of an office building. Failure to act at the right time means that the potential efficiency gains are lost or that they will be much more expensive to acquire later on.

effort to completely change the market. Such programs are likely to be much more cost-effective than are efforts aimed at individual customers because the utility deals with far fewer entities. For example, B.C. Hydro ran an industrial efficient-motors program that largely converted the market in British Columbia from inefficient to efficient motors in three years (Flanigan and Fleming 1993), primarily by working with distributors and manufacturers.

Although such market-transformation programs are likely to be very cost-effective from a broad societal perspective, the very reason for their success—utilities no longer have to deal with individual customers to promote DSM actions—argues against their use in a competitive environment. Why should utilities work with dealers and manufacturers when the benefits will flow to all retail consumers whether or not they purchase their electricity from the utility? This dichotomy between private and public interests argues for a strong PUC role to ensure that the benefits of market transformation are not lost in a retail-competitive world.

Utilities during the past few years shifted their DSM emphasis from the residential sector to the C/I sector. They made this shift in recognition of the larger potentials for improving efficiency in the C/I sectors and the greater cost-effectiveness of such programs. In particular, the transaction costs per kWh or kW saved can be much less than in the residential sector, because each commercial or industrial customer has much greater DSM potential than individual residential customers. Ironically, if retail wheeling becomes widespread and the large C/I customers leave the local utility, the utility may be left with only the less cost-effective DSM opportunities, those in the small-C/I and residential sectors. Here, too, PUCs may want to establish regulations to ensure that the most cost-effective DSM opportunities are captured.

Utilities will have to find suitable niches (combinations of DSM products and services aimed at particular groups of customers) for their DSM activities. These niches might include new services not now widely available, such as building commissioning. They might include services that the utility currently provides for itself but not for customers, such as analysis of distribution transformers. And utilities might continue to offer many of the traditional DSM products they have been offering, although packaged in ways that are more appealing to customers, including computerized home energy audits and commercial-lighting improvements.

High-efficiency transformers provide an interesting example of a service that utilities, perhaps uniquely, could provide their large industrial customers. Such transformers, those bought both by utilities and by large customers, can provide cost-effective savings, with paybacks ranging from 1.5 to 4 years. But "very few utilities treat distribution transformers as a DSM resource or actively help their customers to purchase more efficient models. Absent encouragement from utilities, end users typically overlook efficiency in their transformer purchasing policies, most often buying based solely on lowest first cost" (Howe 1993).

Utilities have unique access to detailed information on the electricity-use patterns of their customers. Utilities might be able to sell information services to their customers, based on these data. For example, utilities could disaggregate residential-customer electric bills to show the costs of operating various appliances. Surveys show that customers do not know how much electricity various end uses consume and would find such information useful (Hopkins, Paquette, and Sioshansi 1994). For large commercial customers, utilities could link real-time price signals with building-automation systems, allowing the building's heating, ventilating, and air-conditioning control systems to respond to hour-by-hour changes in electricity prices. Such a service would benefit both the utility and the customer (Flood, Carmichael, and Vold 1994).

In the future, utility DSM programs are likely to focus more on costumer service, with less emphasis on providing capacity and energy resources to the utility. Utilities will explore, identify, and implement a much wider range of energy-efficiency and load-management programs than exist today as they work with different market segments among their customers.

CONCLUSIONS

A critical issue, not addressed in this report, concerns how today's retail monopoly franchise, cost-of-service regulation will be transformed to a more competitive environment. Utilities are concerned that PUCs may not provide them with enough flexibility to respond appropriately to competitive threats. Others are worried that residential and small commercial customers will subsidize the larger customers that have market power and choices. And environmentalists worry that DSM programs may be a victim of the rush to competition.

Another key issue concerns environmental quality. Environmental concerns may play an important role in determining both the extent of retail wheeling and the nature of utility DSM programs. If, as seems likely, society places stricter controls on emissions of air and water pollutants and remains concerned about possible electromagnetic-field effects of transmission and distribution lines, then state regulators will likely require continuation or expansion of utility DSM programs. The Climate Challenge (part of the *Climate Change Action Plan)*, includes a prominent role for U.S. electric utilities to reduce CO_2 emissions. If the industry is largely fragmented, meeting the CO_2-emission-reduction goals will be very difficult. Utilities can also play an important role in helping their customers meet environmental regulations, as Southern California Edison and others are already doing.

As of early 1994, it is hard to know the extent to which competition will affect utility DSM programs. However, it seems clear that competition at the retail level would have much more effect on DSM programs than would wholesale-level competition. The extent to which retail competition occurs will depend strongly on the role that PUCs play. If PUCs encourage retail wheeling, utility DSM programs that are financed by customers as a whole would be severely restricted. Utilities in that kind of an environment could generally run only those programs that paid for themselves. In other words, individual customers would have to pay for the DSM services that the utility provides. And that payment might have to include both the program costs and the utility's lost revenues.

If, however, PUCs limit retail wheeling or impose a fee on use of the local utility's distribution system, DSM programs could continue much the same as today. Customers covered by such programs would diminish because noncore customers would presumably not be eligible (or want) to participate in the utility's DSM programs. But utilities could continue to rely primarily on the TRC test rather than the RIM test in choosing DSM programs. Even here, though, utilities may likely face pressure to keep rates low. This pressure will require utilities to look for least-cost ways to implement DSM programs; i.e., the rate impact measure will be important here, although less so than in a full blown retail-wheeling scenario.

Utility DSM programs will
• Focus more on customer service and less on the system-resource benefits.

- Emphasize capacity reductions more and energy savings less. This shift will occur as utilities seek to minimize the lost revenues associated with DSM.

- Become more cost-effective as utilities identify better ways to deliver DSM services at lower cost.

- Decrease as utilities increasingly seek to have individual customers pay for their own DSM services.

Regardless of the extent of retail competition, the role of utility DSM will depend in large part on the ability of utilities to deliver DSM, to become more customer driven, and to lower the costs of DSM. Utilities will package DSM with other customer services, including pricing alternatives, reliability levels, power quality, and productivity improvements. All these DSM-program changes will require large changes in utility structure, operation, and culture. The challenge to regulators will be to retain the public benefits of traditional DSM programs in a more competitive environment.

[1]Based on the report "Electric-Utility DSM Programs in a Competitive Market" by Eric Hirst.

Sponsored by Office of Energy Efficiency and Renewable Energy, U.S. Department of Energy

Oak Ridge National Laboratory, Oak Ridge, Tennessee 37831, managed by Martin Marietta Energy Systems, Inc. under contract No. DE-AC05-840R21400 for U.S. Department of Energy

ACKNOWLEDGMENTS

Douglas Bauer, Eric Blank, David Berry, Ralph Cavanagh, John Chamberlin, Shel Feldman, Jane Finleon, James Gallagher, John Hartnett, Patricia Herman, Lynn Hobbie, David Meyer, Diane Pirkey, Ralph Prahl, Dan Quigley, John Rowe, Richard Scheer, Martin Schweitzer, Bruce Tonn, Michael Yokell, Fred O'Hara, and Ethel Schorn.

REFERENCES

E. Blank 1993, "Minimizing Non-Participant DSM Rate Impacts—Without Harming Participation," *The Electricity Journal* 6(4), 32-41.

E. A. Bretz 1994, "Power Marketing: Wave of the Future?" *Electrical World* 208(2), 16, February.

California Public Utilities Commission and California Energy Commission 1987, *Standard Practice Manual, Economic Analysis of Demand-Side Management Programs*, San Francisco and Sacramento, CA, December.

California Public Utilities Commission 1993, *California's Electric Services Industry: Perspectives on the Past, Strategies for the Future*, Division of Strategic Planning, San Francisco, CA, February.

R. Cavanagh 1994, *The Great "Retail Wheeling" Illusion—And More Productive Energy Futures*, Natural Resources Defense Council, New York, February.

J. Chamberlin, P. Herman, and G. Wikler 1993, "Mitigating Rate Impacts of DSM Programs," *The Electricity Journal* 6(9), 46-56, November.

W.J. Clinton and A. Gore, Jr. 1993, *The Climate Change Action Plan*, The White House, Washington, DC, October.

P. Corcoran 1994a, "The Conowingo Power Stranded Investment Case," *Edison Times*, Edison Electric Institute, Washington, DC, January.

P. Corcoran 1994b, "FERC Sets Transit Agency Case for Hearing," *Edison Times*, Edison Electric Institute, Washington, DC, February.

Demand-Side Report 1994, "Noresco Program for University of Rochester Heads Off Plans for Cogeneration Unit," 5, January 6.

Drazen-Brubaker & Associates, Inc. 1993, "Survey of Power Costs in the Southeastern United States," prepared for the Tennessee Valley Industrial Committee, October.

Electric Power Research Institute 1994, *Proceedings: 1994 Innovative Electricity Pricing*, EPRI TR-103629, Palo Alto, CA, February.

16
Appendix

Table 15-1. 10% Interest Factor

Period n	Single-payment compound-amount (F/P) Future value of $1 $(1 + i)^n$	Single-payment present-worth (P/F) Present value of $1 $\dfrac{1}{(1 + i)^n}$	Uniform-series compound-amount (F/A) Future value of uniform series of $1 $\dfrac{(1 + i)^n - 1}{i}$	Sinking-fund payment (A/F) Uniform series whose future value is $1 $\dfrac{i}{(1 + i)^n - 1}$	Capital recovery (A/P Uniform series with present value of $1 $\dfrac{i(1 + i)^n}{(1 + i)^n - 1}$	Uniform-series present-worth (P/A) Present value of uniform series of $1 $\dfrac{(1 + i)^n - 1}{i(1 + i)^n}$
1	1.100	0.9091	1.000	1.00000	1.10000	0.909
2	1.210	0.8264	2.100	0.47619	0.57619	1.736
3	1.331	0.7513	3.310	0.30211	0.40211	2.487
4	1.464	0.6830	4.641	0.21547	0.31547	3.170
5	1.611	0.6209	6.105	0.16380	0.26380	3.791
6	1.772	0.5645	7.716	0.12961	0.22961	4.355
7	1.949	0.5132	9.487	0.10541	0.20541	4.868
8	2.144	0.4665	11.436	0.08744	0.18744	5.335
9	2.358	0.4241	13.579	0.07364	0.17364	5.759
10	2.594	0.3855	15.937	0.06275	0.16275	6.144
11	2.853	0.3505	18.531	0.05396	0.15396	6.495
12	3.138	0.3186	21.384	0.04676	0.14676	6.814
13	3.452	0.2897	24.523	0.04078	0.14078	7.103
14	3.797	0.2633	27.975	0.03575	0.13575	7.367
15	4.177	0.2394	31.772	0.03147	0.13147	7.606
16	4.595	0.2176	35.950	0.02782	0.12782	7.824
17	5.054	0.1978	40.545	0.02466	0.12466	8.022
18	5.560	0.1799	45.599	0.02193	0.12193	8.201
19	6.116	0.1635	51.159	0.01955	0.11955	8.365
20	6.727	0.1486	57.275	0.01746	0.11746	8.514
21	7.400	0.1351	64.002	0.01562	0.11562	8.649
22	8.140	0.1228	71.403	0.01401	0.11401	8.772
23	8.954	0.1117	79.543	0.01257	0.11257	8.883
24	9.850	0.1015	88.497	0.01130	0.11130	8.985
25	10.835	0.0923	98.347	0.01017	0.11017	9.077
26	11.918	0.0839	109.182	0.00916	0.10916	9.161
27	13.110	0.0763	121.100	0.00826	0.10826	9.237
28	14.421	0.0693	134.210	0.00745	0.10745	9.307
29	15.863	0.0630	148.631	0.00673	0.10673	9.370
30	17.449	0.0573	164.494	0.00608	0.10608	9.427
35	28.102	0.0356	271.024	0.00369	0.10369	9.644
40	45.259	0.0221	442.593	0.00226	0.10226	9.779
45	72.890	0.0137	718.905	0.00139	0.10139	9.863
50	117.391	0.0085	1163.909	0.00086	0.10086	9.915
55	189.059	0.0053	1880.591	0.00053	0.10053	9.947
60	304.482	0.0033	3034.816	0.00033	0.10033	9.967
65	490.371	0.0020	4893.707	0.00020	0.10020	9.980
70	789.747	0.0013	7887.470	0.00013	0.10013	9.987
75	1271.895	0.0008	12708.954	0.00008	0.10008	9.992
80	2048.400	0.0005	20474.002	0.00005	0.10005	9.995
85	3298.969	0.0003	32979.690	0.00003	0.10003	9.997
90	5313.023	0.0002	53120.226	0.00002	0.10002	9.998
95	8556.676	0.0001	85556.760	0.00001	0.10001	9.999

Table 15-2. 12% Interest Factor

Period *n*	Single-payment compound-amount (F/P)	Single-payment present-worth (P/F)	Uniform-series compound-amount (F/A)	Sinking-fund payment (A/F)	Capital recovery (A/P	Uniform-series present -worth (P/A)
	Future value of $1 $(1+i)^n$	Present value of $1 $\frac{1}{(1+i)^n}$	Future value of uniform series of $1 $\frac{(1+i)^n-1}{i}$	Uniform series whose future value is $1 $\frac{i}{(1+i)^n-1}$	Uniform series with present value of $1 $\frac{i(1+i)^n}{(1+i)^n-1}$	Present value of uniform series of $1 $\frac{(1+i)^n-1}{i(1+i)^n}$
1	1.120	0.8929	1.000	1.00000	1.12000	0.893
2	1.254	0.7972	2.120	0.47170	0.59170	1.690
3	1.405	0.7118	3.374	0.29635	0.41635	2.402
4	1.574	0.6355	4.779	0.20923	0.32923	3.037
5	1.762	0.5674	6.353	0.15741	0.27741	3.605
6	1.974	0.5066	8.115	0.12323	0.24323	4.111
7	2.211	0.4523	10.089	0.09912	0.21912	4.564
8	2.476	0.4039	12.300	0.08130	0.20130	4.968
9	2.773	0.3606	14.776	0.06768	0.18768	5.328
10	3.106	0.3220	17.549	0.05698	0.17698	5.650
11	3.479	0.2875	20.655	0.04842	0.16842	5.938
12	3.896	0.2567	24.133	0.04144	0.16144	6.194
13	4.363	0.2292	28.029	0.03568	0.15568	6.424
14	4.887	0.2046	32.393	0.03087	0.15087	6.628
15	5.474	0.1827	37.280	0.02682	0.14682	6.811
16	6.130	0.1631	42.753	0.02339	0.14339	6.974
17	6.866	0.1456	48.884	0.02046	0.14046	7.120
18	7.690	0.1300	55.750	0.01794	0.13794	7.250
19	8.613	0.1161	63.440	0.01576	0.13576	7.366
20	9.646	0.1037	72.052	0.01388	0.13388	7.469
21	10.804	0.0926	81.699	0.01224	0.13224	7.562
22	12.100	0.0826	92.503	0.01081	0.13081	7.645
23	13.552	0.0738	104.603	0.00956	0.12956	7.718
24	15.179	0.0659	118.155	0.00846	0.12846	7.784
25	17.000	0.0588	133.334	0.00750	0.12750	7.843
26	19.040	0.0525	150.334	0.00665	0.12665	7.896
27	21.325	0.0469	169.374	0.00590	0.12590	7.943
28	23.884	0.0419	190.699	0.00524	0.12524	7.984
29	26.750	0.0374	214.583	0.00466	0.12466	8.022
30	29.960	0.0334	241.333	0.00414	0.12414	8.055
35	52.800	0.0189	431.663	0.00232	0.12232	8.176
40	93.051	0.0107	767.091	0.00130	0.12130	8.244
45	163.988	0.0061	1358.230	0.00074	0.12074	8.283
50	289.002	0.0035	2400.018	0.00042	0.12042	8.304
55	509.321	0.0020	4236.005	0.00024	0.12024	8.317
60	897.597	0.0011	7471.641	0.00013	0.12013	8.324
65	1581.872	0.0006	13173.937	0.00008	0.12008	8.328
70	2787.800	0.0004	23223.332	0.00004	0.12004	8.330
75	4913.056	0.0002	40933.799	0.00002	0.12002	8.332
80	8658.483	0.0001	72145.692	0.00001	0.12001	8.332

Table 15-3. 15% Interest Factors.

Period n	Single-payment compound-amount (F/P) Future value of $1 $(1 + i)^n$	Single-payment present-worth (P/F) Present value of $1 $\dfrac{1}{(1 + i)^n}$	Uniform-series compound-amount (F/A) Future value of uniform series of $1 $\dfrac{(1 + i)^n - 1}{i}$	Sinking-fund payment (A/F) Uniform series whose future value is $1 $\dfrac{i}{(1 + i)^n - 1}$	Capital recovery (A/P Uniform series with present value of $1 $\dfrac{i(1 + i)^n}{(1 + i)^n - 1}$	Uniform-series present-worth (P/A) Present value of uniform series of $1 $\dfrac{(1 + i)^n - 1}{i(1 + i)^n}$
1	1.150	0.8696	1.000	1.00000	1.15000	0.870
2	1.322	0.7561	2.150	0.46512	0.61512	1.626
3	1.521	0.6575	3.472	0.28798	0.43798	2.283
4	1.749	0.5718	4.993	0.20027	0.35027	2.855
5	2.011	0.4972	6.742	0.14832	0.29832	3.352
6	2.313	0.4323	8.754	0.11424	0.26424	3.784
7	2.660	0.3759	11.067	0.09036	0.24036	4.160
8	3.059	0.3269	13.727	0.07285	0.22285	4.487
9	3.518	0.2843	16.786	0.05957	0.20957	4.772
10	4.046	0.2472	20.304	0.04925	0.19925	5.019
11	4.652	0.2149	24.349	0.04107	0.19107	5.234
12	5.350	0.1869	29.002	0.03448	0.18448	5.421
13	6.153	0.1625	34.352	0.02911	0.17911	5.583
14	7.076	0.1413	40.505	0.02469	0.17469	5.724
15	8.137	0.1229	47.580	0.02102	0.17102	5.847
16	9.358	0.1069	55.717	0.01795	0.16795	5.954
17	10.761	0.0929	65.075	0.01537	0.16537	6.047
18	12.375	0.0808	75.836	0.01319	0.16319	6.128
19	14.232	0.0703	88.212	0.01134	0.16134	6.198
20	16.367	0.0611	102.444	0.00976	0.15976	6.259
21	18.822	0.0531	118.810	0.00842	0.15842	6.312
22	21.645	0.0462	137.632	0.00727	0.15727	6.359
23	24.891	0.0402	159.276	0.00628	0.15628	6.399
24	28.625	0.0349	184.168	0.00543	0.15543	6.434
25	32.919	0.0304	212.793	0.00470	0.15470	6.464
26	37.857	0.0264	245.712	0.00407	0.15407	6.491
27	43.535	0.0230	283.569	0.00353	0.15353	6.514
28	50.066	0.0200	327.104	0.00306	0.15306	6.534
29	57.575	0.0174	377.170	0.00265	0.15265	6.551
30	66.212	0.0151	434.745	0.00230	0.15230	6.566
35	133.176	0.0075	881.170	0.00113	0.15113	6.617
40	267.864	0.0037	1779.090	0.00056	0.15056	6.642
45	538.769	0.0019	3585.128	0.00028	0.15028	6.654
50	1083.657	0.0009	7217.716	0.00014	0.15014	6.661
55	2179.622	0.0005	14524.148	0.00007	0.15007	6.664
60	4383.999	0.0002	29219.992	0.00003	0.15003	6.665
65	8817.787	0.0001	58778.583	0.00002	0.15002	6.666

Table 15-4. 20% Interest Factor

Period n	Single-payment compound-amount (F/P)	Single-payment present-worth (P/F)	Uniform-series compound-amount (F/A)	Sinking-fund payment (A/F)	Capital recovery (A/P	Uniform-series present-worth (P/A)
	Future value of $1 $(1 + i)^n$	Present value of $1 $\dfrac{1}{(1 + i)^n}$	Future value of uniform series of $1 $\dfrac{(1 + i)^n - 1}{i}$	Uniform series whose future value is $1 $\dfrac{i}{(1 + i)^n - 1}$	Uniform series with present value of $1 $\dfrac{i(1 + i)^n}{(1 + i)^n - 1}$	Present value of uniform series of $1 $\dfrac{(1 + i)^n - 1}{i(1 + i)^n}$
1	1.200	0.8333	1.000	1.00000	1.20000	0.833
2	1.440	0.6944	2.200	0.45455	0.65455	1.528
3	1.728	0.5787	3.640	0.27473	0.47473	2.106
4	2.074	0.4823	5.368	0.18629	0.38629	2.589
5	2.488	0.4019	7.442	0.13438	0.33438	2.991
6	2.986	0.3349	9.930	0.10071	0.30071	3.326
7	3.583	0.2791	12.916	0.07742	0.27742	3.605
8	4.300	0.2326	16.499	0.06061	0.26061	3.837
9	5.160	0.1938	20.799	0.04808	0.24808	4.031
10	6.192	0.1615	25.959	0.03852	0.23852	4.192
11	7.430	0.1346	32.150	0.03110	0.23110	4.327
12	8.916	0.1122	39.581	0.02526	0.22526	4.439
13	10.699	0.0935	48.497	0.02062	0.22062	4.533
14	12.839	0.0779	59.196	0.01689	0.21689	4.611
15	15.407	0.0649	72.035	0.01388	0.21388	4.675
16	18.488	0.0541	87.442	0.01144	0.21144	4.730
17	22.186	0.0451	105.931	0.00944	0.20944	4.775
18	26.623	0.0376	128.117	0.00781	0.20781	4.812
19	31.948	0.0313	154.740	0.00646	0.20646	4.843
20	38.338	0.0261	186.688	0.00536	0.20536	4.870
21	46.005	0.0217	225.026	0.00444	0.20444	4.891
22	55.206	0.0181	271.031	0.00369	0.20369	4.909
23	66.247	0.0151	326.237	0.00307	0.20307	4.925
24	79.497	0.0126	392.484	0.00255	0.20255	4.937
25	95.396	0.0105	471.981	0.00212	0.20212	4.948
26	114.475	0.0087	567.377	0.00176	0.20176	4.956
27	137.371	0.0073	681.853	0.00147	0.20147	4.964
28	164.845	0.0061	819.223	0.00122	0.20122	4.970
29	197.814	0.0051	984.068	0.00102	0.20102	4.975
30	237.376	0.0042	1181.882	0.00085	0.20085	4.979
35	590.668	0.0017	2948.341	0.00034	0.20034	4.992
40	1469.772	0.0007	7343.858	0.00014	0.20014	4.997
45	3657.262	0.0003	18281.310	0.00005	0.20005	4.999
50	9100.438	0.0001	45497.191	0.00002	0.20002	4.909

Table 15-5. 25% Interest Factor

Period n	Single-payment compound-amount (F/P) Future value of $1 $(1+i)^n$	Single-payment present-worth (P/F) Present value of $1 $\frac{1}{(1+i)^n}$	Uniform-series compound-amount (F/A) Future value of uniform series of $1 $\frac{(1+i)^n-1}{i}$	Sinking-fund payment (A/F) Uniform series whose future value is $1 $\frac{i}{(1+i)^n-1}$	Capital recovery (A/P) Uniform series with present value of $1 $\frac{i(1+i)^n}{(1+i)^n-1}$	Uniform-series present-worth (P/A) Present value of uniform series of $1 $\frac{(1+i)^n-1}{i(1+i)^n}$
1	1.250	0.8000	1.000	1.00000	1.25000	0.800
2	1.562	0.6400	2.250	0.44444	0.69444	1.440
3	1.953	0.5120	3.812	0.26230	0.51230	1.952
4	2.441	0.4096	5.766	0.17344	0.42344	2.362
5	3.052	0.3277	8.207	0.12185	0.37185	2.689
6	3.815	0.2621	11.259	0.08882	0.33882	2.951
7	4.768	0.2097	15.073	0.06634	0.31634	3.161
8	5.960	0.1678	19.842	0.05040	0.30040	3.329
9	7.451	0.1342	25.802	0.03876	0.28876	3.463
10	9.313	0.1074	33.253	0.03007	0.28007	3.571
11	11.642	0.0859	42.566	0.02349	0.27349	3.656
12	14.552	0.0687	54.208	0.01845	0.26845	3.725
13	18.190	0.0550	68.760	0.01454	0.26454	3.780
14	22.737	0.0440	86.949	0.01150	0.26150	3.824
15	28.422	0.0352	109.687	0.00912	0.25912	3.859
16	35.527	0.0281	138.109	0.00724	0.25724	3.887
17	44.409	0.0225	173.636	0.00576	0.25576	3.910
18	55.511	0.0180	218.045	0.00459	0.25459	3.928
19	69.389	0.0144	273.556	0.00366	0.25366	3.942
20	86.736	0.0115	342.945	0.00292	0.25292	3.954
21	108.420	0.0092	429.681	0.00233	0.25233	3.963
22	135.525	0.0074	538.101	0.00186	0.25186	3.970
23	169.407	0.0059	673.626	0.00148	0.25148	3.976
24	211.758	0.0047	843.033	0.00119	0.25119	3.981
25	264.698	0.0038	1054.791	0.00095	0.25095	3.985
26	330.872	0.0030	1319.489	0.00076	0.25076	3.988
27	413.590	0.0024	1650.361	0.00061	0.25061	3.990
28	516.988	0.0019	2063.952	0.00048	0.25048	3.992
29	646.235	0.0015	2580.939	0.00039	0.25039	3.994
30	807.794	0.0012	3227.174	0.00031	0.25031	3.995
35	2465.190	0.0004	9856.761	0.00010	0.25010	3.998
40	7523.164	0.0001	30088.655	0.00003	0.25003	3.999

Table 15-6. 30% Interest Factor

Period n	Single-payment compound-amount (F/P)	Single-payment present-worth (P/F)	Uniform-series compound-amount (F/A)	Sinking-fund payment (A/F)	Capital recovery (A/P	Uniform-series present-worth (P/A)
	Future value of $1 $(1 + i)^n$	Present value of $1 $\dfrac{1}{(1+i)^n}$	Future value of uniform series of $1 $\dfrac{(1+i)^n - 1}{i}$	Uniform series whose future value is $1 $\dfrac{i}{(1+i)^n - 1}$	Uniform series with present value of $1 $\dfrac{i(1+i)^n}{(1+i)^n - 1}$	Present value of uniform series of $1 $\dfrac{(1+i)^n - 1}{i(1+i)^n}$
1	1.300	0.7692	1.000	1.00000	1.30000	0.769
2	1.690	0.5917	2.300	0.43478	0.73478	1.361
3	2.197	0.4552	3.990	0.25063	0.55063	1.816
4	2.856	0.3501	6.187	0.16163	0.46163	2.166
5	3.713	0.2693	9.043	0.11058	0.41058	2.436
6	4.827	0.2072	12.756	0.07839	0.37839	2.643
7	6.275	0.1594	17.583	0.05687	0.35687	2.802
8	8.157	0.1226	23.858	0.04192	0.34192	2.925
9	10.604	0.0943	32.015	0.03124	0.33124	3.019
10	13.786	0.0725	42.619	0.02346	0.32346	3.092
11	17.922	0.0558	56.405	0.01773	0.31773	3.147
12	23.298	0.0429	74.327	0.01345	0.31345	3.190
13	30.288	0.0330	97.625	0.01024	0.31024	3.223
14	39.374	0.0254	127.913	0.00782	0.30782	3.249
15	51.186	0.0195	167.286	0.00598	0.30598	3.268
16	66.542	0.0150	218.472	0.00458	0.30458	3.283
17	86.504	0.0116	285.014	0.00351	0.30351	3.295
18	112.455	0.0089	371.518	0.00269	0.30269	3.304
19	146.192	0.0068	483.973	0.00207	0.30207	3.311
20	190.050	0.0053	630.165	0.00159	0.30159	3.316
21	247.065	0.0040	820.215	0.00122	0.30122	3.320
22	321.184	0.0031	1067.280	0.00094	0.30094	3.323
23	417.539	0.0024	1388.464	0.00072	0.30072	3.325
24	542.801	0.0018	1806.003	0.00055	0.30055	3.327
25	705.641	0.0014	2348.803	0.00043	0.30043	3.329
26	917.333	0.0011	3054.444	0.00033	0.30033	3.330
27	1192.533	0.0008	3971.778	0.00025	0.30025	3.331
28	1550.293	0.0006	5164.311	0.00019	0.30019	3.331
29	2015.381	0.0005	6714.604	0.00015	0.30015	3.332
30	2619.996	0.0004	8729.985	0.00011	0.30011	3.332
35	9727.860	0.0001	32422.868	0.00003	0.30003	3.333

Table 15-7. 40% Interest Factor

Period n	Single-payment compound-amount (F/P)	Single-payment present-worth (P/F)	Uniform-series compound-amount (F/A)	Sinking-fund payment (A/F)	Capital recovery (A/P	Uniform-series present-worth (P/A)
	Future value of \$1 $(1+i)^n$	Present value of \$1 $\dfrac{1}{(1+i)^n}$	Future value of uniform series of \$1 $\dfrac{(1+i)^n - 1}{i}$	Uniform series whose future value is \$1 $\dfrac{i}{(1+i)^n - 1}$	Uniform series with present value of \$1 $\dfrac{i(1+i)^n}{(1+i)^n - 1}$	Present value of uniform series of \$1 $\dfrac{(1+i)^n - 1}{i(1+i)^n}$
1	1.400	0.7143	1.000	1.00000	1.40000	0.714
2	1.960	0.5102	2.400	0.41667	0.81667	1.224
3	2.744	0.3644	4.360	0.22936	0.62936	1.589
4	3.842	0.2603	7.104	0.14077	0.54077	1.849
5	5.378	0.1859	10.946	0.09136	0.49136	2.035
6	7.530	0.1328	16.324	0.06126	0.46126	2.168
7	10.541	0.0949	23.853	0.04192	0.44192	2.263
8	14.758	0.0678	34.395	0.02907	0.42907	2.331
9	20.661	0.0484	49.153	0.02034	0.42034	2.379
10	28.925	0.0346	69.814	0.01432	0.41432	2.414
11	40.496	0.0247	98.739	0.01013	0.41013	2.438
12	56.694	0.0176	139.235	0.00718	0.40718	2.456
13	79.371	0.0126	195.929	0.00510	0.40510	2.469
14	111.120	0.0090	275.300	0.00363	0.40363	2.478
15	155.568	0.0064	386.420	0.00259	0.40259	2.484
16	217.795	0.0046	541.988	0.00185	0.40185	2.489
17	304.913	0.0033	759.784	0.00132	0.40132	2.492
18	426.879	0.0023	1064.697	0.00094	0.40094	2.494
19	597.630	0.0017	1491.576	0.00067	0.40067	2.496
20	836.683	0.0012	2089.206	0.00048	0.40048	2.497
21	1171.356	0.0009	2925.889	0.00034	0.40034	2.498
22	1639.898	0.0006	4097.245	0.00024	0.40024	2.498
23	2295.857	0.0004	5737.142	0.00017	0.40017	2.499
24	3214.200	0.0003	8032.999	0.00012	0.40012	2.499
25	4499.880	0.0002	11247.199	0.00009	0.40009	2.499
26	6299.831	0.0002	15747.079	0.00006	0.40006	2.500
27	8819.764	0.0001	22046.910	0.00005	0.40005	2.500

Table 15-8.   50% Interest Factor

Period n	Single-payment compound-amount (F/P)	Single-payment present-worth (P/F)	Uniform-series compound-amount (F/A)	Sinking-fund payment (A/F)	Capital recovery (A/P	Uniform-series present-worth (P/A)
	Future value of $1 $(1+i)^n$	Present value of $1 $\dfrac{1}{(1+i)^n}$	Future value of uniform series of $1 $\dfrac{(1+i)^n-1}{i}$	Uniform series whose future value is $1 $\dfrac{i}{(1+i)^n-1}$	Uniform series with present value of $1 $\dfrac{i(1+i)^n}{(1+i)^n-1}$	Present value of uniform series of $1 $\dfrac{(1+i)^n-1}{i(1+i)^n}$
1	1.500	0.6667	1.000	1.00000	1.50000	0.667
2	2.250	0.4444	2.500	0.40000	0.90000	1.111
3	3.375	0.2963	4.750	0.21053	0.71053	1.407
4	5.062	0.1975	8.125	0.12308	0.62308	1.605
5	7.594	0.1317	13.188	0.07583	0.57583	1.737
6	11.391	0.0878	20.781	0.04812	0.54812	1.824
7	17.086	0.0585	32.172	0.03108	0.53108	1.883
8	25.629	0.0390	49.258	0.02030	0.52030	1.922
9	38.443	0.0260	74.887	0.01335	0.51335	1.948
10	57.665	0.0173	113.330	0.00882	0.50882	1.965
11	86.498	0.0116	170.995	0.00585	0.50585	1.977
12	129.746	0.0077	257.493	0.00388	0.50388	1.985
13	194.620	0.0051	387.239	0.00258	0.50258	1.990
14	291.929	0.0034	581.859	0.00172	0.50172	1.993
15	437.894	0.0023	873.788	0.00114	0.50114	1.995
16	656.841	0.0015	1311.682	0.00076	0.50076	1.997
17	985.261	0.0010	1968.523	0.00051	0.50051	1.998
18	1477.892	0.0007	2953.784	0.00034	0.50034	1.999
19	2216.838	0.0005	4431.676	0.00023	0.50023	1.999
20	3325.257	0.0003	6648.513	0.00015	0.50015	1.999
21	4987.885	0.0002	9973.770	0.00010	0.50010	2.000
22	7481.828	0.0001	14961.655	0.00007	0.50007	2.000

Table 15-9. Five-Year Escalation Table

Present Worth of a Series of Escalating Payments Compounded Annually
Discount-Escalation Factors for n = 5 Years

Discount Rate	Annual Escalation Rate					
	0.10	0.12	0.14	0.16	0.18	0.20
0.10	5.000000	5.279234	5.572605	5.880105	6.202627	6.540569
0.11	4.866862	5.136200	5.420152	5.717603	6.029313	6.355882
0.12	4.738562	5.000000	5.274242	5.561868	5.863289	6.179066
0.13	4.615647	4.869164	5.133876	5.412404	5.704137	6.009541
0.14	4.497670	4.742953	5.000000	5.269208	5.551563	5.847029
0.15	4.384494	4.622149	4.871228	5.131703	5.404955	5.691165
0.16	4.275647	4.505953	4.747390	5.000000	5.264441	5.541511
0.17	4.171042	4.394428	4.628438	4.873699	5.129353	5.397964
0.18	4.070432	4.287089	4.513947	4.751566	5.000000	5.259749
0.19	3.973684	4.183921	4.403996	4.634350	4.875619	5.126925
0.20	3.880510	4.084577	4.298207	4.521178	4.755725	5.000000
0.21	3.790801	3.989001	4.196400	4.413341	4.640260	4.877689
0.22	4.704368	3.896891	4.098287	4.308947	4.529298	4.759649
0.23	3.621094	3.808179	4.003835	4.208479	4.422339	4.645864
0.24	3.540773	3.722628	3.912807	4.111612	4.319417	4.536517
0.25	3.463301	3.640161	3.825008	4.018249	4.220158	4.431144
0.26	3.388553	3.560586	3.740376	3.928286	4.124553	4.329514
0.27	3.316408	3.483803	3.658706	3.841442	4.032275	4.231583
0.28	3.246718	3.409649	3.579870	3.757639	3.943295	4.137057
0.29	3.179393	3.338051	3.503722	3.676771	3.857370	4.045902
0.30	3.114338	3.268861	3.430201	3.598653	3.774459	3.957921
0.31	3.051452	3.201978	3.359143	3.523171	3.694328	3.872901
0.32	2.990618	3.137327	3.290436	3.450224	3.616936	3.790808
0.33	2.931764	3.074780	3.224015	3.379722	3.542100	3.711472
0.34	2.874812	3.014281	3.159770	3.311524	3.469775	3.634758

Table 15-10. Ten-Year Escalation Table

Present Worth of a Series of Escalating Payments Compounded Annually
Discount-Escalation Factors for n = 10 Years

Discount Rate	Annual Escalation Rate					
	0.10	0.12	0.14	0.16	0.18	0.20
0.10	10.000000	11.056250	12.234870	13.548650	15.013550	16.646080
0.11	9.518405	10.508020	11.613440	12.844310	14.215140	15.741560
0.12	9.068870	10.000000	11.036530	12.190470	13.474590	14.903510
0.13	8.650280	9.526666	10.498990	11.582430	12.786980	14.125780
0.14	8.259741	9.084209	10.000000	11.017130	12.147890	13.403480
0.15	7.895187	8.672058	9.534301	10.490510	11.552670	12.731900
0.16	7.554141	8.286779	9.099380	10.000000	10.998720	12.106600
0.17	7.234974	7.926784	8.693151	9.542653	10.481740	11.524400
0.18	6.935890	7.589595	8.312960	9.113885	10.000000	10.980620
0.19	6.655455	7.273785	7.957330	8.713262	9.549790	10.472990
0.20	6.392080	6.977461	7.624072	8.338518	9.128122	10.000000
0.21	6.144593	6.699373	7.311519	7.987156	8.733109	9.557141
0.22	5.911755	6.437922	7.017915	7.657542	8.363208	9.141752
0.23	5.692557	6.192047	6.742093	7.348193	8.015993	8.752133
0.24	5.485921	5.960481	6.482632	7.057347	7.690163	8.387045
0.25	5.290990	5.742294	6.238276	6.783767	7.383800	8.044173
0.26	5.106956	5.536463	6.008083	6.526298	7.095769	7.721807
0.27	4.933045	5.342146	5.790929	6.283557	6.824442	7.418647
0.28	4.768518	5.158489	5.585917	6.054608	6.568835	7.133100
0.29	4.612762	4.984826	5.392166	5.838531	6.327682	6.864109
0.30	4.465205	4.820429	5.209000	5.634354	6.100129	6.610435
0.31	4.325286	4.664669	5.035615	5.441257	5.885058	6.370867
0.32	4.192478	4.517015	4.871346	5.258512	5.681746	6.144601
0.33	4.066339	4.376884	4.715648	5.085461	5.489304	5.930659
0.34	3.946452	4.243845	4.567942	4.921409	5.307107	5.728189

Table 15-11. Fifteen-Year Escalation Table

Present Worth of a Series of Escalating Payments Compounded Annually
Discount-Escalation Factors for *n* = 15 years

Discount Rate	Annual Escalation Rate					
	0.10	0.12	0.14	0.16	0.18	0.20
0.10	15.000000	17.377880	20.199780	23.549540	27.529640	32.259620
0.11	13.964150	16.126230	18.690120	21.727370	25.328490	29.601330
0.12	13.026090	15.000000	17.332040	20.090360	23.355070	27.221890
0.13	12.177030	13.981710	16.105770	18.616160	21.581750	25.087260
0.14	11.406510	13.057790	15.000000	17.287320	19.985530	23.169060
0.15	10.706220	12.220570	13.998120	16.086500	18.545150	21.442230
0.16	10.068030	11.459170	13.088900	15.000000	17.244580	19.884420
0.17	9.485654	10.766180	12.262790	14.015480	16.066830	18.477610
0.18	8.953083	10.133630	11.510270	13.118840	15.000000	17.203010
0.19	8.465335	9.555676	10.824310	12.303300	14.030830	16.047480
0.20	8.017635	9.026333	10.197550	11.560150	13.148090	15.000000
0.21	7.606115	8.540965	9.623969	10.881130	12.343120	14.046400
0.22	7.227109	8.094845	9.097863	10.259820	11.608480	13.176250
0.23	6.877548	7.684317	8.614813	9.690559	10.936240	12.381480
0.24	6.554501	7.305762	8.170423	9.167798	10.320590	11.655310
0.25	6.255518	6.956243	7.760848	8.687104	9.755424	10.990130
0.26	5.978393	6.632936	7.382943	8.244519	9.236152	10.379760
0.27	5.721101	6.333429	7.033547	7.836080	8.757889	9.819020
0.28	5.481814	6.055485	6.710042	7.458700	8.316982	9.302823
0.29	5.258970	5.797236	6.410005	7.109541	7.909701	8.827153
0.30	5.051153	5.556882	6.131433	6.785917	7.533113	8.388091
0.31	4.857052	5.332839	5.872303	6.485500	7.184156	7.982019
0.32	4.675478	5.123753	5.630905	6.206250	6.860492	7.606122
0.33	4.505413	4.928297	5.405771	5.946343	6.559743	7.257569
0.34	4.345926	4.745399	5.195502	5.704048	6.280019	6.933897

Table 15-12. Twenty-Year Escalation Table

Present Worth of a Series of Escalating Payments Compounded Annually
Discount-Escalation Factors for n = 20 Years

Discount Rate	Annual Escalation Rate					
	0.10	0.12	0.14	0.16	0.18	0.20
0.10	20.000000	24.295450	29.722090	36.592170	45.308970	56.383330
0.11	18.213210	22.002090	26.776150	32.799710	40.417480	50.067940
0.12	16.642370	20.000000	24.210030	29.505400	36.181240	44.614710
0.13	15.259850	18.243100	21.964990	26.634490	32.502270	39.891400
0.14	14.038630	16.694830	20.000000	24.127100	29.298170	35.789680
0.15	12.957040	15.329770	18.271200	21.929940	26.498510	32.218060
0.16	11.995640	14.121040	16.746150	20.000000	24.047720	29.098950
0.17	11.138940	13.048560	15.397670	18.300390	21.894660	26.369210
0.18	10.373120	12.093400	14.201180	16.795710	20.000000	23.970940
0.19	9.686791	11.240870	13.137510	15.463070	18.326720	21.860120
0.20	9.069737	10.477430	12.188860	14.279470	16.844020	20.000000
0.21	8.513605	9.792256	11.340570	13.224610	15.527270	18.353210
0.22	8.010912	9.175267	10.579620	12.282120	14.355520	16.890730
0.23	7.555427	8.618459	9.895583	11.438060	13.309280	15.589300
0.24	7.141531	8.114476	9.278916	10.679810	12.373300	14.429370
0.25	6.764528	7.657278	8.721467	9.997057	11.533310	13.392180
0.26	6.420316	7.241402	8.216490	9.380883	10.778020	12.462340
0.27	6.105252	6.862203	7.757722	8.823063	10.096710	11.626890
0.28	5.816151	6.515563	7.339966	8.316995	9.480940	10.874120
0.29	5.550301	6.198027	6.958601	7.856833	8.922847	10.194520
0.30	5.305312	5.906440	6.609778	7.437339	8.416060	9.579437
0.31	5.079039	5.638064	6.289875	7.054007	7.954518	9.021190
0.32	4.869585	5.390575	5.995840	6.702967	7.533406	8.513612
0.33	4.675331	5.161809	5.725066	6.380829	7.148198	8.050965
0.34	4.494838	4.949990	5.475180	6.084525	6.795200	7.628322

STEAM TABLES

Table 15-13. Saturated Steam: Temperature

Temp Fahr t	Abs Press. lb per Sq In. p	Specific Volume Sat. Liquid vf	Specific Volume Evap vfg	Specific Volume Sat. Vapor vg	Enthalpy Sat. Liquid hf	Enthalpy Evap hfg	Enthalpy Sat. Vapor hg	Entropy Sat. Liquid sf	Entropy Evap sfg	Entropy Sat. Vapor sg	Temp Fahr t
32.0*	0.08859	0.016022	3304.7	3304.7	0.0179	1075.5	1075.5	0.0000	2.1873	2.1873	32.0*
34.0	0.09600	0.016021	3061.9	3061.9	1.996	1074.4	1076.4	0.0041	2.1762	2.1802	34.0
36.0	0.10395	0.016020	2839.0	2839.0	4.008	1073.2	1077.2	0.0081	2.1651	2.1732	36.0
38.0	0.11249	0.016019	2634.1	2634.2	6.018	1072.1	1078.1	0.0122	2.1541	2.1663	38.0
40.0	0.12163	0.016019	2445.8	2445.8	8.027	1071.0	1079.0	0.0162	2.1432	2.1594	40.0
42.0	0.13143	0.016019	2272.4	2272.4	10.035	1069.8	1079.9	0.0202	2.1325	2.1527	42.0
44.0	0.14192	0.016019	2112.8	2112.8	12.041	1068.7	1080.7	0.0242	2.1217	2.1459	44.0
46.0	0.15314	0.016020	1965.7	1965.7	14.047	1067.6	1081.6	0.0282	2.1111	2.1393	46.0
48.0	0.16514	0.016021	1830.0	1830.0	16.051	1066.4	1082.5	0.0321	2.1006	2.1327	48.0
50.0	0.17796	0.016023	1704.8	1704.8	18.054	1065.3	1083.4	0.0361	2.0901	2.1262	50.0
52.0	0.19165	0.016024	1589.2	1589.2	20.057	1064.2	1084.2	0.0400	2.0798	2.1197	52.0
54.0	0.20625	0.016026	1482.4	1482.4	22.058	1063.1	1085.1	0.0439	2.0695	2.1134	54.0
56.0	0.22183	0.016028	1383.6	1383.6	24.059	1061.9	1086.0	0.0478	2.0593	2.1070	56.0
58.0	0.23843	0.016031	1292.2	1292.2	26.060	1060.8	1086.9	0.0516	2.0491	2.1008	58.0
60.0	0.25611	0.016033	1207.6	1207.6	28.060	1059.7	1087.7	0.0555	2.0391	2.0946	60.0
62.0	0.27494	0.016036	1129.2	1129.2	30.059	1058.5	1088.6	0.0593	2.0291	2.0885	62.0
64.0	0.29497	0.016039	1056.5	1056.5	32.058	1057.4	1089.5	0.0632	2.0192	2.0824	64.0
66.0	0.31626	0.016043	989.0	989.1	34.056	1056.3	1090.4	0.0670	2.0094	2.0764	66.0
68.0	0.33889	0.016046	926.5	926.5	36.054	1055.2	1091.2	0.0708	1.9996	2.0704	68.0
70.0	0.36292	0.016050	868.3	868.4	38.052	1054.0	1092.1	0.0745	1.9900	2.0645	70.0
72.0	0.38844	0.016054	814.3	814.3	40.049	1052.9	1093.0	0.0783	1.9804	2.0587	72.0
74.0	0.41550	0.016058	764.1	764.1	42.046	1051.8	1093.8	0.0821	1.9708	2.0529	74.0
76.0	0.44420	0.016063	717.4	717.4	44.043	1050.7	1094.7	0.0858	1.9614	2.0472	76.0
78.0	0.47461	0.016067	673.8	673.9	46.040	1049.5	1095.6	0.0895	1.9520	2.0415	78.0
80.0	0.50683	0.016072	633.3	633.3	48.037	1048.4	1096.4	0.0932	1.9426	2.0359	80.0
82.0	0.54093	0.016077	595.5	595.5	50.033	1047.3	1097.3	0.0969	1.9334	2.0303	82.0
84.0	0.57702	0.016082	560.3	560.3	52.029	1046.1	1098.2	0.1006	1.9242	2.0248	84.0
86.0	0.61518	0.016087	527.5	527.5	54.026	1045.0	1099.0	0.1043	1.9151	2.0193	86.0
88.0	0.65551	0.016093	496.8	496.8	56.022	1043.9	1099.9	0.1079	1.9060	2.0139	88.0
90.0	0.69813	0.016099	468.1	468.1	58.018	1042.7	1100.8	0.1115	1.8970	2.0086	90.0
92.0	0.74313	0.016105	441.3	441.3	60.014	1041.6	1101.6	0.1152	1.8881	2.0033	92.0
94.0	0.79065	0.016111	416.3	416.3	62.010	1040.5	1102.5	0.1188	1.8792	1.9980	94.0
96.0	0.84072	0.016117	392.8	392.9	64.006	1039.3	1103.3	0.1224	1.8704	1.9928	96.0
98.0	0.89356	0.016123	370.9	370.9	66.003	1038.2	1104.2	0.1260	1.8617	1.9876	98.0

*The states shown are metastable

Temp Fahr t	Abs Press Lb per Sq In p	Specific Volume Sat Liquid v_f	Specific Volume Evap v_{fg}	Specific Volume Sat Vapor v_g	Enthalpy Sat Liquid h_f	Enthalpy Evap h_{fg}	Enthalpy Sat Vapor h_g	Entropy Sat Liquid s_f	Entropy Evap s_{fg}	Entropy Sat Vapor s_g	Temp Fahr t
100.0	0.94924	0.016130	350.4	350.4	67.999	1037.1	1105.1	0.1295	1.8530	1.9825	100.0
102.0	1.00789	0.016137	331.1	331.1	69.995	1035.9	1105.9	0.1331	1.8444	1.9775	102.0
104.0	1.06965	0.016144	313.1	313.1	71.992	1034.8	1106.8	0.1366	1.8358	1.9725	104.0
106.0	1.1347	0.016151	296.16	296.18	73.99	1033.6	1107.6	0.1402	1.8273	1.9675	106.0
108.0	1.2030	0.016158	280.28	280.30	75.98	1032.5	1108.5	0.1437	1.8188	1.9626	108.0
110.0	1.2750	0.016165	265.37	265.39	77.98	1031.4	1109.3	0.1472	1.8105	1.9577	110.0
112.0	1.3505	0.016173	251.21	251.23	79.98	1030.2	1110.2	0.1507	1.8021	1.9528	112.0
114.0	1.4299	0.016180	238.21	238.22	81.97	1029.1	1111.0	0.1542	1.7938	1.9480	114.0
116.0	1.5133	0.016188	225.84	225.85	83.97	1027.9	1111.9	0.1577	1.7856	1.9433	116.0
118.0	1.6009	0.016196	214.20	214.21	85.97	1026.8	1112.7	0.1611	1.7774	1.9386	118.0
120.0	1.6927	0.016204	203.25	203.26	87.97	1025.6	1113.6	0.1646	1.7693	1.9339	120.0
122.0	1.7891	0.016213	192.94	192.95	89.96	1024.5	1114.4	0.1680	1.7613	1.9293	122.0
124.0	1.8901	0.016221	183.23	183.24	91.96	1023.3	1115.3	0.1715	1.7533	1.9247	124.0
126.0	1.9959	0.016229	174.08	174.09	93.96	1022.2	1116.1	0.1749	1.7453	1.9202	126.0
128.0	2.1068	0.016238	165.45	165.47	95.96	1021.0	1117.0	0.1783	1.7374	1.9157	128.0
130.0	2.2230	0.016247	157.32	157.33	97.96	1019.8	1117.8	0.1817	1.7295	1.9112	130.0
132.0	2.3445	0.016256	149.64	149.66	99.95	1018.7	1118.6	0.1851	1.7217	1.9068	132.0
134.0	2.4717	0.016265	142.40	142.41	101.95	1017.5	1119.5	0.1884	1.7140	1.9024	134.0
136.0	2.6047	0.016274	135.55	135.57	103.95	1016.4	1120.3	0.1918	1.7063	1.8980	136.0
138.0	2.7438	0.016284	129.09	129.11	105.95	1015.2	1121.1	0.1951	1.6986	1.8937	138.0
140.0	2.8892	0.016293	122.98	123.00	107.95	1014.0	1122.0	0.1985	1.6910	1.8895	140.0
142.0	3.0411	0.016303	117.20	117.22	109.95	1012.9	1122.8	0.2018	1.6834	1.8852	142.0
144.0	3.1997	0.016312	111.74	111.76	111.95	1011.7	1123.6	0.2051	1.6759	1.8810	144.0
146.0	3.3653	0.016322	106.58	106.59	113.95	1010.5	1124.5	0.2084	1.6684	1.8769	146.0
148.0	3.5381	0.016332	101.68	101.70	115.95	1009.3	1125.3	0.2117	1.6610	1.8727	148.0
150.0	3.7184	0.016343	97.05	97.07	117.95	1008.2	1126.1	0.2150	1.6536	1.8686	150.0
152.0	3.9065	0.016353	92.66	92.68	119.95	1007.0	1126.9	0.2183	1.6463	1.8646	152.0
154.0	4.1025	0.016363	88.50	88.52	121.95	1005.8	1127.7	0.2216	1.6390	1.8606	154.0
156.0	4.3068	0.016374	84.56	84.57	123.95	1004.6	1128.6	0.2248	1.6318	1.8566	156.0
158.0	4.5197	0.016384	80.82	80.83	125.96	1003.4	1129.4	0.2281	1.6245	1.8526	158.0
160.0	4.7414	0.016395	77.27	77.29	127.96	1002.2	1130.2	0.2313	1.6174	1.8487	160.0
162.0	4.9722	0.016406	73.90	73.92	129.96	1001.0	1131.0	0.2345	1.6103	1.8448	162.0
164.0	5.2124	0.016417	70.70	70.72	131.96	999.8	1131.8	0.2377	1.6032	1.8409	164.0
166.0	5.4623	0.016428	67.67	67.68	133.97	998.6	1132.6	0.2409	1.5961	1.8371	166.0
168.0	5.7223	0.016440	64.78	64.80	135.97	997.4	1133.4	0.2441	1.5892	1.8333	168.0
170.0	5.9926	0.016451	62.04	62.06	137.97	996.2	1134.2	0.2473	1.5822	1.8295	170.0
172.0	6.2736	0.016463	59.43	59.45	139.98	995.0	1135.0	0.2505	1.5753	1.8258	172.0
174.0	6.5656	0.016474	56.95	56.97	141.98	993.8	1135.8	0.2537	1.5684	1.8221	174.0
176.0	6.8690	0.016486	54.59	54.61	143.99	992.6	1136.6	0.2568	1.5616	1.8184	176.0
178.0	7.1840	0.016498	52.35	52.36	145.99	991.4	1137.4	0.2600	1.5548	1.8147	178.0

Table 15-13. (Continued)

Temp											Temp
180.0	7.5110	0.016510	50.21	50.22	148.00	990.2	1138.2	0.2631	1.5480	1.8111	180.0
182.0	7.850	0.016522	48.172	18.189	150.01	989.0	1139.0	0.2662	1.5413	1.8075	182.0
184.0	8.203	0.016534	46.232	46.249	152.01	987.8	1139.8	0.2694	1.5346	1.8040	184.0
186.0	8.568	0.016547	44.383	44.400	154.02	986.5	1140.5	0.2725	1.5279	1.8004	186.0
188.0	8.947	0.016559	42.621	42.638	156.03	985.3	1141.3	0.2756	1.5213	1.7969	188.0
190.0	9.340	0.016572	40.941	40.957	158.04	984.1	1142.1	0.2787	1.5148	1.7934	190.0
192.0	9.747	0.016585	39.337	39.354	160.05	982.8	1142.9	0.2818	1.5082	1.7900	192.0
194.0	10.168	0.016598	37.808	37.824	162.05	981.6	1143.7	0.2848	1.5017	1.7865	194.0
196.0	10.605	0.016611	36.348	36.364	164.06	980.4	1144.4	0.2879	1.4952	1.7831	196.0
198.0	11.058	0.016624	34.954	34.970	166.08	979.1	1145.2	0.2910	1.4888	1.7798	198.0
200.0	11.526	0.016637	33.622	33.639	168.09	977.9	1146.0	0.2940	1.4824	1.7764	200.0
204.0	12.512	0.016664	31.135	31.151	172.11	975.4	1147.5	0.3001	1.4697	1.7698	204.0
208.0	13.568	0.016691	28.862	28.878	176.14	972.8	1149.0	0.3061	1.4571	1.7632	208.0
212.0	14.696	0.016719	26.782	26.799	180.17	970.3	1150.5	0.3121	1.4447	1.7568	212.0
216.0	15.901	0.016747	24.878	24.894	184.20	967.8	1152.0	0.3181	1.4323	1.7505	216.0
220.0	17.186	0.016775	23.131	23.148	188.23	965.2	1153.4	0.3241	1.4201	1.7442	220.0
224.0	18.556	0.016805	21.529	21.545	192.27	962.6	1154.9	0.3300	1.4081	1.7380	224.0
228.0	20.015	0.016834	20.056	20.073	196.31	960.0	1156.3	0.3359	1.3961	1.7320	228.0
232.0	21.567	0.016864	18.701	18.718	200.35	957.4	1157.8	0.3417	1.3842	1.7260	232.0
236.0	23.216	0.016895	17.454	17.471	204.40	954.8	1159.2	0.3476	1.3725	1.7201	236.0
240.0	24.968	0.016926	16.304	16.321	208.45	952.1	1160.6	0.3533	1.3609	1.7142	240.0
244.0	26.826	0.016958	15.243	15.260	212.50	949.5	1162.0	0.3591	1.3494	1.7085	244.0
248.0	28.796	0.016990	14.264	14.281	216.56	946.8	1163.4	0.3649	1.3379	1.7028	248.0
252.0	30.883	0.017022	13.358	13.375	220.62	944.1	1164.7	0.3706	1.3266	1.6972	252.0
256.0	33.091	0.017055	12.520	12.538	224.69	941.4	1166.1	0.3763	1.3154	1.6917	256.0
260.0	35.427	0.017089	11.745	11.762	228.76	938.6	1167.4	0.3819	1.3043	1.6862	260.0
264.0	37.894	0.017123	11.025	11.042	232.83	935.9	1168.7	0.3876	1.2933	1.6808	264.0
268.0	40.500	0.017157	10.358	10.375	236.91	933.1	1170.0	0.3932	1.2823	1.6755	268.0
272.0	43.249	0.017193	9.738	9.755	240.99	930.3	1171.3	0.3987	1.2715	1.6702	272.0
276.0	46.147	0.017228	9.162	9.180	245.08	927.5	1172.5	0.4043	1.2607	1.6650	276.0
280.0	49.200	0.017264	8.627	8.644	249.17	924.6	1173.8	0.4098	1.2501	1.6599	280.0
284.0	52.414	0.01730	8.1280	8.1453	253.3	921.7	1175.0	0.4154	1.2395	1.6548	284.0
288.0	55.795	0.01734	7.6634	7.6807	257.4	918.8	1176.2	0.4208	1.2290	1.6498	288.0
292.0	59.350	0.01738	7.2301	7.2475	261.5	915.9	1177.4	0.4263	1.2186	1.6449	292.0
296.0	63.084	0.01741	6.8259	6.8433	265.6	913.0	1178.6	0.4317	1.2082	1.6400	296.0
300.0	67.005	0.01745	6.4483	6.4658	269.7	910.0	1179.7	0.4372	1.1979	1.6351	300.0
304.0	71.119	0.01749	6.0955	6.1130	273.8	907.0	1180.9	0.4426	1.1877	1.6303	304.0
308.0	75.433	0.01753	5.7655	5.7830	278.0	904.0	1182.0	0.4479	1.1776	1.6256	308.0
312.0	79.953	0.01757	5.4566	5.4742	282.1	901.0	1183.1	0.4533	1.1676	1.6209	312.0
316.0	84.688	0.01761	5.1673	5.1849	286.3	897.9	1184.1	0.4586	1.1576	1.6162	316.0
320.0	89.643	0.01766	4.8961	4.9138	290.4	894.8	1185.2	0.4640	1.1477	1.6116	320.0
324.0	94.826	0.01770	4.6418	4.6595	294.6	891.6	1186.2	0.4692	1.1378	1.6071	324.0
328.0	100.245	0.01774	4.4030	4.4208	298.7	888.5	1187.2	0.4745	1.1280	1.6025	328.0
332.0	105.907	0.01779	4.1788	4.1966	302.9	885.3	1188.2	0.4798	1.1183	1.5981	332.0
336.0	111.820	0.01783	3.9681	3.9859	307.1	882.1	1189.1	0.4850	1.1086	1.5936	336.0

Temp Fahr t	Abs Press. Lb per Sq In p	Specific Volume			Enthalpy			Entropy			Temp Fahr t
		Sat Liquid v_f	Evap v_{fg}	Sat Vapor v_g	Sat Liquid h_f	Evap h_{fg}	Sat Vapor h_g	Sat Liquid s_f	Evap s_{fg}	Sat Vapor s_g	
340.0	117.992	0.01787	3.7699	3.7878	311.3	878.8	1190.1	0.4902	1.0990	1.5892	340.0
344.0	124.430	0.01792	3.5834	3.6013	315.5	875.5	1191.0	0.4954	1.0894	1.5849	344.0
348.0	131.142	0.01797	3.4078	3.4258	319.7	872.2	1191.9	0.5006	1.0799	1.5806	348.0
352.0	138.138	0.01801	3.2423	3.2603	323.9	868.9	1192.7	0.5058	1.0705	1.5763	352.0
356.0	145.424	0.01806	3.0863	3.1044	328.1	865.5	1193.6	0.5110	1.0611	1.5721	356.0
360.0	153.010	0.01811	2.9392	2.9573	332.3	862.1	1194.4	0.5161	1.0517	1.5678	360.0
364.0	160.903	0.01816	2.8002	2.8184	336.5	858.6	1195.2	0.5212	1.0424	1.5637	364.0
368.0	169.113	0.01821	2.6691	2.6873	340.8	855.1	1195.9	0.5263	1.0332	1.5595	368.0
372.0	177.648	0.01826	2.5451	2.5633	345.0	851.6	1196.7	0.5314	1.0240	1.5554	372.0
376.0	186.517	0.01831	2.4279	2.4462	349.3	848.1	1197.4	0.5365	1.0148	1.5513	376.0
380.0	195.729	0.01836	2.3170	2.3353	353.6	844.5	1198.0	0.5416	1.0057	1.5473	380.0
384.0	205.294	0.01842	2.2110	2.2304	357.9	840.8	1198.7	0.5466	0.9966	1.5432	384.0
388.0	215.220	0.01847	2.1126	2.1311	362.2	837.2	1199.3	0.5516	0.9876	1.5392	388.0
392.0	225.516	0.01853	2.0184	2.0369	366.5	833.4	1199.9	0.5567	0.9786	1.5352	392.0
396.0	236.193	0.01858	1.9291	1.9477	370.8	829.7	1200.4	0.5617	0.9696	1.5313	396.0
400.0	247.259	0.01864	1.8444	1.8630	375.1	825.9	1201.0	0.5667	0.9607	1.5274	400.0
404.0	258.725	0.01870	1.7640	1.7827	379.4	822.0	1201.5	0.5717	0.9518	1.5234	404.0
408.0	270.600	0.01875	1.6877	1.7064	383.8	818.2	1201.9	0.5766	0.9429	1.5195	408.0
412.0	282.894	0.01881	1.6152	1.6340	388.1	814.2	1202.4	0.5816	0.9341	1.5157	412.0
416.0	295.617	0.01887	1.5463	1.5651	392.5	810.2	1202.8	0.5866	0.9253	1.5118	416.0
420.0	308.780	0.01894	1.4808	1.4997	396.9	806.2	1203.1	0.5915	0.9165	1.5080	420.0
424.0	322.391	0.01900	1.4184	1.4374	401.3	802.2	1203.5	0.5964	0.9077	1.5042	424.0
428.0	336.463	0.01906	1.3591	1.3782	405.7	798.0	1203.7	0.6014	0.8990	1.5004	428.0
432.0	351.00	0.01913	1.30266	1.32179	410.1	793.9	1204.0	0.6063	0.8903	1.4966	432.0
436.0	366.03	0.01919	1.24887	1.26806	414.6	789.7	1204.2	0.6112	0.8816	1.4928	436.0
440.0	381.54	0.01926	1.19761	1.21687	419.0	785.4	1204.4	0.6161	0.8729	1.4890	440.0
444.0	397.56	0.01933	1.14874	1.16806	423.5	781.1	1204.6	0.6210	0.8643	1.4853	444.0
448.0	414.09	0.01940	1.10212	1.12152	428.0	776.7	1204.7	0.6259	0.8557	1.4815	448.0
452.0	431.14	0.01947	1.05764	1.07711	432.5	772.3	1204.8	0.6308	0.8471	1.4778	452.0
456.0	448.73	0.01954	1.01518	1.03472	437.0	767.8	1204.8	0.6356	0.8385	1.4741	456.0
460.0	466.87	0.01961	0.97463	0.99424	441.5	763.2	1204.8	0.6405	0.8299	1.4704	460.0
464.0	485.56	0.01969	0.93588	0.95557	446.1	758.6	1204.7	0.6454	0.8213	1.4667	464.0
468.0	504.83	0.01976	0.89885	0.91862	450.7	754.0	1204.6	0.6502	0.8127	1.4629	468.0
472.0	524.67	0.01984	0.86345	0.88329	455.2	749.3	1204.5	0.6551	0.8042	1.4592	472.0
476.0	545.11	0.01992	0.82958	0.84950	459.9	744.5	1204.3	0.6599	0.7956	1.4555	476.0

Table 15-13. (Continued)

T	P										T
480.0	566.15	0.02000	0.79716	0.81717	464.5	739.6	1204.1	0.6648	0.7871	1.4518	480.0
484.0	587.81	0.02009	0.76613	0.78622	469.1	734.7	1203.8	0.6696	0.7785	1.4481	484.0
488.0	610.10	0.02017	0.73641	0.75658	473.8	729.7	1203.5	0.6745	0.7700	1.4444	488.0
492.0	633.03	0.02026	0.70794	0.72820	478.5	724.6	1203.1	0.6793	0.7614	1.4407	492.0
496.0	656.61	0.02034	0.68065	0.70100	483.2	719.5	1202.7	0.6842	0.7528	1.4370	496.0
500.0	680.86	0.02043	0.65448	0.67492	487.9	714.3	1202.2	0.6890	0.7443	1.4333	500.0
504.0	705.78	0.02053	0.62938	0.64991	492.7	709.0	1201.7	0.6939	0.7357	1.4296	504.0
508.0	731.40	0.02062	0.60530	0.62592	497.5	703.7	1201.1	0.6987	0.7271	1.4258	508.0
512.0	757.72	0.02072	0.58218	0.60289	502.3	698.2	1200.5	0.7036	0.7185	1.4221	512.0
516.0	784.76	0.02081	0.55997	0.58079	507.1	692.7	1199.8	0.7085	0.7099	1.4183	516.0
520.0	812.53	0.02091	0.53864	0.55956	512.0	687.0	1199.0	0.7133	0.7013	1.4146	520.0
524.0	841.04	0.02102	0.51814	0.53916	516.9	681.3	1198.2	0.7182	0.6926	1.4108	524.0
528.0	870.31	0.02112	0.49843	0.51955	521.8	675.5	1197.3	0.7231	0.6839	1.4070	528.0
532.0	900.34	0.02123	0.47947	0.50070	526.8	669.6	1196.4	0.7280	0.6752	1.4032	532.0
536.0	931.17	0.02134	0.46123	0.48257	531.7	663.6	1195.4	0.7329	0.6665	1.3993	536.0
540.0	962.79	0.02146	0.44367	0.46513	536.8	657.5	1194.3	0.7378	0.6577	1.3954	540.0
544.0	995.22	0.02157	0.42677	0.44834	541.8	651.3	1193.1	0.7427	0.6489	1.3915	544.0
548.0	1028.49	0.02169	0.41048	0.43217	546.9	645.0	1191.9	0.7476	0.6400	1.3876	548.0
552.0	1062.59	0.02182	0.39479	0.41660	552.0	638.5	1190.6	0.7525	0.6311	1.3837	552.0
556.0	1097.55	0.02194	0.37966	0.40160	557.2	632.0	1189.2	0.7575	0.6222	1.3797	556.0
560.0	1133.38	0.02207	0.36507	0.38714	562.4	625.3	1187.7	0.7625	0.6132	1.3757	560.0
564.0	1170.10	0.02221	0.35099	0.37320	567.6	618.5	1186.1	0.7674	0.6041	1.3716	564.0
568.0	1207.72	0.02235	0.33741	0.35975	572.9	611.5	1184.5	0.7725	0.5950	1.3675	568.0
572.0	1246.26	0.02249	0.32429	0.34678	578.3	604.5	1182.7	0.7775	0.5859	1.3634	572.0
576.0	1285.74	0.02264	0.31162	0.33426	583.7	597.2	1180.9	0.7825	0.5766	1.3592	576.0
580.0	1326.17	0.02279	0.29937	0.32216	589.1	589.9	1179.0	0.7876	0.5673	1.3550	580.0
584.0	1367.7	0.02295	0.28753	0.31048	594.6	582.4	1176.9	0.7927	0.5580	1.3507	584.0
588.0	1410.0	0.02311	0.27608	0.29919	600.1	574.7	1174.8	0.7978	0.5485	1.3464	588.0
592.0	1453.3	0.02328	0.26499	0.28827	605.7	566.8	1172.6	0.8030	0.5390	1.3420	592.0
596.0	1497.8	0.02345	0.25425	0.27770	611.4	558.8	1170.2	0.8082	0.5293	1.3375	596.0
600.0	1543.2	0.02364	0.24384	0.26747	617.1	550.6	1167.7	0.8134	0.5196	1.3330	600.0
604.0	1589.7	0.02382	0.23374	0.25757	622.9	542.2	1165.1	0.8187	0.5097	1.3284	604.0
608.0	1637.3	0.02402	0.22394	0.24796	628.8	533.6	1162.4	0.8240	0.4997	1.3238	608.0
612.0	1686.1	0.02422	0.21442	0.23865	634.8	524.7	1159.5	0.8294	0.4896	1.3190	612.0
616.0	1735.9	0.02444	0.20516	0.22960	640.8	515.6	1156.4	0.8348	0.4794	1.3141	616.0
620.0	1786.9	0.02466	0.19615	0.22081	646.9	506.3	1153.2	0.8403	0.4689	1.3092	620.0
624.0	1839.0	0.02489	0.18737	0.21226	653.1	496.6	1149.8	0.8458	0.4583	1.3041	624.0
628.0	1892.4	0.02514	0.17880	0.20394	659.5	486.7	1146.1	0.8514	0.4474	1.2988	628.0
632.0	1947.0	0.02539	0.17044	0.19583	665.9	476.4	1142.2	0.8571	0.4364	1.2934	632.0
636.0	2002.8	0.02566	0.16226	0.18792	672.4	465.7	1138.1	0.8628	0.4251	1.2879	636.0
640.0	2059.9	0.02595	0.15427	0.18021	679.1	454.6	1133.7	0.8686	0.4134	1.2821	640.0
644.0	2118.3	0.02625	0.14644	0.17269	685.9	443.1	1129.0	0.8746	0.4015	1.2761	644.0
648.0	2178.1	0.02657	0.13876	0.16534	692.9	431.1	1124.0	0.8806	0.3893	1.2699	648.0
652.0	2239.2	0.02691	0.13124	0.15816	700.0	418.7	1118.7	0.8868	0.3767	1.2634	652.0
656.0	2301.7	0.02728	0.12387	0.15115	707.4	405.7	1113.1	0.8931	0.3637	1.2567	656.0

Temp Fahr t	Abs Press. lb per Sq In p	Specific Volume			Enthalpy			Entropy			Temp Fahr t
		Sat. Liquid v_f	Evap v_{fg}	Sat. Vapor v_g	Sat. Liquid h_f	Evap h_{fg}	Sat. Vapor h_g	Sat. Liquid s_f	Evap s_{fg}	Sat. Vapor s_g	
660.0	2365.7	0.02768	0.11663	0.14431	714.9	392.1	1107.0	0.8995	0.3502	1.2498	660.0
664.0	2431.1	0.02811	0.10947	0.13757	722.9	377.7	1100.6	0.9064	0.3361	1.2425	664.0
668.0	2498.1	0.02858	0.10229	0.13087	731.5	362.1	1093.5	0.9137	0.3210	1.2347	668.0
672.0	2566.6	0.02911	0.09514	0.12424	740.2	345.7	1085.9	0.9212	0.3054	1.2266	672.0
676.0	2636.8	0.02970	0.08799	0.11769	749.2	328.5	1077.6	0.9287	0.2892	1.2179	676.0
680.0	2708.6	0.03037	0.08080	0.11117	758.5	310.1	1068.5	0.9365	0.2720	1.2086	680.0
684.0	2782.1	0.03114	0.07349	0.10463	768.2	290.2	1058.4	0.9447	0.2537	1.1984	684.0
688.0	2857.4	0.03204	0.06595	0.09799	778.8	268.2	1047.0	0.9535	0.2337	1.1872	688.0
692.0	2934.5	0.03313	0.05797	0.09110	790.5	243.1	1033.6	0.9634	0.2110	1.1744	692.0
696.0	3013.4	0.03455	0.04916	0.08371	804.4	212.8	1017.2	0.9749	0.1841	1.1591	696.0
700.0	3094.3	0.03662	0.03857	0.07519	822.4	172.7	995.2	0.9901	0.1490	1.1390	700.0
702.0	3135.5	0.03824	0.03173	0.06997	835.0	144.7	979.7	1.0006	0.1246	1.1252	702.0
704.0	3177.2	0.04108	0.02192	0.06300	854.2	102.0	956.2	1.0169	0.0876	1.1046	704.0
705.0	3198.3	0.04427	0.01304	0.05730	873.0	61.4	934.4	1.0329	0.0527	1.0856	705.0
705.47*	3208.2	0.05078	0.00000	0.05078	906.0	0.0	906.0	1.0612	0.0000	1.0612	705.47*

*Critical temperature

Table 15-14. Saturated Steam: Pressure Table

Abs Press. Lb/Sq In. p	Temp Fahr t	Specific Volume Sat. Liquid v_f	Specific Volume Evap v_{fg}	Specific Volume Sat. Vapor v_g	Enthalpy Sat. Liquid h_f	Enthalpy Evap h_{fg}	Enthalpy Sat. Vapor h_g	Entropy Sat. Liquid s_f	Entropy Evap s_{fg}	Entropy Sat. Vapor s_g	Abs Press. Lb/Sq In. p
0.08865	32.018	0.016022	3302.4	3302.4	0.0003	1075.5	1075.5	0.0000	2.1872	2.1872	0.08865
0.25	59.323	0.016032	1235.5	1235.5	27.382	1060.1	1087.4	0.0542	2.0425	2.0967	0.25
0.50	79.586	0.016071	641.5	641.5	47.623	1048.6	1096.2	0.0925	1.9446	2.0370	0.50
1.0	101.74	0.016136	333.59	333.60	69.73	1036.1	1105.8	0.1326	1.8455	1.9781	1.0
5.0	162.24	0.016407	73.515	73.532	130.20	1000.9	1131.1	0.2349	1.6094	1.8443	5.0
10.0	193.21	0.016592	38.404	38.420	161.26	982.1	1143.3	0.2836	1.5043	1.7879	10.0
14.696	212.00	0.016719	26.782	26.799	180.17	970.3	1150.5	0.3121	1.4447	1.7568	14.696
15.0	213.03	0.016726	26.274	26.290	181.17	969.7	1150.9	0.3137	1.4415	1.7552	15.0
20.0	227.96	0.016834	20.070	20.087	196.27	960.1	1156.3	0.3358	1.3962	1.7320	20.0
30.0	250.34	0.017009	13.7266	13.7436	218.9	945.2	1164.1	0.3682	1.3313	1.6995	30.0
40.0	267.25	0.017151	10.4794	10.4965	236.1	933.6	1169.8	0.3921	1.2844	1.6765	40.0
50.0	281.02	0.017274	8.4967	8.5140	250.2	923.9	1174.1	0.4112	1.2474	1.6586	50.0
60.0	292.71	0.017383	7.1562	7.1736	262.2	915.4	1177.6	0.4273	1.2167	1.6440	60.0
70.0	302.93	0.017482	6.1875	6.2050	272.7	907.8	1180.6	0.4411	1.1905	1.6316	70.0
80.0	312.04	0.017573	5.4536	5.4711	282.1	900.9	1183.1	0.4534	1.1675	1.6208	80.0
90.0	320.28	0.017659	4.8779	4.8953	290.7	894.6	1185.3	0.4643	1.1470	1.6113	90.0
100.0	327.82	0.017740	4.4133	4.4310	298.5	888.6	1187.2	0.4743	1.1284	1.6027	100.0
110.0	334.79	0.01782	4.0306	4.0484	305.8	883.1	1188.9	0.4834	1.1115	1.5950	110.0
120.0	341.27	0.01789	3.7097	3.7275	312.6	877.8	1190.4	0.4919	1.0960	1.5879	120.0
130.0	347.33	0.01796	3.4364	3.4544	319.0	872.8	1191.7	0.4998	1.0815	1.5813	130.0
140.0	353.04	0.01803	3.2010	3.2190	325.0	868.0	1193.0	0.5071	1.0681	1.5752	140.0
150.0	358.43	0.01809	2.9958	3.0139	330.6	863.4	1194.1	0.5141	1.0554	1.5695	150.0
160.0	363.55	0.01815	2.8155	2.8336	336.1	859.0	1195.1	0.5206	1.0435	1.5641	160.0
170.0	368.42	0.01821	2.6556	2.6738	341.2	854.8	1196.0	0.5269	1.0322	1.5591	170.0
180.0	373.08	0.01827	2.5129	2.5312	346.2	850.7	1196.9	0.5328	1.0215	1.5543	180.0
190.0	377.53	0.01833	2.3847	2.4030	350.9	846.7	1197.6	0.5384	1.0113	1.5498	190.0
200.0	381.80	0.01839	2.2689	2.2873	355.5	842.8	1198.3	0.5438	1.0016	1.5454	200.0
210.0	385.91	0.01844	2.16373	2.18217	359.9	839.1	1199.0	0.5490	0.9923	1.5413	210.0
220.0	389.88	0.01850	2.06779	2.08629	364.2	835.4	1199.6	0.5540	0.9834	1.5374	220.0
230.0	393.70	0.01855	1.97991	1.99846	368.3	831.8	1200.1	0.5588	0.9748	1.5336	230.0
240.0	397.39	0.01860	1.89909	1.91769	372.3	828.4	1200.6	0.5634	0.9665	1.5299	240.0
250.0	400.97	0.01865	1.82452	1.84317	376.1	825.0	1201.1	0.5679	0.9585	1.5264	250.0
260.0	404.44	0.01870	1.75548	1.77418	379.9	821.6	1201.5	0.5722	0.9508	1.5230	260.0
270.0	407.80	0.01875	1.69137	1.71013	383.6	818.3	1201.9	0.5764	0.9433	1.5197	270.0
280.0	411.07	0.01880	1.63169	1.65049	387.1	815.1	1202.3	0.5805	0.9361	1.5166	280.0
290.0	414.25	0.01885	1.57597	1.59482	390.6	812.0	1202.6	0.5844	0.9291	1.5135	290.0
300.0	417.35	0.01889	1.52384	1.54274	394.0	808.9	1202.9	0.5882	0.9223	1.5105	300.0
350.0	431.73	0.01912	1.30642	1.32554	409.8	794.2	1204.0	0.6059	0.8909	1.4968	350.0
400.0	444.60	0.01934	1.14162	1.16095	424.2	780.4	1204.6	0.6217	0.8630	1.4847	400.0

Abs Press. Lb/Sq In. p	Temp Fahr t	Specific Volume			Enthalpy			Entropy			Abs Press. Lb/Sq In. p
		Sat. Liquid v_f	Evap v_{fg}	Sat. Vapor v_g	Sat. Liquid h_f	Evap h_{fg}	Sat. Vapor h_g	Sat. Liquid s_f	Evap s_{fg}	Sat. Vapor s_g	
450.0	456.28	0.01954	1.01224	1.03179	437.3	767.5	1204.8	0.6360	0.8378	1.4738	450.0
500.0	467.01	0.01975	0.90787	0.92762	449.5	755.1	1204.7	0.6490	0.8148	1.4639	500.0
550.0	476.94	0.01994	0.82183	0.84177	460.9	743.3	1204.3	0.6611	0.7936	1.4547	550.0
600.0	486.20	0.02013	0.74962	0.76975	471.7	732.0	1203.7	0.6723	0.7738	1.4461	600.0
650.0	494.89	0.02032	0.68811	0.70843	481.9	720.9	1202.8	0.6828	0.7552	1.4381	650.0
700.0	503.08	0.02050	0.63505	0.65556	491.6	710.2	1201.8	0.6928	0.7377	1.4304	700.0
750.0	510.84	0.02069	0.58880	0.60949	500.9	699.8	1200.7	0.7022	0.7210	1.4232	750.0
800.0	518.21	0.02087	0.54809	0.56896	509.8	689.6	1199.4	0.7111	0.7051	1.4163	800.0
850.0	525.24	0.02105	0.51197	0.53302	518.4	679.5	1198.0	0.7197	0.6899	1.4096	850.0
900.0	531.95	0.02123	0.47968	0.50091	526.7	669.7	1196.4	0.7279	0.6753	1.4032	900.0
950.0	538.39	0.02141	0.45064	0.47205	534.7	660.0	1194.7	0.7358	0.6612	1.3970	950.0
1000.0	544.58	0.02159	0.42436	0.44596	542.6	650.4	1192.9	0.7434	0.6476	1.3910	1000.0
1050.0	550.53	0.02177	0.40047	0.42224	550.1	640.9	1191.0	0.7507	0.6344	1.3851	1050.0
1100.0	556.28	0.02195	0.37863	0.40058	557.5	631.5	1189.1	0.7578	0.6216	1.3794	1100.0
1150.0	561.82	0.02214	0.35859	0.38073	564.8	622.2	1187.0	0.7647	0.6091	1.3738	1150.0
1200.0	567.19	0.02232	0.34013	0.36245	571.9	613.0	1184.8	0.7714	0.5969	1.3683	1200.0
1250.0	572.38	0.02250	0.32306	0.34556	578.8	603.8	1182.6	0.7780	0.5850	1.3630	1250.0
1300.0	577.42	0.02269	0.30722	0.32991	585.6	594.6	1180.2	0.7843	0.5733	1.3577	1300.0
1350.0	582.32	0.02288	0.29250	0.31537	592.3	585.4	1177.8	0.7906	0.5620	1.3525	1350.0
1400.0	587.07	0.02307	0.27871	0.30178	598.8	576.5	1175.3	0.7966	0.5507	1.3474	1400.0
1450.0	591.70	0.02327	0.26584	0.28911	605.3	567.4	1172.8	0.8026	0.5397	1.3423	1450.0
1500.0	596.20	0.02346	0.25372	0.27719	611.7	558.4	1170.1	0.8085	0.5288	1.3373	1500.0
1550.0	600.59	0.02366	0.24235	0.26601	618.0	549.4	1167.4	0.8142	0.5182	1.3324	1550.0
1600.0	604.87	0.02387	0.23159	0.25545	624.2	540.3	1164.5	0.8199	0.5076	1.3274	1600.0
1650.0	609.05	0.02407	0.22143	0.24551	630.4	531.3	1161.6	0.8254	0.4971	1.3225	1650.0
1700.0	613.13	0.02428	0.21178	0.23607	636.5	522.2	1158.6	0.8309	0.4867	1.3176	1700.0
1750.0	617.12	0.02450	0.20263	0.22713	642.5	513.1	1155.6	0.8363	0.4765	1.3128	1750.0
1800.0	621.02	0.02472	0.19390	0.21861	648.5	503.8	1152.3	0.8417	0.4662	1.3079	1800.0
1850.0	624.83	0.02495	0.18558	0.21052	654.5	494.6	1149.0	0.8470	0.4561	1.3030	1850.0
1900.0	628.56	0.02517	0.17761	0.20278	660.4	485.2	1145.6	0.8522	0.4459	1.2981	1900.0
1950.0	632.22	0.02541	0.16999	0.19540	666.3	475.8	1142.0	0.8574	0.4358	1.2931	1950.0
2000.0	635.80	0.02565	0.16266	0.18831	672.1	466.2	1138.3	0.8625	0.4256	1.2881	2000.0
2100.0	642.76	0.02615	0.14885	0.17501	683.8	446.7	1130.5	0.8727	0.4053	1.2780	2100.0
2200.0	649.45	0.02669	0.13603	0.16272	695.5	426.7	1122.2	0.8828	0.3848	1.2676	2200.0
2300.0	655.89	0.02727	0.12406	0.15133	707.2	406.0	1113.2	0.8929	0.3640	1.2569	2300.0
2400.0	662.11	0.02790	0.11287	0.14076	719.0	384.8	1103.7	0.9031	0.3430	1.2460	2400.0
2500.0	668.11	0.02859	0.10209	0.13068	731.7	361.6	1093.3	0.9139	0.3206	1.2345	2500.0
2600.0	673.91	0.02938	0.09172	0.12110	744.5	337.6	1082.0	0.9247	0.2977	1.2225	2600.0
2700.0	679.53	0.03029	0.08165	0.11194	757.3	312.3	1069.7	0.9356	0.2741	1.2097	2700.0
2800.0	684.96	0.03134	0.07171	0.10305	770.7	285.1	1055.8	0.9468	0.2491	1.1958	2800.0
2900.0	690.22	0.03262	0.06158	0.09420	785.1	254.7	1039.8	0.9588	0.2215	1.1803	2900.0
3000.0	695.33	0.03428	0.05073	0.08500	801.8	218.4	1020.3	0.9728	0.1891	1.1619	3000.0
3100.0	700.28	0.03681	0.03771	0.07452	824.0	169.3	993.3	0.9914	0.1460	1.1373	3100.0
3200.0	705.08	0.04472	0.01191	0.05663	875.5	56.1	931.6	1.0351	0.0482	1.0832	3200.0
3208.2*	705.47	0.05078	0.00000	0.05078	906.0	0.0	906.0	1.0612	0.0000	1.0612	3208.2*

*Critical pressure

Table 15-15. Superheated Steam

Temperature (°F)

Abs. Press. (psi) (Sat. Temp.)		Sat. Water	Sat. Steam	200	250	300	350	400	450	500	600	700	800	900	1000	1100	1200
1 (101.74)	Sh			98.26	148.26	198.26	248.26	298.26	348.26	398.26	498.26	598.26	698.26	798.26	898.26	998.26	1098.26
	v	0.01614	333.6	392.5	422.4	452.3	482.1	511.9	541.7	571.5	631.1	690.7	750.3	809.8	869.4	929.0	988.6
	h	69.73	1105.8	1150.2	1172.9	1195.7	1218.7	1241.8	1265.1	1288.6	1336.1	1384.5	1433.7	1483.8	1534.9	1586.8	1639.7
	s	0.1326	.9781	2.0509	2.0841	2.1152	2.1445	2.1722	2.1985	2.2237	2.2708	2.3144	2.3551	2.3934	2.4296	2.4640	2.4969
5 (162.24)	Sh			37.76	87.76	137.76	187.76	237.76	287.76	337.76	437.76	537.76	637.76	737.76	837.76	937.76	1037.76
	v	0.01641	73.53	78.14	84.21	90.24	96.25	102.24	108.23	114.21	126.15	138.08	150.01	161.94	173.86	185.78	197.70
	h	130.20	1131.1	1148.6	1171.7	1194.8	1218.0	1241.3	1264.7	1288.2	1335.9	1384.3	1433.6	1483.7	1534.7	1586.8	1639.6
	s	0.2349	1.8443	1.8716	1.9054	1.9369	1.9664	1.9943	2.0208	2.0460	2.0932	2.1369	2.1776	2.2159	2.2521	2.2866	2.3194
10 (193.21)	Sh			6.79	56.79	106.79	156.79	206.79	256.79	306.79	406.79	506.79	606.79	706.79	806.79	906.79	1006.79
	v	0.01659	38.42	38.84	41.93	44.98	48.02	51.03	54.04	57.04	63.03	69.00	74.98	80.94	86.91	92.87	98.84
	h	161.26	1143.3	1146.6	1170.2	1193.7	1217.1	1240.6	1264.1	1287.8	1335.5	1384.0	1433.4	1483.5	1534.6	1586.6	1639.5
	s	0.2836	1.7879	1.7928	1.8273	1.8592	1.8892	1.9173	1.9439	1.9692	2.0166	2.0603	2.1011	2.1394	2.1757	2.2101	2.2430
14.696 (212.00)	Sh				38.00	88.00	138.00	188.00	238.00	288.00	388.00	488.00	588.00	688.00	788.00	888.00	988.00
	v	.0167	26.799		28.42	30.52	32.60	34.67	36.72	38.77	42.86	46.93	51.00	55.06	59.13	63.19	67.25
	h	180.17	1150.5		1168.8	1192.6	1216.3	1239.9	1263.6	1287.4	1335.2	1383.8	1433.2	1483.4	1534.5	1586.5	1639.4
	s	.3121	1.7568		1.7833	1.8158	1.8459	1.8743	1.9010	1.9265	1.9739	2.0177	2.0585	2.0969	2.1332	2.1676	2.2005
15 (213.03)	Sh				36.97	86.97	136.97	186.97	236.97	286.97	386.97	486.97	586.97	686.97	786.97	886.97	986.97
	v	0.01673	26.290		27.837	29.899	31.939	33.963	35.977	37.985	41.986	45.978	49.964	53.946	57.926	61.905	65.882
	h	181.21	1150.9		1168.7	1192.5	1216.2	1239.9	1263.6	1287.3	1335.2	1383.8	1433.2	1483.4	1534.5	1586.5	1639.4
	s	.3137	1.7552		1.7809	1.8134	1.8437	1.8720	1.8988	1.9242	1.9717	2.0155	2.0563	2.0946	2.1309	2.1653	2.1982
20 (227.96)	Sh				22.04	72.04	122.04	172.04	222.04	272.04	372.04	472.04	572.04	672.04	772.04	872.04	972.04
	v	0.01683	20.087		20.788	22.356	23.900	25.428	26.946	28.457	31.466	34.465	37.458	40.447	43.435	46.420	49.405
	h	196.27	1156.3		1167.1	1191.4	1215.4	1239.2	1263.0	1286.9	1334.9	1383.5	1432.9	1483.2	1534.3	1586.3	1639.3
	s	.3358	1.7320		1.7475	1.7805	1.8111	1.8397	1.8666	1.8921	1.9397	1.9836	2.0244	2.0628	2.0991	2.1336	2.1982
25 (240.07)	Sh				9.93	59.93	109.93	159.93	209.93	259.93	359.93	459.93	559.93	659.93	759.93	859.93	959.93
	v	0.01693	16.301		16.558	17.829	19.076	20.307	21.527	22.740	25.153	27.557	29.954	32.348	34.740	37.130	39.518
	h	208.52	1160.6		1165.6	1190.2	1214.5	1238.5	1262.5	1286.4	1334.6	1383.3	1432.7	1483.0	1534.2	1586.2	1639.2
	s	.3535	1.7141		1.7212	1.7547	1.7856	1.8145	1.8415	1.8672	1.9149	1.9588	1.9997	2.0381	2.0744	2.1089	2.1418
30 (250.34)	Sh					49.66	99.66	149.66	199.66	249.66	349.66	449.66	549.66	649.66	749.66	849.66	949.66
	v	0.01701	13.744			14.810	15.859	16.892	17.914	18.929	20.945	22.951	24.952	26.949	28.943	30.936	32.927
	h	218.93	1164.1			1189.0	1213.6	1237.8	1261.9	1286.0	1334.2	1383.0	1432.5	1482.8	1534.0	1586.1	1639.0
	s	.3682	1.6995			1.7334	1.7647	1.7937	1.8210	1.8467	1.8946	1.9386	1.9795	2.0179	2.0543	2.0888	2.1217

Superheated steam table (values of v, h, s at superheat Sh).

p (psia) (Tsat)		Sat. liq.	Sat. vap.												
35 (259.29)	Sh			40.71	90.71	140.71	190.71	240.71	340.71	440.71	540.71	640.71	740.71	840.71	940.71
	v	0.01708	11.896	12.654	13.562	14.453	15.334	16.207	17.939	19.662	21.379	23.092	24.803	26.512	28.220
	h	228.03	1167.1	1187.8	1212.7	1237.1	1261.3	1285.5	1333.9	1382.8	1432.3	1482.7	1533.9	1586.0	1638.0
	s	0.3809	1.6872	1.7152	1.7468	1.7761	1.8035	1.8294	1.8774	1.9214	1.9624	2.0009	2.0372	2.0717	2.1046
40 (267.25)	Sh			32.75	82.75	132.75	182.75	232.75	332.75	432.75	532.75	632.75	732.75	832.75	932.75
	v	0.01715	10.497	11.036	11.838	12.624	13.398	14.165	15.685	17.195	18.699	20.199	21.697	23.194	24.689
	h	236.14	1169.8	1186.6	1211.7	1236.4	1260.8	1285.0	1333.6	1382.5	1432.1	1482.5	1533.7	1585.8	1638.8
	s	0.3921	1.6765	1.6992	1.7312	1.7608	1.7883	1.8143	1.8624	1.9065	1.9476	1.9860	2.0224	2.0569	2.0899
45 (274.44)	Sh			25.56	75.56	125.56	175.56	225.56	325.56	425.56	525.56	625.56	725.56	825.56	925.56
	v	0.01721	9.399	9.777	10.497	11.201	11.892	12.577	13.932	15.276	16.614	17.950	19.282	20.613	21.943
	h	243.49	1172.1	1185.4	1210.4	1235.7	1260.2	1284.6	1333.3	1382.3	1431.9	1482.3	1533.6	1585.7	1638.7
	s	0.4021	1.6671	1.6849	1.7173	1.7471	1.7748	1.8010	1.8492	1.8934	1.9345	1.9730	2.0093	2.0439	2.0768
50 (281.02)	Sh			18.98	68.98	118.98	168.98	218.98	318.98	418.98	518.98	618.98	718.98	818.98	918.98
	v	0.01727	8.514	8.769	9.424	10.062	10.688	11.306	12.529	13.741	14.947	16.150	17.350	18.549	19.746
	h	250.21	1174.1	1184.1	1209.9	1234.9	1259.6	1284.1	1332.9	1382.0	1431.7	1482.2	1533.4	1585.6	1638.6
	s	0.4112	1.6586	1.6720	1.7048	1.7349	1.7628	1.7890	1.8374	1.8816	1.9227	1.9613	1.9977	2.0322	2.0652
55 (287.07)	Sh			12.93	62.93	112.93	162.93	212.93	312.93	412.93	512.93	612.93	712.93	812.93	912.93
	v	0.01733		7.945	8.546	9.130	9.702	10.267	11.381	12.485	13.583	14.677	15.769	16.859	17.948
	h	256.43		1182.9	1208.9	1234.2	1259.1	1283.6	1332.6	1381.8	1431.5	1482.0	1533.3	1585.5	1638.5
	s	0.4196		1.6601	1.6913	1.7237	1.7518	1.7781	1.8266	1.8710	1.9121	1.9507	1.9871	2.022	2.055
60 (292.71)	Sh			7.29	57.29	107.29	157.29	207.29	307.29	407.29	507.29	607.29	707.29	807.29	907.29
	v	0.01738	7.174	7.257	7.815	8.354	8.881	9.400	10.425	11.438	12.446	13.450	14.452	15.452	16.450
	h	262.21	1177.6	1181.6	1208.0	1233.5	1258.5	1283.2	1332.3	1381.5	1431.3	1481.8	1533.2	1585.3	1638.4
	s	0.4273	1.6440	1.6492	1.6934	1.7134	1.7417	1.7681	1.8168	1.8612	1.9024	1.9410	1.9774	2.0120	2.0450
65 (297.98)	Sh			2.02	52.02	102.02	152.02	202.02	302.02	402.02	502.02	602.02	702.02	802.02	902.02
	v	0.01743	6.653	6.675	7.195	7.697	8.186	8.667	9.615	10.552	11.484	12.412	13.337	14.261	15.183
	h	267.63	1179.1	1180.3	1207.0	1232.7	1257.9	1282.7	1331.9	1381.3	1431.1	1481.6	1533.0	1585.2	1638.3
	s	0.4344	1.6375	1.6390	1.6731	1.7040	1.7324	1.7590	1.8077	1.8522	1.8935	1.9321	1.9685	2.0031	2.0361
70 (302.93)	Sh				47.07	97.07	147.07	197.07	297.07	397.07	497.07	597.07	697.07	797.07	897.07
	v	0.01748	6.205		6.664	7.133	7.590	8.039	8.922	9.793	10.659	11.522	12.382	13.240	14.097
	h	272.74	1180.6		1206.0	1232.0	1257.3	1282.2	1331.6	1381.0	1430.9	1481.5	1532.9	1585.1	1638.2
	s	0.4411	1.6316		1.6640	1.6951	1.7237	1.7504	1.7993	1.8439	1.8852	1.9238	1.9603	1.9949	2.0279
75 (307.61)	Sh				42.39	92.39	142.39	192.39	292.39	392.39	492.39	592.39	692.39	792.39	892.39
	v	0.01753	5.814		6.204	6.645	7.074	7.494	8.320	9.135	9.945	10.750	11.553	12.355	13.155
	h	277.56	1181.9		1205.0	1231.2	1256.7	1281.7	1331.3	1380.7	1430.7	1481.3	1532.7	1585.0	1638.1
	s	0.4474	1.6260		1.6554	1.6868	1.7156	1.7424	1.7915	1.8361	1.8774	1.9161	1.9526	1.9872	2.0202

Table 15-15. Continued

Abs. Press. (psi) (Sat. Temp.)		Sat. Water	Sat. Steam	350	400	450	500	550	600	700	800	900	1000	1100	1200	1300	1400
80 (312.04)	Sh			37.96	87.96	137.96	187.96	237.96	287.96	387.96	487.96	587.96	687.96	787.96	887.96	987.96	1087.96
	v	0.01757	5.471	5.801	6.218	6.622	7.018	7.408	7.794	8.560	9.319	10.075	10.829	11.581	12.331	13.081	13.829
	h	282.15	1183.1	1204.0	1230.5	1256.1	1281.3	1306.2	1330.9	1380.5	1430.5	1481.;	1532.6	1584.9	1638.0	1692.0	1746.8
	s	0.4534	1.6208	1.6473	1.6790	1.7080	1.7349	1.7602	1.7842	1.8289	1.8702	1.9089	1.9454	1.9800	2.0131	2.0446	2.0750
85 (316.26)	Sh			33.74	83.74	133.74	183.74	233.74	283.74	383.74	483.74	583.74	683.74	783.74	883.74	983.74	1083.74
	v	0.01762	5.167	5.445	5.840	6.223	6.597	6.966	7.330	8.052	8.768	9.480	10.190	10.898	11.604	12.310	13.014
	h	286.52	1184.2	1203.0	1229.7	1255.5	1280.8	1305.8	1330.6	1380.2	1430.3	1481.0	1532.4	1584.7	1637.9	1691.9	1746.8
	s	0.4590	1.6159	1.6396	1.6716	1.7008	1.7279	1.7532	1.7772	1.8220	1.8634	1.9021	1.9386	1.9733	2.0063	2.0379	2.0682
90 (320.28)	Sh			29.72	79.72	129.72	179.72	229.72	279.72	379.72	479.72	579.72	679.72	779.72	879.72	979.72	1079.72
	v	0.01766	4.895	5.128	5.505	5.869	6.223	6.572	6.917	7.600	8.277	8.950	9.621	10.290	10.958	11.625	12.290
	h	290.69	1185.3	1202.0	1228.9	1254.9	1280.3	1305.4	1330.2	1380.0	1430.1	1480.8	1532.3	1584.6	1637.8	1691.8	1746.7
	s	0.4643	1.6113	1.6323	1.6646	1.6940	1.7212	1.7467	1.7707	1.8156	1.8570	1.8957	1.9323	1.9669	2.0000	2.0316	2.0619
95 (324.13)	Sh			25.87	75.87	125.87	175.87	225.87	275.87	375.87	475.87	575.87	675.87	775.87	875.87	975.87	1075.87
	v	0.01770	4.651	4.845	5.205	5.551	5.889	6.221	6.548	7.196	7.838	8.477	9.113	9.747	10.380	11.012	11.643
	h	294.70	1186.2	1200.9	1228.1	1254.3	1279.8	1305.0	1329.9	1379.7	1429.9	1480.6	1532.1	1584.5	1637.7	1691.7	1746.6
	s	0.4694	1.6069	1.6253	1.6580	1.6876	1.7149	1.7404	1.7645	1.8094	1.8509	1.8897	1.9262	1.9609	1.9940	2.0256	2.0559
100 (327.82)	Sh			22.18	72.18	122.18	172.18	222.18	272.18	372.18	472.18	572.18	672.18	772.18	872.18	972.18	1072.18
	v	0.01774	4.431	4.590	4.935	5.266	5.588	5.904	6.216	6.833	7.443	8.050	8.655	9.258	9.860	10.460	11.060
	h	298.54	1187.2	1199.9	1227.4	1253.7	1279.3	1304.6	1329.6	1379.5	1429.7	1480.4	1532.0	1584.4	1637.6	1691.6	1746.5
	s	0.4743	1.6027	1.6187	1.6516	1.6814	1.7088	1.7344	1.7586	1.8036	1.8451	1.8839	1.9205	1.9552	1.9883	2.0199	2.0502
105 (331.37)	Sh			18.63	68.63	118.63	168.63	218.63	268.63	368.63	468.63	568.63	668.63	768.63	868.63	968.63	1068.63
	v	0.01778	4.231	4.359	4.690	5.007	5.315	5.617	5.915	6.504	7.086	7.665	8.241	8.816	9.389	9.961	10.532
	h	302.24	1188.0	1198.8	1226.6	1253.1	1278.8	1304.2	1329.2	1379.2	1429.4	1480.3	1531.8	1584.2	1637.5	1691.5	1746.4
	s	0.4790	1.5988	1.6122	1.6455	1.6755	1.7031	1.7288	1.7530	1.7981	1.8396	1.8785	1.9151	1.9498	1.9823	2.0145	2.0448
110 (334.79)	Sh			15.21	65.21	115.21	165.21	215.21	265.21	365.21	465.21	565.21	665.21	765.21	865.21	965.21	1065.21
	v	0.01782	4.048	4.149	4.468	4.772	5.068	5.357	5.642	6.205	6.761	7.314	7.865	8.413	8.961	9.507	10.053
	h	305.80	1188.9	1197.7	1225.8	1252.5	1278.3	1303.8	1328.9	1379.0	1429.2	1480.1	1531.7	1584.1	1637.4	1691.4	1746.4
	s	0.4834	1.5950	1.6061	1.6396	1.6698	1.6975	1.7233	1.7476	1.7928	1.8344	1.8732	1.9099	1.9446	1.9777	2.0093	2.0397
115 (338.08)	Sh			11.92	61.92	111.92	161.92	211.92	261.92	361.92	461.92	561.92	661.92	761.92	861.92	961.92	1061.92
	v	0.01785	3.881	3.957	4.265	4.558	4.841	5.119	5.392	5.932	6.465	6.994	7.521	8.046	8.570	9.093	9.615
	h	309.25	1189.6	1196.7	1225.0	1251.8	1277.7	1303.3	1328.6	1378.7	1429.0	1479.9	1531.6	1584.0	1637.2	1691.4	1746.3
	s	0.4877	1.5913	1.6001	1.6340	1.6644	1.6922	1.7181	1.7425	1.7877	1.8294	1.8682	1.9049	1.9396	1.9727	2.0044	2.0347

Temperature (°F)

120 (341.27)

Sh	Sat. liq	Sat. vap	8.73	58.73	108.73	158.73	208.73	258.73	358.73	458.73	558.73	658.73	758.73	858.73	958.73	1058.73
v	0.01789	3.7275	3.7815	4.0786	4.3610	4.6341	4.9009	5.1637	5.6813	6.1928	6.7006	7.2060	7.7096	8.2119	8.7130	9.2134
h	312.58	1190.4	1195.6	1224.1	1251.2	1277.4	1302.9	1328.2	1378.4	1428.8	1479.8	1531.4	1583.9	1637.1	1691.3	1746.2
s	0.4919	1.5879	1.5943	1.6284	1.6597	1.6872	1.7132	1.7376	1.7829	1.8246	1.8635	1.9001	1.9349	1.9680	1.9996	2.0300

130 (347.33)

Sh	Sat. liq	Sat. vap	2.67	52.67	102.67	152.67	202.67	252.67	352.67	452.67	552.67	652.67	752.67	852.67	952.67	1052.67
v	0.01796	3.4544	3.4699	3.7489	4.0129	4.2672	4.5151	4.7589	5.2384	5.7118	6.1814	6.6486	7.1140	7.5781	8.0411	8.5033
h	318.95	1191.7	1193.4	1222.5	1249.9	1276.4	1302.1	1327.5	1377.9	1428.4	1479.4	1531.1	1583.6	1636.9	1691.1	1746.1
s	0.4998	1.5813	1.5833	1.6182	1.6493	1.6775	1.7037	1.7283	1.7737	1.8155	1.8545	1.8911	1.9259	1.9591	1.9907	2.0211

140 (353.04)

Sh	Sat. liq	Sat. vap	46.96	96.96	146.96	196.96	246.96	346.96	446.96	546.96	646.96	746.96	846.96	946.96	1046.96
v	0.01803	3.2190	3.4661	3.7143	3.9526	4.1844	4.4119	4.8588	5.2995	5.7364	6.1709	6.6036	7.0349	7.4652	7.8946
h	324.96	1193.0	1220.8	1248.7	1275.3	1301.3	1326.8	1377.4	1428.0	1479.1	1530.8	1583.4	1636.7	1690.9	1745.9
s	0.5071	1.5752	1.6085	1.6400	1.6686	1.6949	1.7196	1.7652	1.8071	1.8461	1.8828	1.9176	1.9508	1.9825	2.0129

150 (358.43)

Sh	Sat. liq	Sat. vap	41.57	91.57	141.57	191.57	241.57	341.57	441.57	541.57	641.57	741.57	841.57	941.57	1041.57
v	0.01809	3.0139	3.2208	3.4555	3.6799	3.8978	4.1112	4.5295	4.9421	5.3507	5.7568	6.1612	6.5642	6.9661	7.3671
h	330.65	1194.1	1219.1	1247.4	1274.3	1300.5	1326.1	1376.9	1427.6	1478.7	1530.5	1583.1	1636.5	1690.7	1745.7
s	0.5141	1.5695	1.5993	1.6313	1.6602	1.6867	1.7115	1.7573	1.7992	1.8383	1.8751	1.9099	1.9431	1.9748	2.0052

160 (363.55)

Sh	Sat. liq	Sat. vap	36.45	86.45	136.45	186.45	236.45	336.45	436.45	536.45	636.45	736.45	836.45	936.45	1036.45
v	0.01815	2.8336	3.0060	3.2288	3.4413	3.6469	3.8480	4.2420	4.6295	5.0132	5.3945	5.7741	6.1522	6.5293	6.9055
h	336.07	1195.1	1217.4	1246.0	1273.3	1299.6	1325.4	1376.4	1427.2	1478.4	1530.3	1582.9	1636.3	1690.5	1745.6
s	0.5206	1.5641	1.5906	1.6231	1.6522	1.6790	1.7039	1.7499	1.7919	1.8310	1.8678	1.9027	1.9359	1.9676	1.9980

170 (368.42)

Sh	Sat. liq	Sat. vap	31.58	81.58	131.58	181.58	231.58	331.58	431.58	531.58	631.58	731.58	831.58	931.58	1031.58
v	0.01821	2.6738	2.8162	3.0288	3.2306	3.4255	3.6158	3.9879	4.3536	4.7155	5.0749	5.4325	5.7888	6.1440	6.4983
h	341.24	1196.0	1215.6	1244.7	1272.2	1298.8	1324.7	1375.8	1426.8	1478.0	1529.9	1582.6	1636.1	1690.4	1745.4
s	0.5269	1.5591	1.5823	1.6152	1.6447	1.6717	1.6968	1.7428	1.7850	1.8241	1.8610	1.8959	1.9291	1.9608	1.9913

180 (373.08)

Sh	Sat. liq	Sat. vap	26.92	76.92	126.92	176.92	226.92	326.92	426.92	526.92	626.92	726.92	826.92	926.92	1026.92
v	0.01827	2.5312	2.6474	2.8508	3.0433	3.2286	3.4093	3.7621	4.1084	4.4508	4.7907	5.1289	5.4657	5.8014	6.1363
h	346.19	1196.9	1213.8	1243.4	1271.2	1297.9	1324.0	1375.3	1426.3	1477.7	1529.7	1582.4	1635.9	1690.2	1745.3
s	0.5328	1.5543	1.5743	1.6078	1.6376	1.6647	1.6900	1.7362	1.7784	1.8176	1.8545	1.8894	1.9227	1.9545	1.9849

190 (377.53)

Sh	Sat. liq	Sat. vap	22.47	72.47	122.47	172.47	222.47	322.47	422.47	522.47	622.47	722.47	822.47	922.47	1022.47
v	0.01833	2.4030	2.4961	2.6915	2.8756	3.0525	3.2246	3.5601	3.8889	4.2140	4.5365	4.8572	5.1766	5.4949	5.8124
h	350.94	1197.6	1212.0	1242.0	1270.1	1297.1	1323.3	1374.8	1425.9	1477.4	1529.4	1582.1	1635.7	1690.0	1745.1
s	0.5384	1.5498	1.5667	1.6006	1.6307	1.6581	1.6835	1.7299	1.7722	1.8115	1.8484	1.8834	1.9166	1.9484	1.9789

200 (381.80)

Sh	Sat. liq	Sat. vap	18.20	68.20	118.20	168.20	218.20	318.20	418.20	518.20	618.20	718.20	818.20	918.20	1018.20
v	0.01839	2.2873	2.3598	2.5480	2.7247	2.8939	3.0583	3.3783	3.6915	4.0008	4.3077	4.6128	4.9165	5.2191	5.5209
h	355.51	1198.3	1210.1	1240.6	1269.0	1296.2	1322.6	1374.3	1425.5	1477.0	1529.1	1581.9	1635.4	1689.8	1745.0
s	0.5438	1.5454	1.5593	1.5938	1.6242	1.6518	1.6773	1.7239	1.7663	1.8057	1.8426	1.8776	1.9109	1.9427	1.9732

Table 15-15. Continued

Abs. Press. (psi) (Sat. Temp.)		Sat. Water	Sat. Steam	Temperature (°F) 400	450	500	550	600	700	800	900	1000	1100	1200	1300	1400	1500
210 (385.91)	Sh			14.09	64.09	114.09	164.09	214.09	314.09	414.09	514.09	614.09	714.09	814.09	914.09	1014.09	1114.09
	v	0.01844	2.1822	2.2364	2.4181	2.5880	2.7504	2.9078	3.2137	3.5128	3.8080	4.1007	4.3915	4.6811	4.9695	5.2571	5.5440
	h	359.91	1199.0	1208.02	1239.2	1268.0	1295.3	1321.9	1373.7	1425.1	1476.7	1528.8	1581.6	1635.2	1689.6	1744.8	1800.8
	s	0.5490	1.5413	1.5522	1.5872	1.6180	1.6458	1.6715	1.7182	1.7607	1.8001	1.8371	1.8721	1.9054	1.9372	1.9677	1.9970
220 (389.88)	Sh			10.12	60.12	110.12	160.12	210.12	310.12	410.12	510.12	610.12	710.12	810.12	910.12	1010.12	1110.12
	v	0.01850	2.0863	2.1240	2.2999	2.4638	2.6199	2.7710	3.0642	3.3504	3.6327	3.9125	4.1905	4.4671	4.7426	5.0173	5.2913
	h	364.17	1199.6	1206.3	1237.8	1266.9	1294.5	1321.2	1373.2	1424.7	1476.3	1528.5	1581.4	1635.0	1689.4	1744.7	1800.6
	s	0.5540	1.5374	1.5453	1.5808	1.6120	1.6400	1.6658	1.7128	1.7553	1.7948	1.8318	1.8668	1.9002	1.9320	1.9625	1.9919
230 (393.70)	Sh			6.30	56.30	106.30	156.30	206.30	306.30	406.30	506.30	606.30	706.30	806.30	906.30	1006.30	1106.30
	v	0.01855	1.9985	2.0212	2.1919	2.3503	2.5008	2.6461	2.9276	3.2020	3.4726	3.7406	4.0068	4.2717	4.5355	4.7984	5.0606
	h	368.28	1200.1	1204.4	1236.3	1265.7	1293.6	1320.4	1372.7	1424.2	1476.0	1528.2	1581.1	1634.8	1689.3	1744.5	1800.5
	s	0.5588	1.5336	1.5385	1.5747	1.6062	1.6344	1.6604	1.7075	1.7502	1.7897	1.8268	1.8618	1.8952	1.9270	1.9576	1.9869
240 (397.39)	Sh			2.61	52.61	102.61	152.61	202.61	302.61	402.61	502.61	602.61	702.61	802.61	902.61	1002.61	1102.61
	v	0.01860	1.9177	1.9268	2.0923	2.2462	2.3915	2.5316	2.8024	3.0661	3.3259	3.5831	3.8385	4.0926	4.3456	4.5977	4.8492
	h	372.27	1200.6	1202.4	1234.9	1264.6	1292.7	1319.7	1372.1	1423.8	1475.6	1527.9	1580.9	1634.6	1689.1	1744.3	1800.4
	s	0.5634	1.5299	1.5320	1.5687	1.6006	1.6291	1.6552	1.7025	1.7452	1.7848	1.8219	1.8570	1.8904	1.9223	1.9528	1.9822
250 (400.97)	Sh				49.03	99.03	149.03	199.03	299.03	399.03	499.03	599.03	699.03	799.03	899.03	999.03	1099.03
	v	0.01865	1.8432		2.0016	2.1504	2.2909	2.4262	2.6872	2.9410	3.1909	3.4382	3.6837	3.9278	4.1709	4.4131	4.6546
	h	376.14	1201.1		1233.4	1263.5	1291.8	1319.0	1371.6	1423.4	1475.3	1527.6	1580.6	1634.4	1688.9	1744.2	1800.2
	s	0.5679	1.5264		1.5629	1.5951	1.6239	1.6502	1.6976	1.7405	1.7801	1.8173	1.8524	1.8858	1.9177	1.9482	1.9776
260 (404.44)	Sh				45.56	95.56	145.56	195.56	295.56	395.56	495.56	595.56	695.56	795.56	895.56	995.56	1095.56
	v	0.01870	1.7742		1.9173	2.0619	2.1981	2.3289	2.5808	2.8256	3.0663	3.3044	3.5408	3.7758	4.0097	4.2427	4.4750
	h	379.90	1201.5		1231.9	1262.4	1290.9	1318.2	1371.1	1423.0	1474.9	1527.3	1580.4	1634.2	1688.7	1744.0	1800.1
	s	0.5722	1.5230		1.5573	1.5899	1.6189	1.6453	1.6930	1.7359	1.7756	1.8128	1.8480	1.8814	1.9133	1.9439	1.9732
270 (407.80)	Sh				42.20	92.20	142.20	192.20	292.20	392.20	492.20	592.20	692.20	792.20	892.20	992.20	1092.20
	v	0.01875	1.7101		1.8391	1.9799	2.1121	2.2388	2.4824	2.7186	2.9509	3.1806	3.4084	3.6349	3.8603	4.0849	4.3087
	h	383.56	1201.9		1230.4	1261.2	1290.0	1317.5	1370.5	1422.6	1474.6	1527.1	1580.1	1634.0	1688.5	1743.9	1800.0
	s	0.5764	1.5197		1.5518	1.5848	1.6140	1.6406	1.6885	1.7315	1.7713	1.8085	1.8437	1.8771	1.9090	1.9396	1.9690
280 (411.07)	Sh				38.93	88.93	138.93	188.93	288.93	388.93	488.93	588.93	688.93	788.93	888.93	988.93	1088.93
	v	0.01880	1.6505		1.7665	1.9037	2.0332	2.1551	2.3909	2.6194	2.8437	3.0655	3.2855	3.5042	3.7217	3.9384	4.1543
	h	387.12	1202.3		1228.8	1260.0	1289.1	1316.8	1370.0	1422.1	1474.2	1526.8	1579.9	1633.8	1688.4	1743.7	1799.8
	s	0.5805	1.5166		1.5464	1.5798	1.6093	1.6361	1.6841	1.7273	1.7671	1.8043	1.8395	1.8730	1.9050	1.9356	1.9649

Superheated Steam Tables (pressure in psia, Sh = superheat °F; v = specific volume, h = enthalpy, s = entropy)

290 (414.25)

	Sat. water	Sat. steam	35.75	85.75	135.75	185.75	285.75	385.75	485.75	585.75	685.75	785.75	885.75	985.75	1085.75
Sh															
v	0.01885	1.5948	1.6988	1.8327	1.9578	2.0772	2.3058	2.5269	2.7440	2.9585	3.1711	3.3824	3.5926	3.8019	4.0106
h	390.60	1202.6	1227.3	1258.9	1288.1	1316.0	1369.5	1421.7	1473.9	1526.5	1579.6	1633.5	1688.2	1743.6	1799.7
s	0.5844	1.5135	1.5412	1.5750	1.6048	1.6317	1.6799	1.7232	1.7630	1.8003	1.8356	1.8690	1.9010	1.9316	1.9610

300 (417.35)

	Sat. water	Sat. steam	32.65	82.65	132.65	182.65	282.65	382.65	482.65	582.65	682.65	782.65	882.65	982.65	1082.65
Sh															
v	0.01889	1.5427	1.6356	1.7665	1.8883	2.0044	2.2263	2.4407	2.6509	2.8585	3.0643	3.2688	3.4721	3.6746	3.8764
h	393.99	1202.9	1225.7	1257.7	1287.2	1315.2	1368.9	1421.3	1473.6	1526.2	1579.4	1633.3	1688.0	1743.4	1799.6
s	0.5882	1.5105	1.5361	1.5703	1.6003	1.6274	1.6758	1.7192	1.7591	1.7964	1.8317	1.8652	1.8972	1.9278	1.9572

310 (420.36)

	Sat. water	Sat. steam	29.64	79.64	129.64	179.64	279.64	379.64	479.64	579.64	679.64	779.64	879.64	979.64	1079.64
Sh															
v	0.01894	1.4939	1.5763	1.7044	1.8233	1.9363	2.1520	2.3600	2.5638	2.7650	2.9644	3.1625	3.3594	3.5555	3.7509
h	397.30	1203.2	1224.1	1256.5	1286.3	1314.5	1368.4	1420.9	1473.2	1525.9	1579.2	1633.1	1687.8	1743.3	1799.4
s	0.5920	1.5076	1.5311	1.5657	1.5960	1.6233	1.6719	1.7153	1.7553	1.7927	1.8280	1.8615	1.8935	1.9241	1.9536

320 (423.31)

	Sat. water	Sat. steam	26.69	76.69	126.69	176.69	276.69	376.69	476.69	576.69	676.69	776.69	876.69	976.69	1076.69
Sh															
v	0.01899	1.4480	1.5207	1.6462	1.7623	1.8725	2.0823	2.2843	2.4821	2.6774	2.8708	3.0628	3.2538	3.4438	3.6332
h	400.53	1203.4	1222.5	1255.2	1285.3	1313.7	1367.8	1420.5	1472.9	1525.6	1578.9	1632.9	1687.6	1743.1	1799.3
s	0.5956	1.5048	1.5261	1.5612	1.5918	1.6192	1.6680	1.7116	1.7516	1.7890	1.8243	1.8579	1.8899	1.9206	1.9500

330 (426.18)

	Sat. water	Sat. steam	23.82	73.82	123.82	173.82	273.82	373.82	473.82	573.82	673.82	773.82	873.82	973.82	1073.82
Sh															
v	0.01903	1.4048	1.4684	1.5915	1.7050	1.8125	2.0168	2.2132	2.4054	2.5950	2.7828	2.9692	3.1545	3.3389	3.5227
h	403.70	1203.6	1220.9	1254.0	1284.4	1313.0	1367.3	1420.0	1472.5	1525.3	1578.7	1632.7	1687.5	1742.9	1799.2
s	0.5991	1.5021	1.5213	1.5568	1.5876	1.6153	1.6643	1.7079	1.7480	1.7855	1.8208	1.8544	1.8864	1.9171	1.9466

340 (428.99)

	Sat. water	Sat. steam	21.01	71.01	121.01	171.01	271.01	371.01	471.01	571.01	671.01	771.01	871.01	971.01	1071.01
Sh															
v	0.01908	1.3640	1.4191	1.5399	1.6511	1.7561	1.9552	2.1463	2.3333	2.5175	2.7000	2.8811	3.0611	3.2402	3.4186
h	406.80	1203.8	1219.2	1252.8	1283.4	1312.2	1366.7	1419.6	1472.2	1525.0	1578.4	1632.5	1687.3	1742.8	1799.0
s	0.6026	1.4994	1.5165	1.5525	1.5836	1.6114	1.6606	1.7044	1.7445	1.7820	1.8174	1.8510	1.8831	1.9138	1.9432

350 (431.73)

	Sat. water	Sat. steam	18.27	68.27	118.27	168.27	268.27	368.27	468.27	568.27	668.27	768.27	868.27	968.27	1068.27
Sh															
v	0.01912	1.3255	1.3725	1.4913	1.6002	1.7028	1.8970	2.0832	2.2652	2.4445	2.6219	2.7980	2.9730	3.1471	3.3205
h	409.83	1204.0	1217.5	1251.5	1282.4	1311.4	1366.2	1419.2	1471.8	1524.7	1578.2	1632.3	1687.1	1742.6	1798.9
s	0.6059	1.4968	1.5119	1.5483	1.5797	1.6077	1.6571	1.7009	1.7411	1.7787	1.8141	1.8477	1.8798	1.9105	1.9400

360 (434.41)

	Sat. water	Sat. steam	15.59	65.59	115.59	165.59	265.59	365.59	465.59	565.59	665.59	765.59	865.59	965.59	1065.59
Sh															
v	0.01917	1.2891	1.3285	1.4454	1.5521	1.6525	1.8421	2.0237	2.2009	2.3755	2.5482	2.7196	2.8898	3.0592	3.2279
h	412.81	1204.1	1215.8	1250.3	1281.5	1310.6	1365.6	1418.7	1471.5	1524.4	1577.9	1632.1	1686.9	1742.5	1798.8
s	0.6092	1.4943	1.5073	1.5441	1.5758	1.6040	1.6536	1.6976	1.7379	1.7754	1.8109	1.8445	1.8766	1.9073	1.9368

380 (439.61)

	Sat. water	Sat. steam	10.39	60.39	110.39	160.39	260.39	360.39	460.39	560.39	660.39	760.39	860.39	960.39	1060.39
Sh															
v	0.01925	1.2218	1.2472	1.3606	1.4635	1.5598	1.7410	1.9139	2.0825	2.2484	2.4124	2.5750	2.7366	2.8973	3.0572
h	418.59	1204.4	1212.4	1247.7	1279.5	1309.0	1364.5	1417.9	1470.8	1523.8	1577.4	1631.6	1686.5	1742.2	1798.5
s	0.6156	1.4894	1.4982	1.5360	1.5683	1.5969	1.6470	1.6911	1.7315	1.7692	1.8047	1.8384	1.8705	1.9012	1.9307

Table 15-15. Continued

Abs. Press. (psi) (Sat. Temp.)		Sat. Water	Sat. Steam	450	500	550	600	650	700	800	900	1000	1100	1200	1300	1400	1500
										Temperature (°F)							
400 (444.60)	Sh	0.01934	1.1610	5.40	55.40	105.40	155.40	205.40	255.40	355.40	455.40	555.40	655.40	755.40	855.40	955.40	1055.40
	v	424.17	1204.6	1.1738	1.2841	1.3836	1.4763	1.5646	1.6499	1.8151	1.9759	2.1339	2.2901	2.4450	2.5987	2.7515	2.9037
	h	0.6217	1.4847	1208.8	1245.1	1277.5	1307.4	1335.9	1363.4	1417.0	1470.1	1523.3	1576.9	1631.2	1686.2	1741.9	1798.2
	s			1.4894	1.5282	1.5611	1.5901	1.6163	1.6406	1.6850	1.7255	1.7632	1.7988	1.8325	1.8647	1.8955	1.9250
420 (449.40)	Sh	0.01942	1.1057	.60	50.60	100.60	150.60	200.60	250.60	350.60	450.60	550.60	650.60	750.60	850.60	950.60	1050.60
	v	429.56	1204.7	1.1071	1.2148	1.3113	1.4007	1.4856	1.5676	1.7258	1.8795	2.0304	2.1795	2.3273	2.4739	2.6196	2.7647
	h	0.6276	1.4802	1205.2	1242.4	1275.4	1305.8	1334.5	1362.3	1416.2	1469.4	1522.7	1576.4	1630.8	1685.8	1741.6	1798.0
	s			1.4808	1.5206	1.5542	1.5835	1.6100	1.6345	1.6791	1.7197	1.7575	1.7932	1.8269	1.8591	1.8899	1.9195
440 (454.03)	Sh	0.01950	1.0554		45.97	95.97	145.97	195.97	245.97	345.97	445.97	545.97	645.97	745.97	845.97	945.97	1045.97
	v	434.77	1204.8		1.1517	1.2454	1.3319	1.4138	1.4926	1.6445	1.7918	1.9363	2.0790	2.2203	2.3605	2.4998	2.6384
	h	0.6332	1.4759		1239.7	1273.4	1304.2	1333.2	1361.1	1415.3	1468.7	1522.1	1575.9	1630.4	1685.5	1741.2	1797.7
	s				1.5132	1.5474	1.5772	1.6040	1.6286	1.6734	1.7142	1.7521	1.7878	1.8216	1.8538	1.8847	1.9143
460 (458.50)	Sh	0.01959	1.0092		41.50	91.50	141.50	191.50	241.50	341.50	441.50	541.50	641.50	741.50	841.50	941.50	1041.50
	v	439.83	1204.8		1.0939	1.1852	1.2691	1.3482	1.4242	1.5703	1.7117	1.8504	1.9872	2.1226	2.2569	2.3903	2.5230
	h	0.6387	1.4718		1236.9	1271.3	1302.5	1331.8	1360.0	1414.4	1468.0	1521.5	1575.4	1629.9	1685.1	1740.9	1797.4
	s				1.5060	1.5409	1.5711	1.5982	1.6230	1.6680	1.7089	1.7469	1.7826	1.8165	1.8488	1.8797	1.9093
480 (462.82)	Sh	0.01967	0.9668		37.18	87.18	137.18	187.18	237.18	337.18	437.18	537.18	637.18	737.18	837.18	937.18	1037.18
	v	444.75	1204.8		1.0409	1.1130	1.2115	1.2881	1.3615	1.5023	1.6384	1.7716	1.9030	2.0330	2.1619	2.2900	2.4173
	h	0.6439	1.4677		1234.1	1269.1	1300.8	1330.5	1358.8	1413.6	1467.3	1520.9	1574.9	1629.5	1684.7	1740.6	1797.2
	s				1.4990	1.5346	1.5652	1.5925	1.6176	1.6628	1.7038	1.7419	1.7777	1.8116	1.8439	1.8748	1.9045
500 (467.01)	Sh	0.01975	0.9276		32.99	82.99	132.99	182.99	232.99	332.99	432.99	532.99	632.99	732.99	832.99	932.99	1032.99
	v	449.52	1204.7		0.9919	1.0791	1.1584	1.2327	1.3037	1.4397	1.5708	1.6992	1.8256	1.9507	2.0746	2.1977	2.3200
	h	0.6490	1.4639		1231.2	1267.0	1299.1	1329.1	1357.7	1412.7	1466.6	1520.3	1574.4	1629.1	1684.4	1740.3	1796.9
	s				1.4921	1.5284	1.5595	1.5871	1.6123	1.6578	1.6990	1.7371	1.7730	1.8069	1.8393	1.8702	1.8998
520 (471.07)	Sh	0.01982	0.8914		28.93	78.93	128.93	178.93	228.93	328.93	428.93	528.93	628.93	728.93	828.93	928.93	1028.93
	v	454.18	1204.5		0.9466	1.0321	1.1094	1.1816	1.2504	1.3819	1.5085	1.6323	1.7542	1.8746	1.9940	2.1125	2.2302
	h	0.6540	1.4601		1228.3	1264.8	1297.4	1327.7	1356.5	1411.8	1465.5	1519.7	1573.7	1628.7	1684.0	1740.0	1796.7
	s				1.4853	1.5223	1.5539	1.5818	1.6072	1.6530	1.6943	1.7325	1.7684	1.8024	1.8348	1.8657	1.8954
540 (475.01)	Sh	0.01990	0.8577		24.99	74.99	124.99	174.99	224.99	324.99	424.99	524.99	624.99	724.99	824.99	924.99	1024.99
	v	458.71	1204.4		0.9045	0.9884	1.0640	1.1342	1.2010	1.3284	1.4508	1.5704	1.6880	1.8042	1.9193	2.0336	2.1471
	h	0.6587	1.4565		1225.3	1262.5	1295.7	1326.3	1355.3	1410.9	1465.1	1519.1	1573.4	1628.2	1683.6	1739.7	1796.4
	s				1.4786	1.5164	1.5485	1.5767	1.6023	1.6483	1.6897	1.7280	1.7640	1.7981	1.8305	1.8615	1.8911

560 (478.84) — Sat.: v = 0.01998 / 0.8264, h = 463.14 / 1204.2, s = 0.6634 / 1.4529

Sh	21.16	71.16	121.16	171.16	221.16	321.16	421.16	521.16	621.16	721.16	821.16	921.16	1021.16
v	0.8653	0.9479	1.0217	1.0902	1.1552	1.2787	1.3972	1.5129	1.6266	1.7388	1.8500	1.9603	2.0699
h	1222.2	1260.3	1293.9	1324.9	1354.2	1410.0	1464.4	1518.6	1572.9	1627.8	1683.3	1739.4	1796.1
s	1.4720	1.5106	1.5431	1.5717	1.5975	1.6438	1.6853	1.7237	1.7598	1.7939	1.8263	1.8573	1.8870

580 (482.57) — Sat.: v = 0.02006 / 0.7971, h = 467.47 / 1203.9, s = 0.6679 / 1.4495

Sh	17.43	67.43	117.43	167.43	217.43	317.43	417.43	517.43	617.43	717.43	817.43	917.43	1017.43
v	0.8287	0.9100	0.9824	1.0492	1.1125	1.2324	1.3473	1.4593	1.5693	1.6780	1.7855	1.8921	1.9980
h	1219.1	1258.0	1292.1	1323.4	1353.0	1409.2	1463.7	1518.0	1572.4	1627.4	1682.9	1739.1	1795.9
s	1.4654	1.5049	1.5380	1.5668	1.5929	1.6394	1.6811	1.7196	1.7556	1.7898	1.8223	1.8533	1.8831

600 (486.20) — Sat.: v = 0.02013 / 0.7697, h = 471.70 / 1203.7, s = 0.6723 / 1.4461

Sh	13.80	63.80	113.80	163.80	213.80	313.80	413.80	513.80	613.80	713.80	813.80	913.80	1013.80
v	0.7944	0.8746	0.9456	1.0109	1.0726	1.1892	1.3008	1.4093	1.5160	1.6211	1.7252	1.8284	1.9309
h	1215.9	1255.6	1290.3	1322.0	1351.8	1408.3	1463.0	1517.4	1571.9	1627.0	1682.6	1738.8	1795.6
s	1.4590	1.4993	1.5329	1.5621	1.5884	1.6351	1.6769	1.7155	1.7517	1.7859	1.8184	1.8494	1.8792

650 (494.89) — Sat.: v = 0.02032 / 0.7084, h = 481.89 / 1202.8, s = 0.6828 / 1.4381

Sh	5.11	55.11	105.11	155.11	205.11	305.11	405.11	505.11	605.11	705.11	805.11	905.11	1005.11
v	0.7173	0.7954	0.8634	0.9254	0.9835	1.0929	1.1969	1.2979	1.3969	1.4944	1.5909	1.6864	1.7813
h	1207.6	1249.6	1285.7	1318.3	1348.7	1406.0	1461.2	1515.9	1570.4	1625.9	1681.6	1738.0	1794.9
s	1.4430	1.4858	1.5207	1.5507	1.5775	1.6249	1.6671	1.7059	1.7422	1.7765	1.8092	1.8403	1.8701

700 (503.08) — Sat.: v = 0.02050 / 0.6556, h = 491.60 / 1201.8, s = 0.6928 / 1.4304

Sh	46.92	96.92	146.92	196.92	296.92	396.92	496.92	596.92	696.92	796.92	896.92	996.92
v	0.7271	0.7928	0.8520	0.9072	1.0102	1.1078	1.2023	1.2948	1.3858	1.4757	1.5647	1.6530
h	1243.4	1281.0	1314.6	1345.6	1403.7	1459.4	1514.4	1569.4	1624.8	1680.7	1737.2	1794.3
s	1.4726	1.5090	1.5399	1.5673	1.6154	1.6580	1.6970	1.7335	1.7679	1.8006	1.8318	1.8617

750 (510.84) — Sat.: v = 0.02069 / 0.6095, h = 500.89 / 1200.7, s = 0.7022 / 1.4232

Sh	39.16	89.16	139.16	189.16	289.16	389.16	489.16	589.16	689.16	789.16	889.16	989.16
v	0.6676	0.7313	0.7882	0.8409	0.9386	1.0306	1.1195	1.2063	1.2916	1.3759	1.4592	1.5419
h	1236.9	1276.1	1310.7	1342.5	1401.5	1457.6	1512.9	1568.2	1623.8	1679.8	1736.4	1793.6
s	1.4598	1.4977	1.5296	1.5577	1.6065	1.6494	1.6886	1.7252	1.7598	1.7926	1.8239	1.8538

800 (518.21) — Sat.: v = 0.02087 / 0.5690, h = 509.81 / 1199.4, s = 0.7111 / 1.4163

Sh	31.79	81.79	131.79	181.79	281.79	381.79	481.79	581.79	681.79	781.79	881.79	981.79
v	0.6151	0.6774	0.7323	0.7828	0.8759	0.9631	1.0470	1.1289	1.2093	1.2885	1.3669	1.4446
h	1230.1	1271.1	1306.8	1339.3	1399.1	1455.8	1511.4	1566.9	1622.7	1678.9	1735.7	1792.9
s	1.4472	1.4869	1.5198	1.5484	1.5980	1.6413	1.6807	1.7175	1.7522	1.7851	1.8164	1.8464

850 (525.24) — Sat.: v = 0.02105 / 0.5330, h = 518.40 / 1198.0, s = 0.7197 / 1.4096

Sh	24.76	74.76	124.76	174.76	274.76	374.76	474.76	574.76	674.76	774.76	874.76	974.76
v	0.5683	0.6296	0.6829	0.7315	0.8205	0.9034	0.9830	1.0606	1.1366	1.2115	1.2855	1.3588
h	1223.0	1265.5	1302.8	1336.0	1396.8	1454.0	1510.0	1565.7	1621.6	1678.0	1734.9	1792.3
s	1.4347	1.4763	1.5102	1.5396	1.5899	1.6336	1.6733	1.7102	1.7450	1.7780	1.8094	1.8395

900 (531.95) — Sat.: v = 0.02123 / 0.5009, h = 526.70 / 1196.4, s = 0.7279 / 1.4032

Sh	18.05	68.05	118.05	168.05	268.05	368.05	468.05	568.05	668.05	768.05	868.05	968.05
v	0.5263	0.5869	0.6388	0.6858	0.7713	0.8504	0.9262	0.9998	1.0720	1.1430	1.2131	1.2825
h	1215.5	1260.6	1298.6	1332.7	1394.4	1452.2	1508.5	1564.4	1620.6	1677.1	1734.1	1791.6
s	1.4223	1.4659	1.5010	1.5311	1.5822	1.6263	1.6662	1.7033	1.7382	1.7713	1.8028	1.8329

Table 15-15. Continued

Abs. Press. (psi) (Sat. Temp.)		Sat. Water	Sat. Steam	Temperature (°F)													
				550	600	650	700	750	800	850	900	1000	1100	1200	1300	1400	1500
950 (538.39)	Sh			11.61	61.61	111.61	161.61	211.61	261.61	311.61	361.61	461.61	561.61	661.61	761.61	861.61	961.61
	v	0.02141	0.4721	0.4883	0.5485	0.5993	0.6449	0.6871	0.7272	0.7656	0.8030	0.8753	0.9455	1.0142	1.0817	1.1484	1.2143
	h	534.74	1194.7	1207.6	1255.1	1294.4	1329.3	1361.5	1392.0	1421.5	1450.3	1507.0	1563.2	1619.5	1676.2	1733.3	1791.0
	s	0.7358	1.3970	1.4098	1.4557	1.4921	1.5228	1.5500	1.5748	1.5977	1.6193	1.6595	1.6967	1.7317	1.7649	1.7965	1.8267
1000 (544.58)	Sh			5.42	55.42	105.42	155.42	205.42	255.42	305.42	355.42	455.42	555.42	655.42	755.42	855.42	955.42
	v	0.02159	0.4460	0.4535	0.5137	0.5636	0.6080	0.6489	0.6875	0.7245	0.7603	0.8295	0.8966	0.9622	1.0266	1.0901	1.1529
	h	542.55	1192.9	1199.3	1249.3	1290.1	1325.9	1358.7	1389.6	1419.4	1448.5	1505.4	1561.9	1618.4	1675.3	1732.5	1790.3
	s	0.7434	1.3910	1.3973	1.4457	1.4833	1.5149	1.5426	1.5677	1.5908	1.6126	1.6530	1.6905	1.7256	1.7589	1.7905	1.8207
1050 (550.53)	Sh				49.47	99.47	149.47	199.47	249.47	299.47	349.47	449.47	549.47	649.47	749.47	849.47	949.47
	v	0.02177	0.4222		0.4821	0.5312	0.5745	0.6142	0.6515	0.6872	0.7216	0.7881	0.8524	0.9151	0.9767	1.0373	1.0973
	h	550.15	1191.0		1243.4	1285.7	1322.4	1355.8	1387.2	1417.3	1446.6	1503.9	1560.7	1617.4	1674.4	1731.8	1789.6
	s	0.7507	1.3851		1.4358	1.4748	1.5072	1.5354	1.5608	1.5842	1.6062	1.6469	1.6845	1.7197	1.7551	1.7848	1.8151
1100 (556.28)	Sh				43.72	93.72	143.72	193.72	243.72	293.72	343.72	443.72	543.72	643.72	743.72	843.72	943.72
	v	0.02195	0.4006		0.4531	0.5017	0.5440	0.5826	0.6188	0.6533	0.6865	0.7505	0.8121	0.8723	0.9313	0.9894	1.0468
	h	557.55	1189.1		1237.3	1281.2	1318.8	1352.9	1384.7	1415.2	1444.7	1502.4	1559.4	1616.3	1673.5	1731.0	1789.0
	s	0.7578	1.3794		1.4259	1.4664	1.4996	1.5284	1.5542	1.5779	1.6000	1.6410	1.6787	1.7141	1.7475	1.7793	1.8097
1150 (561.82)	Sh				39.18	89.18	139.18	189.18	239.18	289.18	339.18	439.18	539.18	639.18	739.18	839.18	939.18
	v	0.02214	0.3807		0.4263	0.4746	0.5162	0.5538	0.5889	0.6223	0.6544	0.7161	0.7754	0.8332	0.8899	0.9456	1.0007
	h	564.78	1187.0		1230.9	1276.6	1315.2	1349.9	1382.2	1413.0	1442.8	1500.9	1558.1	1615.2	1672.6	1730.2	1788.3
	s	0.7647	1.3738		1.4160	1.4582	1.4923	1.5216	1.5478	1.5717	1.5941	1.6353	1.6732	1.7087	1.7422	1.7741	1.8045
1200 (567.19)	Sh				32.81	82.81	132.81	182.81	232.81	282.81	332.81	432.81	532.81	632.81	732.81	832.81	932.81
	v	0.02232	0.3624		0.4016	0.4497	0.4905	0.5273	0.5615	0.5939	0.6250	0.6845	0.7418	0.7974	0.8519	0.9055	0.9584
	h	571.85	1184.8		1224.2	1271.8	1311.5	1346.9	1379.7	1410.8	1440.9	1499.4	1556.9	1614.2	1671.6	1729.4	1787.6
	s	0.7714	1.3683		1.4061	1.4501	1.4851	1.5150	1.5415	1.5658	1.5883	1.6298	1.6679	1.7035	1.7371	1.7691	1.7996
1300 (577.42)	Sh				22.58	72.58	122.58	172.58	222.58	272.58	322.58	422.58	522.58	622.58	722.58	822.58	922.58
	v	0.02269	0.3299		0.3570	0.4052	0.4451	0.4804	0.5129	0.5436	0.5729	0.6287	0.6822	0.7341	0.7847	0.8345	0.8836
	h	585.58	1180.2		1209.9	1261.9	1303.9	1340.8	1374.6	1406.4	1437.1	1496.3	1554.3	1612.0	1669.8	1727.7	1786.1
	s	0.7843	1.3577		1.3860	1.4340	1.4711	1.5022	1.5296	1.5544	1.5773	1.6194	1.6578	1.6937	1.7275	1.7596	1.7902
1400 (587.07)	Sh				12.93	62.93	112.93	162.93	212.93	262.93	312.93	412.93	512.93	612.93	712.93	812.93	912.93
	v	0.02307	0.3018		0.3176	0.3667	0.4059	0.4400	0.4712	0.5004	0.5282	0.5809	0.6311	0.6798	0.7272	0.7737	0.8195
	h	598.83	1175.3		1194.1	1251.4	1296.1	1334.5	1369.3	1402.0	1433.2	1493.2	1551.8	1609.9	1668.0	1726.3	1785.0
	s	0.7966	1.3474		1.3652	1.4181	1.4575	1.4900	1.5182	1.5436	1.5670	1.6096	1.6484	1.6845	1.7185	1.7508	1.7815

Superheated steam — properties (Sh = degrees of superheat, v, h, s) at temperature (°F)

P (psia) (Tsat °F)		600	650	700	750	800	850	900	1000	1100	1200	1300	1400	1500
1500 (596.20)	Sh	3.80	53.80	103.80	153.80	203.80	253.80	303.80	403.80	503.80	603.80	703.80	803.80	903.80
	v	0.2820	0.3328	0.3717	0.4049	0.4350	0.4629	0.4894	0.5394	0.5869	0.6327	0.6773	0.7210	0.7639
	h	1176.3	1240.2	1287.9	1328.0	1364.0	1397.4	1429.2	1490.1	1549.2	1607.7	1666.2	1724.8	1783.7
	s	1.3431	1.4022	1.4443	1.4782	1.5073	1.5333	1.5572	1.6004	1.6395	1.6759	1.7101	1.7425	1.7734
1600 (604.87)	Sh		45.13	95.13	145.13	195.13	245.13	295.13	395.13	495.13	595.13	695.13	795.13	895.13
	v		0.3026	0.3415	0.3741	0.4032	0.4301	0.4555	0.5031	0.5482	0.5915	0.6336	0.6748	0.7153
	h		1228.3	1279.4	1321.4	1358.5	1392.8	1425.2	1486.9	1546.6	1605.6	1664.3	1723.2	1782.3
	s		1.3861	1.4312	1.4667	1.4968	1.5235	1.5478	1.5916	1.6312	1.6678	1.7022	1.7347	1.7657
1700 (613.13)	Sh		36.87	86.87	136.87	186.87	236.87	286.87	386.87	486.87	586.87	686.87	786.87	886.87
	v		0.2754	0.3147	0.3468	0.3751	0.4011	0.4255	0.4711	0.5140	0.5552	0.5951	0.6341	0.6724
	h		1215.3	1270.5	1314.5	1352.9	1388.1	1421.2	1483.8	1544.0	1603.4	1662.5	1721.7	1781.0
	s		1.3697	1.4183	1.4555	1.4867	1.5140	1.5388	1.5833	1.6232	1.6601	1.6947	1.7274	1.7585
1800 (621.02)	Sh		28.98	78.98	128.98	178.98	228.98	278.98	378.98	478.98	578.98	678.98	778.98	878.98
	v		0.2505	0.2906	0.3223	0.3500	0.3752	0.3988	0.4426	0.4836	0.5229	0.5609	0.5980	0.6343
	h		1201.2	1261.1	1307.4	1347.2	1383.3	1417.1	1480.6	1541.4	1601.2	1660.7	1720.1	1779.7
	s		1.3526	1.4054	1.4446	1.4768	1.5049	1.5302	1.5753	1.6156	1.6528	1.6876	1.7204	1.7516
1900 (628.56)	Sh		21.44	71.44	121.44	171.44	221.44	271.44	371.44	471.44	571.44	671.44	771.44	871.44
	v		0.2274	0.2687	0.3004	0.3275	0.3521	0.3749	0.4171	0.4565	0.4940	0.5303	0.5656	0.6002
	h		1185.7	1251.3	1300.2	1341.4	1378.4	1412.9	1477.4	1538.8	1599.1	1658.8	1718.6	1778.4
	s		1.3346	1.3925	1.4338	1.4672	1.4960	1.5219	1.5677	1.6084	1.6458	1.6808	1.7138	1.7451
2000 (635.80)	Sh		14.20	64.20	114.20	164.20	214.20	264.20	364.20	464.20	564.20	664.20	764.20	864.20
	v		0.2056	0.2488	0.2805	0.3072	0.3312	0.3534	0.3942	0.4320	0.4680	0.5027	0.5365	0.5695
	h		1168.3	1240.9	1292.6	1335.4	1373.5	1408.7	1474.1	1536.2	1596.9	1657.0	1717.0	1777.1
	s		1.3154	1.3794	1.4231	1.4578	1.4874	1.5138	1.5603	1.6014	1.6391	1.6743	1.7075	1.7389
2100 (642.76)	Sh		7.24	57.24	107.24	157.24	207.24	257.24	357.24	457.24	557.24	657.24	757.24	857.24
	v		0.1847	0.2304	0.2624	0.2888	0.3123	0.3339	0.3734	0.4099	0.4445	0.4778	0.5101	0.5418
	h		1148.5	1229.8	1284.9	1329.3	1368.4	1404.4	1470.9	1533.6	1594.7	1655.2	1715.4	1775.7
	s		1.2942	1.3661	1.4125	1.4486	1.4790	1.5060	1.5532	1.5948	1.6327	1.6681	1.7014	1.7330
2200 (649.45)	Sh		0.55	50.55	100.55	150.55	200.55	250.55	350.55	450.55	550.55	650.55	750.55	850.55
	v		0.1636	0.2134	0.2458	0.2720	0.2950	0.3161	0.3545	0.3897	0.4231	0.4551	0.4862	0.5165
	h		1123.9	1218.0	1276.8	1323.1	1363.3	1400.0	1467.6	1530.9	1592.5	1653.3	1713.9	1774.4
	s		1.2691	1.3523	1.4020	1.4395	1.4708	1.4984	1.5463	1.5883	1.6266	1.6622	1.6956	1.7273
2300 (655.89)	Sh			44.11	94.11	144.11	194.11	244.11	344.11	444.11	544.11	644.11	744.11	844.11
	v			0.1975	0.2305	0.2566	0.2793	0.2999	0.3372	0.3714	0.4035	0.4344	0.4643	0.4935
	h			1205.3	1268.4	1316.7	1358.1	1395.7	1464.2	1528.3	1590.3	1651.5	1712.3	1773.1
	s			1.3381	1.3914	1.4305	1.4628	1.4910	1.5397	1.5821	1.6207	1.6565	1.6901	1.7219

Saturation properties

P (psia)	Tsat (°F)	v_f	h_f	s_f	v_g	h_g	s_g
1500	596.20	0.02346	611.68	0.8085	0.2772	1170.1	1.3373
1600	604.87	0.02387	624.20	0.8199	0.2555	1164.5	1.3274
1700	613.13	0.02428	636.45	0.8309	0.2361	1158.6	1.3176
1800	621.02	0.02472	648.49	0.8417	0.2186	1152.3	1.3079
1900	628.56	0.02517	660.36	0.8522	0.2028	1145.6	1.2981
2000	635.80	0.02565	672.11	0.8625	0.1883	1138.3	1.2881
2100	642.76	0.02615	683.79	0.8727	0.1750	1130.5	1.2780
2200	649.45	0.02669	695.46	0.8828	0.1627	1122.2	1.2676
2300	655.89	0.02727	707.18	0.8929	0.1513	1113.2	1.2569

Table 15-15. Continued

Abs. Press. (psi) (Sat. Temp.)		Sat. Water	Sat. Steam	\multicolumn{14}{l}{Temperature (°F)}													
				700	750	800	850	900	950	1000	1050	1100	1150	1200	1300	1400	1500
2400 (662.11)	Sh			37.89	87.89	137.89	187.89	237.89	287.89	337.89	387.89	437.89	487.89	537.89	637.89	737.89	837.89
	v	0.02790	0.1408	0.1834	0.2164	0.2424	0.2648	0.2850	0.3037	0.3214	0.3382	0.3545	0.3703	0.3856	0.4155	0.4443	0.4724
	h	718.95	1103.7	1191.6	1259.7	1310.1	1352.8	1391.2	1426.9	1460.9	1493.7	1525.6	1557.0	1588.1	1649.6	1710.8	1771.8
	s	0.9031	1.2460	1.3232	1.3808	1.4217	1.4549	1.4837	1.5095	1.5332	1.5553	1.5761	1.5959	1.6149	1.6509	1.6847	1.7167
2500 (668.11)	Sh			31.89	81.89	131.89	181.89	231.89	281.89	331.89	381.89	431.89	481.89	531.89	631.89	731.89	831.89
	v	0.02859	0.1307	0.1681	0.2032	0.2293	0.2514	0.2712	0.2896	0.3068	0.3232	0.3390	0.3543	0.3692	0.3980	0.4259	0.4529
	h	731.71	1093.3	1176.7	1250.6	1303.4	1347.4	1386.7	1423.1	1457.5	1490.7	1522.9	1554.6	1585.9	1647.8	1709.2	1770.4
	s	0.9139	1.2345	1.3076	1.3701	1.4129	1.4472	1.4766	1.5029	1.5269	1.5492	1.5703	1.5903	1.6094	1.6456	1.6796	1.7116
2600 (673.91)	Sh			26.09	76.09	126.09	176.09	226.09	276.09	326.09	376.09	426.09	476.09	526.09	626.09	726.09	826.09
	v	0.02938	0.1211	0.1544	0.1909	0.2171	0.2390	0.2585	0.2765	0.2933	0.3093	0.3247	0.3395	0.3540	0.3819	0.4088	0.4350
	h	744.47	1082.0	1160.2	1241.1	1296.5	1341.9	1382.1	1419.2	1454.1	1487.7	1520.2	1552.2	1583.7	1646.0	1707.7	1769.1
	s	0.9247	1.2225	1.2908	1.3592	1.4042	1.4395	1.4696	1.4964	1.5208	1.5434	1.5646	1.5848	1.6040	1.6405	1.6746	1.7068
2700 (679.53)	Sh			20.47	70.47	120.47	170.47	220.47	270.47	320.47	370.47	420.47	470.47	520.47	620.47	720.47	820.47
	v	0.03029	0.1119	0.1411	0.1794	0.2058	0.2275	0.2468	0.2644	0.2809	0.2965	0.3114	0.3259	0.3399	0.3670	0.3931	0.4184
	h	757.34	1069.7	1142.0	1231.1	1289.5	1336.3	1377.5	1415.2	1450.7	1484.6	1517.5	1549.8	1581.5	1644.1	1706.1	1767.8
	s	0.9356	1.2097	1.2727	1.3481	1.3954	1.4319	1.4628	1.4900	1.5148	1.5376	1.5591	1.5794	1.5988	1.6355	1.6697	1.7021
2800 (684.96)	Sh			15.04	65.04	115.04	165.04	215.04	265.04	315.04	365.04	415.04	465.04	515.04	615.04	715.04	815.04
	v	0.03134	0.1030	0.1278	0.1685	0.1952	0.2168	0.2358	0.2531	0.2693	0.2845	0.2991	0.3132	0.3268	0.3532	0.3785	0.4030
	h	770.69	1055.8	1121.2	1220.6	1282.2	1330.7	1372.8	1411.2	1447.2	1481.6	1514.8	1547.3	1579.3	1642.2	1704.5	1766.5
	s	0.9468	1.1958	1.2527	1.3368	1.3867	1.4245	1.4561	1.4838	1.5089	1.5321	1.5537	1.5742	1.5938	1.6306	1.6651	1.6975
2900 (690.22)	Sh			9.78	59.78	109.78	159.78	209.78	259.78	309.78	359.78	409.78	459.78	509.78	609.78	709.78	809.78
	v	0.03262	0.0942	0.1138	0.1581	0.1853	0.2068	0.2256	0.2427	0.2585	0.2734	0.2877	0.3014	0.3147	0.3403	0.3649	0.3887
	h	785.13	1039.8	1095.3	1209.6	1274.7	1324.9	1368.0	1407.2	1443.7	1478.5	1512.1	1544.9	1577.0	1640.4	1703.0	1765.2
	s	0.9588	1.1803	1.2283	1.3251	1.3780	1.4171	1.4494	1.4777	1.5032	1.5266	1.5485	1.5692	1.5889	1.6259	1.6605	1.6931
3000 (695.33)	Sh			4.67	54.67	104.67	154.67	204.67	254.67	304.67	354.67	404.67	454.67	504.67	604.67	704.67	804.67
	v	0.03428	0.0850	0.0982	0.1483	0.1759	0.1975	0.2161	0.2329	0.2484	0.2630	0.2770	0.2904	0.3033	0.3282	0.3522	0.3753
	h	801.84	1020.3	1060.5	1197.9	1267.0	1319.0	1363.2	1403.1	1440.2	1475.4	1509.4	1542.4	1574.8	1638.5	1701.4	1763.8
	s	0.9728	1.1619	1.1966	1.3131	1.3692	1.4097	1.4429	1.4717	1.4976	1.5213	1.5434	1.5642	1.5841	1.6214	1.6561	1.6888
3100 (700.28)	Sh				49.72	99.72	149.72	199.72	249.72	299.72	349.72	399.72	449.72	499.72	599.72	699.72	799.72
	v	0.03681	0.0745		0.1389	0.1671	0.1887	0.2071	0.2237	0.2390	0.2533	0.2670	0.2800	0.2927	0.3170	0.3403	0.3628
	h	823.97	993.3		1185.4	1259.1	1313.0	1358.4	1399.0	1436.7	1472.3	1506.6	1539.9	1572.6	1636.7	1699.8	1762.5
	s	0.9914	1.1373		1.3007	1.3604	1.4024	1.4364	1.4658	1.4920	1.5161	1.5384	1.5594	1.5794	1.6169	1.6518	1.6847

Pressure				44.92	94.92	144.92	194.92	244.92	294.92	344.92	394.92	444.92	494.92	594.92	694.92	794.92
3200 (705.08)	Sh															
	v	0.04472	0.0566	0.1300	0.1588	0.1804	0.1987	0.2151	0.2301	0.2442	0.2576	0.2704	0.2827	0.3065	0.3291	0.3510
	h	875.54	931.6	1172.3	1250.9	1306.9	1353.4	1394.9	1433.1	1469.2	1503.8	1537.4	1570.3	1634.8	1698.3	1761.2
	s	1.0351	1.0832	1.2877	1.3515	1.3951	1.4300	1.4600	1.4866	1.5110	1.5335	1.5547	1.5749	1.6126	1.6477	1.6806
3300	Sh															
	v			0.1213	0.1510	0.1727	0.1908	0.2070	0.2218	0.2357	0.2488	0.2613	0.2734	0.2966	0.3187	0.3400
	h			1158.2	1242.5	1300.7	1348.4	1390.7	1429.5	1466.1	1501.0	1534.9	1568.1	1632.9	1696.7	1759.9
	s			1.2742	1.3425	1.3879	1.4237	1.4542	1.4813	1.5059	1.5287	1.5501	1.5704	1.6084	1.6436	1.6767
3400	Sh															
	v			0.1129	0.1435	0.1653	0.1834	0.1994	0.2140	0.2276	0.2405	0.2528	0.2646	0.2872	0.3088	0.3296
	h			1143.2	1233.7	1294.3	1343.4	1386.4	1425.9	1462.9	1498.3	1532.4	1565.8	1631.1	1695.1	1758.2
	s			1.2600	1.3334	1.3807	1.4174	1.4486	1.4761	1.5010	1.5240	1.5456	1.5660	1.6042	1.6396	1.6728
3500	Sh															
	v			0.1048	0.1364	0.1583	0.1764	0.1922	0.2066	0.2200	0.2326	0.2447	0.2563	0.2784	0.2995	0.3198
	h			1127.1	1224.6	1287.8	1338.2	1382.2	1422.2	1459.7	1495.5	1529.9	1563.6	1629.2	1693.6	1757.2
	s			1.2450	1.3242	1.3734	1.4112	1.4430	1.4709	1.4962	1.5194	1.5412	1.5618	1.6002	1.6358	1.6691
3600	Sh															
	v			0.0966	0.1296	0.1517	0.1697	0.1854	0.1996	0.2128	0.2252	0.2371	0.2485	0.2702	0.2908	0.3106
	h			1108.6	1215.3	1281.2	1333.0	1377.9	1418.6	1456.5	1492.6	1527.4	1561.3	1627.3	1692.0	1755.9
	s			1.2281	1.3148	1.3662	1.4050	1.4374	1.4658	1.4914	1.5149	1.5369	1.5576	1.5962	1.6320	1.6654
3800	Sh															
	v			0.0799	0.1169	0.1395	0.1574	0.1729	0.1868	0.1996	0.2116	0.2231	0.2340	0.2549	0.2746	0.2936
	h			1064.2	1195.5	1267.6	1322.4	1369.1	1411.2	1450.1	1487.0	1522.4	1556.8	1623.6	1688.9	1753.2
	s			1.1888	1.2955	1.3517	1.3928	1.4265	1.4558	1.4821	1.5061	1.5284	1.5495	1.5886	1.6247	1.6584
4000	Sh															
	v			0.0631	0.1052	0.1284	0.1463	0.1616	0.1752	0.1877	0.1994	0.2105	0.2210	0.2411	0.2601	0.2783
	h			1007.4	1174.3	1253.4	1311.6	1360.2	1403.6	1443.6	1481.3	1517.3	1552.2	1619.8	1685.7	1750.6
	s			1.1396	1.2754	1.3371	1.3807	1.4158	1.4461	1.4730	1.4976	1.5203	1.5417	1.5812	1.6177	1.6516
4200	Sh															
	v			0.0498	0.0945	0.1183	0.1362	0.1513	0.1647	0.1769	0.1883	0.1991	0.2093	0.2287	0.2470	0.2645
	h			950.1	1151.6	1238.6	1300.4	1351.2	1396.0	1437.1	1475.5	1512.2	1547.6	1616.1	1682.6	1748.0
	s			1.0905	1.2544	1.3223	1.3686	1.4053	1.4366	1.4642	1.4893	1.5124	1.5341	1.5742	1.6109	1.6452
4400	Sh															
	v			0.0421	0.0846	0.1090	0.1270	0.1420	0.1552	0.1671	0.1782	0.1887	0.1986	0.2174	0.2351	0.2519
	h			909.5	1127.3	1223.3	1289.0	1342.0	1388.3	1430.4	1469.7	1507.1	1543.0	1612.3	1679.4	1745.3
	s			1.0556	1.2325	1.3073	1.3566	1.3949	1.4272	1.4556	1.4812	1.5048	1.5268	1.5673	1.6044	1.6389

Table 15-15. Continued

Abs. Press. (psi) (Sat. Temp.)		Sat. Water	Sat. Steam	750	800	850	900	950	1000	1050	1100	1150	1200	1250	1300	1400	1500
4600	Sh																
	v			0.0380	0.0751	0.1005	0.1186	0.1335	0.1465	0.1582	0.1691	0.1792	0.1889	0.1982	0.2071	0.2242	0.2404
	h			883.8	1100.0	1207.3	1277.2	1332.6	1380.5	1423.7	1463.9	1501.9	1538.4	1573.8	1608.5	1676.3	1742.7
	s			1.0331	1.2084	1.2922	1.3446	1.3847	1.4181	1.4472	1.4734	1.4974	1.5197	1.5407	1.5607	1.5982	1.6330
4800	Sh																
	v			0.0355	0.0665	0.0927	0.1109	0.1257	0.1385	0.1500	0.1606	0.1706	0.1800	0.1890	0.1977	0.2142	0.2299
	h			866.9	1071.2	1190.7	1265.2	1323.1	1372.6	1417.0	1458.0	1496.7	1533.8	1569.7	1604.7	1673.1	1740.0
	s			1.0180	1.1835	1.2768	1.3327	1.3745	1.4090	1.4390	1.4657	1.4901	1.5128	1.5341	1.5543	1.5921	1.6272
5000	Sh																
	v			0.0338	0.591	0.0855	0.1038	0.1185	0.1312	0.1425	0.1529	0.1626	0.1718	0.1806	0.1890	0.2050	0.2203
	h			854.9	1042.9	1173.6	1252.9	1313.5	1364.6	1410.2	1452.1	1491.5	1529.1	1565.5	1600.9	1670.0	1737.4
	s			1.0070	1.1593	1.2612	1.3207	1.3645	1.4001	1.4309	1.4582	1.4831	1.5061	1.5277	1.5481	1.5863	1.6216
5200	Sh																
	v			0.0326	0.0531	0.0789	0.0973	0.1119	0.1244	0.1356	0.1458	0.1553	0.1642	0.1728	0.1810	0.1966	0.2114
	h			845.8	1016.9	1156.0	1240.4	1303.7	1356.6	1403.4	1446.2	1486.3	1524.5	1561.3	1597.2	1666.8	1734.7
	s			0.9985	1.1370	1.2455	1.3088	1.3545	1.3914	1.4229	1.4509	1.4762	1.4995	1.5214	1.5420	1.5806	1.6161
5400	Sh																
	v			0.0317	0.0483	0.0728	0.0912	0.1058	0.1182	0.1292	0.1392	0.1485	0.1572	0.1656	0.1736	0.1888	0.2031
	h			838.5	994.3	1138.1	1227.7	1293.7	1348.4	1396.5	1440.3	1481.1	1519.8	1557.1	1593.4	1663.7	1732.1
	s			0.9915	1.1175	1.2296	1.2969	1.3446	1.3827	1.4151	1.4437	1.4694	1.4931	1.5153	1.5362	1.5750	1.6109
5600	Sh																
	v			0.0309	0.0447	0.0672	0.0856	0.1001	0.1124	0.1232	0.1331	0.1422	0.1508	0.1589	0.1667	0.1815	0.1954
	h			832.4	975.0	1119.9	1214.8	1283.7	1340.2	1389.6	1434.3	1475.9	1515.2	1552.9	1589.6	1660.5	1729.5
	s			0.9855	1.1008	1.2137	1.2850	1.3348	1.3742	1.4075	1.4366	1.4628	1.4869	1.5093	1.5304	1.5697	1.6058
5800	Sh																
	v			0.0303	0.0419	0.0622	0.0805	0.0949	0.1070	0.1177	0.1274	0.1363	0.1447	0.1527	0.1603	0.1747	0.1883
	h			827.3	958.8	1101.8	1201.8	1273.6	1332.0	1382.6	1428.3	1470.6	1510.5	1548.7	1585.8	1657.4	1726.8
	s			0.9803	1.0867	1.1981	1.2732	1.3250	1.3658	1.3999	1.4297	1.4564	1.4808	1.5035	1.5248	1.5644	1.6008
6000	Sh																
	v			0.0298	0.0397	0.0579	0.0757	0.0900	0.1020	0.1126	0.1221	0.1309	0.1391	0.1469	0.1544	0.1684	0.1817
	h			822.9	945.1	1084.6	1188.8	1263.4	1323.6	1375.7	1422.3	1465.4	1505.9	1544.6	1582.0	1654.2	1724.2
	s			0.9758	1.0746	1.1833	1.2615	1.3154	1.3574	1.3925	1.4229	1.4500	1.4748	1.4978	1.5194	1.5593	1.5960

Temperature (°F)

6500	Sh														
	v	0.1669	0.1544	0.1411	0.1340	0.1266	0.1188	0.1104	0.1012	0.0909	0.0793	0.0655	0.0495	0.0358	0.0287
	h	1717.6	1646.4	1572.5	1534.1	1494.2	1452.2	1407.3	1358.1	1302.7	1237.8	1156.3	1046.7	919.5	813.9
	s	1.5844	1.5471	1.5062	1.4841	1.4604	1.4347	1.4064	1.3743	1.3370	1.2917	1.2328	1.1506	1.0515	0.9661
7000	Sh														
	v	0.1542	0.1424	0.1298	0.1231	0.1160	0.1085	0.1004	0.0915	0.0816	0.0704	0.0573	0.0438	0.0334	0.0279
	h	1711.1	1638.6	1563.1	1523.7	1482.6	1439.1	1392.2	1340.5	1281.7	1212.6	1124.9	1016.5	901.8	806.9
	s	1.5735	1.5355	1.4938	1.4710	1.4466	1.4200	1.3904	1.3567	1.3171	1.2689	1.2055	1.1243	1.0350	0.9582
7500	Sh														
	v	0.1433	0.1321	0.1200	0.1136	0.1068	0.0996	0.0918	0.0833	0.0737	0.0631	0.0512	0.0399	0.0318	0.0272
	h	1704.6	1630.8	1553.7	1513.3	1471.0	1426.0	1377.2	1322.9	1261.0	1188.3	1097.7	992.9	889.0	801.3
	s	1.5632	1.5245	1.4819	1.4586	1.4335	1.4059	1.3751	1.3397	1.2980	1.2473	1.1818	1.1033	1.0224	0.9514
8000	Sh														
	v	0.1338	0.1230	0.1115	0.1054	0.0989	0.0920	0.0845	0.0762	0.0671	0.0571	0.0465	0.0371	0.0306	0.0267
	h	1698.1	1623.1	1544.5	1503.1	1459.6	1413.0	1362.2	1305.5	1241.0	1165.4	1074.3	974.4	879.1	796.6
	s	1.5533	1.5140	1.4705	1.4467	1.4208	1.3924	1.3603	1.3233	1.2798	1.2271	1.1613	1.0864	1.0122	0.9455
8500	Sh														
	v	0.1254	0.1151	0.1041	0.0982	0.0919	0.0853	0.0780	0.0701	0.0615	0.0522	0.0429	0.0350	0.0296	0.0262
	h	1691.7	1615.4	1535.3	1492.9	1448.2	1400.2	1347.5	1288.5	1221.9	1144.0	1054.5	959.8	871.2	792.7
	s	1.5439	1.5040	1.4597	1.4352	1.4087	1.3793	1.3460	1.3076	1.2627	1.2084	1.1437	1.0727	1.0037	0.9402
9000	Sh														
	v	0.1179	0.1081	0.0975	0.0918	0.0858	0.0794	0.0724	0.0649	0.0568	0.0483	0.0402	0.0335	0.0288	0.0258
	h	1685.3	1607.9	1526.3	1482.9	1437.1	1387.5	1333.0	1272.1	1204.1	1125.4	1037.6	948.0	864.7	789.3
	s	1.5349	1.4944	1.4492	1.4243	1.3970	1.3667	1.3323	1.2926	1.2468	1.1918	1.1285	1.0613	0.9964	0.9354
9500	Sh														
	v	0.1113	0.1019	0.0917	0.0862	0.0804	0.0742	0.0675	0.0603	0.0528	0.0451	0.0380	0.0322	0.0282	0.0254
	h	1679.0	1600.4	1517.3	1473.1	1426.1	1375.1	1318.9	1256.6	1187.7	1108.9	1023.4	938.3	859.2	786.4
	s	1.5263	1.4851	1.4392	1.4137	1.3858	1.3546	1.3191	1.2785	1.2320	1.1771	1.1153	1.0516	0.9900	0.9310
10000	Sh														
	v	0.1054	0.0963	0.0865	0.0812	0.0757	0.0697	0.0633	0.0565	0.0495	0.0425	0.0362	0.0312	0.0276	0.0251
	h	1672.8	1593.1	1508.6	1463.4	1415.3	1362.9	1305.3	1242.0	1172.6	1094.2	1011.3	930.2	854.5	783.8
	s	1.5180	1.4763	1.4295	1.4035	1.3749	1.3429	1.3065	1.2652	1.2185	1.1638	1.1039	1.0432	0.9842	0.9270
10500	Sh														
	v	0.1001	0.0913	0.0818	0.0768	0.0714	0.0656	0.0595	0.0532	0.0467	0.0404	0.0347	0.0303	0.0271	0.0248
	h	1666.7	1585.8	1500.0	1453.9	1404.7	1351.1	1292.4	1228.4	1158.9	1081.3	1001.0	923.4	850.5	781.5
	s	1.5100	1.4677	1.4202	1.3937	1.3644	1.3371	1.2946	1.2529	1.2060	1.1519	1.0939	1.0358	0.9790	0.9232

Table 15-15. Continued

Abs. Press. (psi) (Sat. Temp.)		Sat. Water	Sat. Steam	Temperature (°F)													
				750	800	850	900	950	1000	1050	1100	1150	1200	1250	1300	1400	1500
11000	v			0.0245	0.0267	0.0296	0.0335	0.0386	0.0443	0.0503	0.0562	0.0620	0.0676	0.0727	0.0776	0.0868	0.0952
	h			779.5	846.9	917.5	992.1	1069.9	1146.3	1215.9	1280.2	1339.7	1394.4	1444.6	1491.5	1578.7	1660.6
	s			0.9196	0.9742	1.0292	1.0851	1.1412	1.1945	1.2414	1.2833	1.3209	1.3544	1.3842	1.4112	1.4595	1.5023
11500	v			0.0243	0.0263	0.0290	0.0325	0.0370	0.0423	0.0478	0.0534	0.0588	0.0641	0.0691	0.0739	0.0827	0.0909
	h			777.7	843.8	912.4	984.5	1059.8	1134.9	1204.3	1268.7	1328.8	1384.4	1435.5	1483.2	1571.8	1654.7
	s			0.9163	0.9698	1.0232	1.0772	1.1316	1.1840	1.2308	1.2727	1.3107	1.3446	1.3750	1.4025	1.4515	1.4949
12000	v			0.0241	0.0260	0.0284	0.0317	0.0357	0.0405	0.0456	0.0508	0.0560	0.0610	0.0659	0.0704	0.0790	0.0869
	h			776.1	841.0	907.9	977.8	1050.9	1124.5	1193.7	1258.0	1318.5	1374.7	1426.6	1475.1	1564.9	1648.8
	s			0.9131	0.9657	1.0177	1.0701	1.1229	1.1742	1.2209	1.2627	1.3010	1.3355	1.3662	1.3941	1.4438	1.4877
12500	v			0.0238	0.0256	0.0279	0.0309	0.0346	0.0390	0.0437	0.0486	0.0535	0.0583	0.0629	0.0673	0.0756	0.0832
	h			774.7	838.6	903.9	971.9	1043.1	1115.2	1184.1	1247.9	1308.8	1365.4	1418.0	1467.2	1558.2	1643.1
	s			0.9101	0.9618	1.0127	1.0637	1.1151	1.1653	1.2117	1.2534	1.2918	1.3264	1.3576	1.3860	1.4363	1.4808
13000	v			0.0236	0.0253	0.0275	0.0302	0.0336	0.0376	0.0420	0.0466	0.0512	0.0558	0.0602	0.0645	0.0725	0.0799
	h			773.5	836.3	900.4	966.8	1036.2	1106.7	1174.8	1238.5	1299.6	1356.5	1409.6	1459.4	1551.6	1637.4
	s			0.9073	0.9582	1.0080	1.0578	1.1079	1.1571	1.2030	1.2445	1.2831	1.3179	1.3494	1.3781	1.4291	1.4741
13500	v			0.0235	0.0251	0.0271	0.0297	0.0328	0.0364	0.0405	0.0448	0.0492	0.0535	0.0577	0.0619	0.0696	0.0768
	h			772.3	834.4	897.2	962.2	1030.0	1099.1	1166.0	1229.7	1291.0	1348.1	1401.5	1451.8	1545.2	1631.9
	s			0.9045	0.9548	1.0037	1.0524	1.1014	1.1495	1.1948	1.2361	1.2749	1.3098	1.3415	1.3705	1.4221	1.4675
14000	v			0.0233	0.0248	0.0267	0.0291	0.0320	0.0354	0.0392	0.0432	0.0474	0.0515	0.0555	0.0595	0.0670	0.0740
	h			771.3	832.6	894.3	958.0	1024.5	1092.3	1158.5	1221.4	1283.0	1340.2	1393.8	1444.4	1538.8	1626.5
	s			0.9019	0.9515	0.9996	1.0473	1.0953	1.1426	1.1872	1.2282	1.2671	1.3021	1.3339	1.3631	1.4153	1.4612
14500	v			0.0231	0.0246	0.0264	0.0287	0.0314	0.0345	0.0380	0.0418	0.0458	0.0496	0.0534	0.0573	0.0646	0.0714
	h			770.4	831.0	891.7	954.3	1019.6	1086.2	1151.4	1213.8	1275.4	1332.9	1386.4	1437.3	1532.6	1621.1
	s			0.8994	0.9484	0.9957	1.0426	1.0897	1.1362	1.1801	1.2208	1.2597	1.2949	1.3266	1.3560	1.4087	1.4551
15000	v			0.0230	0.0244	0.0261	0.0282	0.0308	0.0337	0.0369	0.0405	0.0443	0.0479	0.0516	0.0552	0.0624	0.0690
	h			769.6	829.5	889.3	950.9	1015.1	1080.6	1144.9	1206.8	1268.1	1326.0	1379.4	1430.3	1526.4	1615.9
	s			0.8970	0.9455	0.9920	1.0382	1.0846	1.1302	1.1735	1.2139	1.2525	1.2880	1.3197	1.3491	1.4022	1.4491
15500	v			0.0228	0.0242	0.0258	0.0278	0.0302	0.0329	0.0360	0.0393	0.0429	0.0464	0.0499	0.0534	0.0603	0.0668
	h			768.9	828.2	887.2	947.8	1011.1	1075.7	1139.0	1200.3	1261.1	1319.6	1372.8	1423.6	1520.4	1610.8
	s			0.8946	0.9427	0.9886	1.0340	1.0797	1.1247	1.1674	1.2073	1.2457	1.2815	1.3131	1.3424	1.3959	1.4433

[a] Sh-superheat. °F; v specific volume. ft³/lb; h enthalpy, Btu/lb; s entropy. Btu/°F · lb.

Source: Copyright 1967 ASME (Abridged): reprinted by permission.

References

1. Verderber, R.R. and O. Morse. Lawrence Berkeley Laboratory, Applied Science Division. "Preformance of Electronic Ballasts and Other New Lighting Equipment (Phase II: The 34-Watt F40 Rapid Start T-12 Fluorescent Lamp)." University of California, February, 1988.
2. Morse, O. and R. Verderber. Lawrence Berkeley Laboratory, Applied Science Division. "Cost Effective Lighting." University of California, July, 1987.
3. Hollister, D.D. Lawrence Berkeley Laboratory, Applied Science Division. "Overview of Advances in Light Sources." University of California, June, 1986.
4. Verderber, R. and O. Morse. Lawrence Berkeley Laboratory, Applied Science Division. "Performance of Electronic Ballasts and Other New Lighting Equipment: Final Report." University of California, October, 1985.
5. Lovins, Amory B. "The State of the Art: Space Cooling." Rocky Mountain Institute, Old Snowmass, CO, November, 1986.
6. U.S. Department of Energy, Assistant Secretary, Conservation and Renewable Energy, Office of Buildings and Community Systems. "Energy Conservation Goals for Buildings." Washington, DC, May, 1988.
7. Rubinstein, F., T. Clark, M. Simonovitch, and R. Verderber. Lawrence Berkeley Laboratory, Applied Science Division. "The Effect of Lighting System Components on Lighting Quality, Energy Use, and Life-Cycle Cost." University of California, July, 1986.
8. Manufacturing Energy Consumption Survey: Consumption of Energy 1985 Energy Information Administration, November 18, 1988.
9. Heery Energy Consultants, Inc. "Georgia Cogeneration Handbook." The Governor's Office of Energy Resources, August, 1988.
10. Thermodynamics, New York, F. Sears, Reading, MA, Addison-Wesley Publishing Company, Inc., 1952.
11. Climatic Atlas of the U.S., reprinted under the title: Weather Atlas of the United States, Gale Research Co., 1975.
12. Thumann, Albert, "Plant Engineers and Managers Guide to Energy Conservation - 3rd Edition." Fairmont Press, 1987.
13. Thumann, Albert. "Handbook of Energy Audits - 2nd Edition." Fairmont Press, 1983.
14. Kreider and McNeil. "Waste Heat Management Guidebook." NBS Handbook 121.
15. U.S. Department of Energy, Office of the Assistant Secretary for Conservation and Solar, Office of Building and Community Systems, Oak Ridge National Laboratory, Oak Ridge, TN, University of Massachusetts Cooperative Extensive Service Energy Education Center, Amherst, MA, The Solar Energy Research Institute, Golden, CO. "Residential Energy Audit Manual." Fairmont Press, 1981.

16. American Society of Heating, Refrigerating and Air-Conditioning Engineers, Inc. "ASHRAE Handbook, 1982 Fundamentals." Chapter 2 - Heat Transfer; Chapter 11 - Applied Heat Pump Systems; Chapter 23 - Design Heat Transmission Coefficients; Chapter 25 - Heating Load; Chapter 26 - Air Conditioning Cooling Load. Atlanta, GA, 1982.
17. American Society of Heating, Refrigerating and Air-Conditioning Engineers, Inc. "ASHRAE Handbook, 1982 Applications." Chapter 57 - Solar Energy Utilization for Heating and Cooling. Atlanta, GA, 1982.
18. American Society of Heating, Refrigerating and Air-Conditioning Engineers, Inc. "ASHRAE Handbook & Product Directory, 1982 Systems." Chapter 7 - Heat Recovery Systems; Chapter 43 - Energy Estimating Methods. Atlanta, GA, 1982.
19. Trane Air Conditioning Manual. The Trane Company, LaCrosse WI.

Index